Asian

NSQF Level - 4 D.G.T. Syllabus

1st & 2nd Year

Latest Revised Edition

Q-Bank Series

ELECTRICIAN

Industrial Training Institute

Skill India
कौशल भारत - कुशल भारत

Trade Theory

Workshop Calculation & Science

Engineering Drawing

Sector - Power

As per NSQF Level-4 Compliant Competency based Curriculum prescribed by D.G.T. Govt of India For 1st and 2nd Year

3 SETS Latest Bharat Skills Question Bank (MCQs) 1st and 2nd Year

Multiple Choice Questions (with key)

Latest online CBT Solved Papers

MCQs For All Competitive Exams

K. Mehta
G.V. Ramana Murthy

For Competitive Exams / Job Interview of RRB - ALP, IOF, SAIL, AIL, ONGC, BHEL, NTPC, DRDO, ISRO, HAL, CCL, DMRC, etc and useful for preparation of technician recruitment in other PSUs

Asian

COMPUTECH PUBLICATIONS LIMITED
An ISO 9001 : 2008 Certified Company
4591/15, Aggarwal Road (Near Happy School),
Darya Ganj, New Delhi - 110002 (India)
Division - AP Asian Publishers

Azadi Ka Amrit Mahotsav

Latest Edition : **2025-26**

Price : ₹ **349.50**

ISBN 978-81-7317-765-1

Published by :

Computech Publications Ltd. (*Division* – **Asian Publishers**)
4591/15, Aggarwal Road (Near Happy School)
Darya Ganj, New Delhi – 110002 (India)
Phone : 011-43080162, 43080163, 43080164, 43080165
WhatsApp : 9716836703, 9716836717, 9716836701
E-mail : nap@newasianpub.in ; info@computechpublications.in
Buy books online : www.computechpublications.in

Printed at : Glorious Printers, Delhi.

Preface to NSQF Level-4 Compliant

Directorate General of Training (DGT) under the Ministry of Skill Development and Entrepreneurship, being the apex organization for development and coordination of vocational training at the national level, undertakes steps from time to time to improve its various aspects in line with changing market and industry requirements.

DGT vide its communication No. MSDE(GT)-19/03(02)/2022-CD dated 5th April 2022 conveyed its decision to simplify the syllabus and trade-testing procedure related to Workshop Calculation & Science (WCS) and Engineering Drawing (ED) by merging these subjects with Trade Theory subject of all engineering trades.

Taking note of this latest development, this Q – Bank has been prepared which covers besides trade theory of Electrician and complete syllabus of WCS and ED as per the latest notified syllabus for Electrician and all other 2-year Electrical Group trades.

This book covers:

- **Trade Theory**
- **Workshop Calculation and Science**
- **Engineering Drawing**
- **Q-Bank (3 Sets each for 1st and 2nd Year)**
- **Online CBT Solved Papers**

The subject matter has been sequentially arranged through different Modules, each containing large number of MCQs. Further, it has been supplemented by Latest Q-Banks – 3 sets each for 1st year and 2nd year (taking cognizance of Bharat Skills Sample Papers), along with solved AITT CBT (Computer Based Test) question papers.

Main features of this book

- Covers Latest Syllabus
- CBT exam pattern as per Bharat Skills MCQs
- Latest CBT exam papers
- Useful for I.T.I. and Competitive Exams / Job Interviews for RRB, various PSUs and technical posts

We sincerely feel that this concise Q – Bank will be of immense help to the trainees to prepare for Annual All India Trade Test and candidates preparing for test / interview for various competitive exams and employment. Undoubtedly, we would welcome suggestions from our esteemed readers for further improvement of this Q – Bank in our subsequent editions.

–Authors

Contents

1st Year

2nd Year

ELECTRICIAN TRADE THEORY

(As Per NSQF Level-4 Compliant Competency Based Curriculum)

1st YEAR

Module 1

Safety Practice and Hand Tools

MULTIPLE CHOICE QUESTIONS

1. An accident is
   (a) unplanned event
   (b) non-controlled event
   (c) undesirable event
   (d) all of the above
   (I.T.I. February 2019, Electrician, Electroplator, Wireman)
2. The safe way of working is
   (a) an effective and right way of working
   (b) an ancient way of working
   (c) a way of handling the work in a hurry
   (d) a way of normal working
   (I.T.I. February 2019, Electrician, Electroplator, Wireman)
3. An oily floor should be cleaned by
   (a) cotton waste
   (b) putting water
   (c) putting saw dust
   (d) spraying carbon dioxide or sand
4. Fire is a combination of
   (a) fuel, light and oxygen
   (b) fuel, heat and oxygen
   (c) fuel, heat and carbon dioxide
   (d) fuel, light and nitrogen
5. The safety sign given below identifies which of the following hazards?
   (a) Hard hats must be worn
   (b) Danger electric shock
   (c) Wear eye protection
   (d) No smoking

6. The following safety sign (white symbol on a blue background) is an example of a:

   (a) Prohibition sign (b) Mandatory sign
   (c) Warning sign (d) Information sign
7. On finding a fire in a building a carpenter's first action should be
   (a) Raise an alarm
   (b) Fight the fire
   (c) Leave the building
   (d) Carry on working
8. A person being electrocuted should be removed from the live supply by means of
   (a) A length of metal pipe
   (b) A length of dry timber
   (c) A length of wet timber
   (d) A scaffold pole

9. If a person faints due to suffocation, the first aid to be given is
   (a) Give him a warm drink
   (b) Sprinkle cold water on his face
   (c) Place him on a well ventilated location
   (d) Give him artificial respiration
10. Long hair must be tied back because
   (a) It can get dirty with the oil and dust from the workshop
   (b) It gets in the way when you are working
   (c) It looks better tied up
   (d) It can get caught in the machines and cause serious injury
11. Which one of the following fire extinguishers is suitable for the live electrical fire?
   (a) Halon
   (b) Water
   (c) Foam
   (d) Liquified chemical
12. In ABC of first aid, A stands for Airway, B stands for breathing and C stands for ....
   (a) Cardio problem
   (b) Circulation
   (c) Cardio arrest
   (d) Cleaning and dressing
13. What is the benefit of 5S system?
   (a) Increase in manpower
   (b) Increase in economy
   (c) Increase in productivity
   (d) Reduction in pollution
14. What is biological hazard?
   (a) Virus (b) Fatigue
   (c) Alcoholism (d) Old age
15. What is the first action if a colleague (victim) received a heavy electric shock?
   (a) Pull the victim from the live conductor
   (b) Pour water to avoid burning of victim
   (c) Fetch the first aid box
   (d) Switch off the current immediately
16. In case of arc welding, one should protect his eyes by using
   (a) dark glass screen
   (b) sun goggles
   (c) mask
   (d) clear glasses
17. While grinding one must use
   (a) dark glass screen
   (b) mask
   (c) safety goggles
   (d) sun goggles
18. The cutting edge of a flat cold chisel is slightly convexed so that :
   (a) less power is required in cutting a metal and its edge doesn't fasten in the job
   (b) metal can be cut easily
   (c) chipping can be done in a straight line
   (d) its edge will not break during chipping
19. Chiesels are made of :
   (a) high carbon steel
   (b) high speed steel
   (c) cast steel
   (d) tungsten steel
20. The length of a hack-saw blade is measured:
   (a) between the inner ends of the two pin-holes
   (b) between the outer ends of the two pin-holes
   (c) between centres of the pin-holes
   (d) between the inner end of pin-hole and tho outer end of the other pin-hole
21. A file is made of :
   (a) cast iron
   (b) cast steel or special tool steel
   (c) mild steel
   (d) nickle steel
22. A piece of aluminium has to be filed. What type of file cut is most suitable?
   (a) Rasp cut (b) Double cut
   (c) Single cut (d) Curved cut
23. Which type of file you would select for filing a wooden plank?
   (a) Rough file (b) Rasp cut file
   (c) Bastard file (d) Single cut file

(I.T.I. February 2014, Wireman; I.T.I. February 2015, Electrician)

24. The figures shows the part of a file. What is the name of the part indicated as 'X' in the figure?

(a) Tang (b) Handle
(c) Heel (d) Shoulder

(I.T.I. August 2019, Electrician, Wireman, Electroplator, Lift & Escalator Mechanic)

25. Which type of the file given in figure?

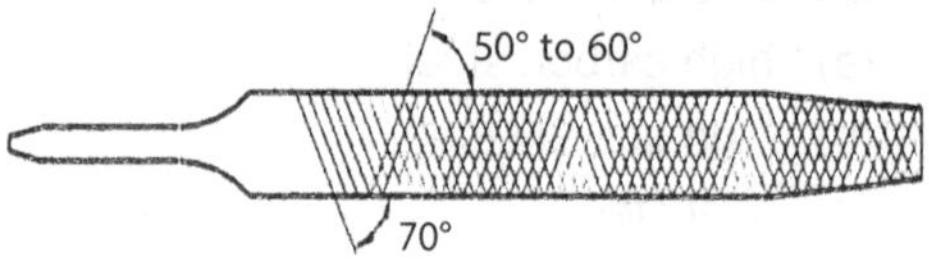

(a) Rasp cut file
(b) Single cut file
(c) Double cut file
(d) Curved cut file

(I.T.I. August 2014, Wireman)

26. Hack saw blades are made up of ....
(a) mild steel
(b) stainless steel
(c) high carbon steel
(d) high speed steel

(I.T.I. February 2014, Electrician)

27. The distance between two adjacent teeth of hacksaw blades are called as ....
(a) specification of the blades
(b) pitch of the blades
(c) designation of the blades
(d) size of the blades

(I.T.I. February 2015, Electrician)

28. Which pitch of hacksaw blade is suitable to cut the sheet metal?
(a) 14 TPI (b) 18 TPI
(c) 24 TPI (d) 32 TPI

(I.T.I. August 2014, Electrician)

29. A wire gauge is used to measure ....
(a) length of a wire
(b) area of a wire
(c) diameter of a wire
(d) thickness of insulation layer of a wire

30. Name the tool which is used for making wire hooks and loops.
(a) Flat nose pliers
(b) Side cutting pliers
(c) Round nose pliers
(d) Long nose pliers

(I.T.I. February 2015, Electrician)

31. Which tool is used for cutting, twisting, holding and gripping of a small job?
(a) Side cutting plier
(b) Flat nose plier
(c) Combination plier
(d) Round nose plier

(I.T.I. August 2014, Wireman)

32. The size of the hammer is specified by its ....
(a) length of the peen
(b) diameter of the face
(c) weight of the hammer
(d) height of the head

(I.T.I. February 2019, Electrician Electroplator, Wireman)

33. Which type of spanner is used to tight the nuts in places, where the space is restricted and high leverage is required?
(a) Ring spanner
(b) Socket (box) spanner
(c) Double ended spanner
(d) Single ended open jaw adjustable spanner

(I.T.I. February 2015, Wireman)

34. Name the tool used for loosening or tightening of nuts and bolts, which are located in narrow space or depth?
(a) Double ended spanner
(b) Ring spanner

(c) Socket (box) spanner
(d) Single ended open jaw adjustable spanner

(I.T.I. August 2014, Wireman)

35. Scribers are made of ....
(a) Mild steel
(b) High carbon steel
(c) Brass (d) Cast iron

(I.T.I. February 2014, Wireman)

36. A tool which is used for boring small hole on wooden article is ......
(a) Bradawl (b) Gimlet
(c) Pocker (d) Rawl plug tool

(I.T.I. February 2015, Wireman)

37. The Rawl tool is made of ....
(a) Mild steel (b) Carbon steel
(c) Stainless steel (d) Aluminium

(I.T.I. February 2014, Wireman)

38. Which type of lifting device is used to pull heavy loads along the ground?
(a) Cranes and slings
(b) Winches
(c) Machines moving platform
(d) Layers and rollers

(I.T.I. August 2014, Electrician)

39. What is the type of chisel you would select for squaring the corners?
(a) Cross cut chisel
(b) Half round nose chisel
(c) Diamond point chisel
(d) Web chisel

(I.T.I. February 2015, Electrician)

40. Name the material by which the blade of screw driver is made of?
(a) Tempered stainless steel
(b) Hardened and tempered mild steel
(c) Hardened and tempered soft steel
(d) Hardened and tempered carbon steel

(I.T.I. February 2015, Electrician)

41. What type of screw driver is used by Electrician?
(a) Heavy duty screw driver
(b) Light duty screw driver
(c) Star tip head screw driver set
(d) Flexible screw driver

(I.T.I. February 2014, Electrician)

42. What is the first step you would like to take to rescure a person who is in contact with live wire?
(a) Call to doctor immediately
(b) Pull him away from live wire holding his hand
(c) Isolate him from the live wire
(d) Shift him to hospital

(I.T.I. February 2014, Wireman)

43. There are different methods of extinguishing of fire smothering is the method of ....
(a) removing the fuel of the fire
(b) pouring of water to the fire
(c) isolating supply of oxygen to the fire
(d) lowering the temperature of the fire

(I.T.I. August 2014, Wireman)

44. Which artificial resuscitation method is suitable to the victim in the case of burns/ injury in the back?
(a) Mouth to mouth method
(b) Schafer's method
(c) Nelson's method
(d) Holgen Nelson method

(I.T.I. February 2015, Wireman)

45. Which type of fire extinguisher is to be used on electrical equipment or installation under fire?
(a) Foam type (b) Halon type
(c) Gas cartridge water filled type
(d) Stored pressure water filled type

(I.T.I. February 2015, Electrician)

46. Which artificial respiration method is suitable in the case of injuries to the victim, when his mouth will not open?
(a) Nelson's artificial respiration method
(b) Schafer's artificial respiration method
(c) Mouth to Mouth artificial respiration method
(d) Mouth to nose artificial respiration method

(I.T.I. August 2014, Wireman)

47. In which type fire, both foam and dry powder fire extinguishers can be used?

(a) Class "A" fire (b) Class "B" fire
(c) Class "C" fire (d) Class "D" fire

(I.T.I. February 2014, Electrician)

48. The colour used for displaying the prohibition sign is ....

(a) red border and cross bar, black symbol and white background
(b) red border and cross bar
(c) black symbol and white background
(d) black symbol

(I.T.I. February 2015, Wireman)

49. Which is the most suitable fire extinguisher for the class "D" fire shown in figure?

(a) Carbon tetra chloride (CTC) extinguisher
(b) Foam Extinguisher
(c) Water Filled Extinguisher
(d) Gas Cartridge Halon Extinguisher

(I.T.I. February 2014, Electrician)

50. Name the class of fire for the fuel shown in the figure?

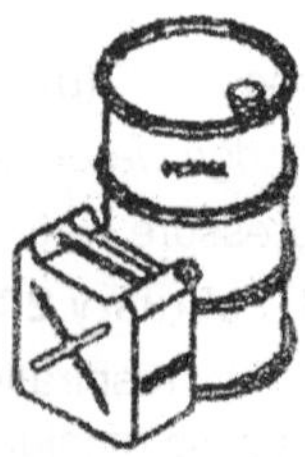

(a) Class "D" fire (b) Class "C" fire
(c) Class "B" fire (d) Class "A" fire

(I.T.I. August 2015, Wireman)

51. Which type of fire extinguisher must NOT be used on electrical equipment?

(a) Foam Type
(b) Carbon dioxide type
(c) Vapourising liquid (CTC)
(d) Dry powder

(I.T.I. August 2015, Electrician, Wireman)

52. Which artificial resuscitation method is suitable to the victim, whose mouth is closed tightly?

(a) Schafer's method
(b) Mouth to mouth method
(c) Nelson's method
(d) Holgen-Nelson method

(I.T.I. August 2015, Wireman)

53. What class of fire extinguisher is suitable to extinguish the fire cuased by liquefied gases?

(a) Foam type extinguisher
(b) Carbon Tetra Chloride (CTC) extinguisher
(c) Dry powder extinguisher
(d) Halon type fire extinguisher

(I.T.I. February 2016, Electrician, Wireman)

54. If a person receiving electric shock and getting severe bleeding due to electric burns; the immediate reaction to stop the bleeding is to ....

(a) cover the bleeding area with a cloth
(b) apply the pressure to the wound
(c) put a dressing over the wound
(d) send for medical treatment

(I.T.I. August 2015, Electrician)

55. Which type of safety sign indicates 'what must be done'?

(a) Prohibition signs
(b) Mandatory signs
(c) Warning signs
(d) Information signs

(I.T.I. February 2016, Electrician, Wireman)

56. Which categories of safety sign is represented in the figure shown?

(a) Prohibition signs
(b) Mandatory signs
(c) Warning signs
(d) Information signs

(I.T.I. August 2015, Electrician)

57. Which tool is used to check the squareness of filed surfaces?
(a) Straight edge
(b) Try square
(c) Universal suface gauge
(d) Steel rule

(I.T.I. February 2016, Electrician, Wireman)

58. The rawl plug tool is specified by ....
(a) length (b) thickness
(c) material (d) number

(I.T.I. February 2016, Wireman)

59. Which material is used to make the head of the hammer?
(a) Drop-forged carbons steel
(b) High carbon steel
(c) Mild steel
(d) Cast iron

(I.T.I. February 2016, Electrician, Wireman)

60. Which type of tool is used for holding small objects in places whore fingers cannot reach?
(a) Flat nose pliers
(b) Long nose pliers
(c) Sido cutting pliers
(d) Round nose pliers

(I.T.I. August 2015, Electrician, Wireman)

61. Which type of file is shown in the figure?

(a) Single cut file (b) Bastard file
(c) Curved cut file (d) Rasp cut file

(I.T.I. August 2015, Electrician)

62. What type of file is shown in the figure?

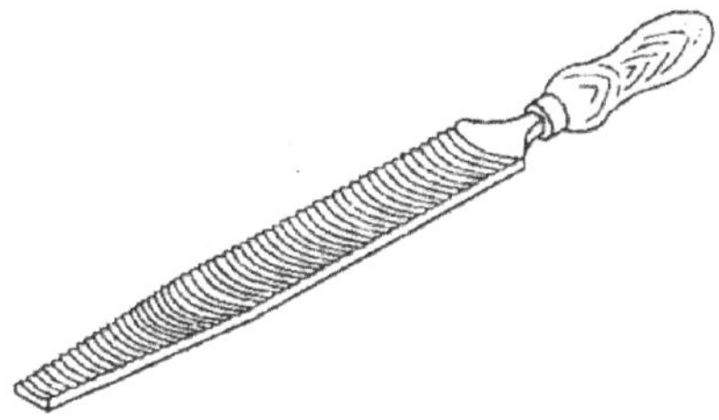

(a) Single cut file (b) Double cut file
(c) Rasp cut file (d) Curved cut file

(I.T.I. February 2016, Electrician, Wireman)

63. While filing soft metals, the file teeth are clogged with minute particles of the material being filed. The file should be cleaned with...
(a) Washing it with water
(b) Washing it with dilute acid
(c) Rubbing it on wood
(d) Using file card

(I.T.I. February 2017, Electrician Electroplator, Wireman)

64. The cold chisel is made by...
(a) Drawing (b) Friging
(c) Rolling (d) Piercing

(I.T.I. February 2017, Electrician Electroplator, Wireman)

65. What is the name of the tool?

(a) Diagonal cutting pliers
(b) Crimping tool
(c) Combination pliers
(d) Wire stripper

(I.T.I. August 2019 Electrician, Electroplator, Wircman, Lift and Escalator Mechanic)

66. What is the name of the tool?

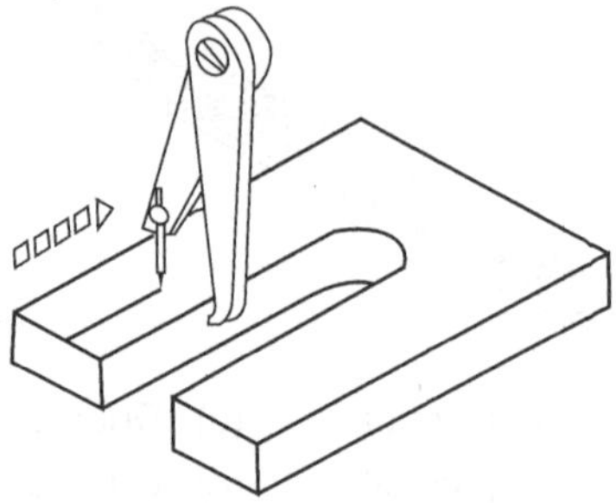

(a) Firm joint calliper
(b) Inside calliper
(c) Outside calliper
(d) Jenny calliper

(I.T.I. August 2019 Electrician, Electroplator, Wireman, Lift & Escalator Mechanic)

67. What is the name of the angle marked "X" of the chisel?

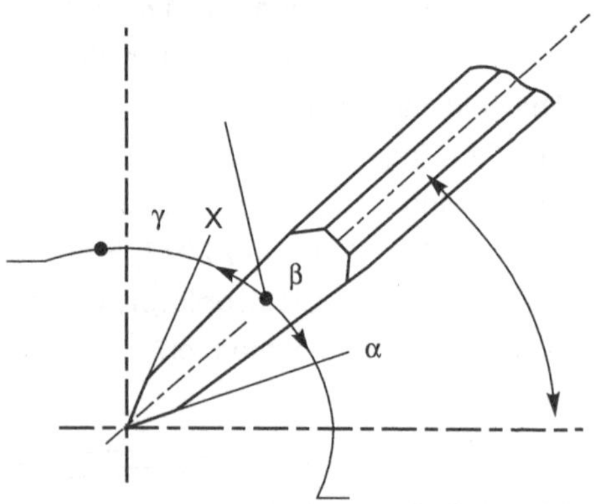

(a) Rake angle (b) Inclination angle
(c) Clearance angle (d) Point angle

(I.T.I. August 2019 Electrician, Electroplator, Wireman, Lift & Escalator Mechanic)

68. A steel hand tool with small sharp teeth used for sharpening or shaping as shown figure is.........

(a) Metal vice (b) Steel file
(c) Chalk line (d) Speed square

(I.T.I. August 2019 Electrician)

69. The cutting tool wears due to....
(a) Edge wear (b) Crater wear
(c) Flank wear (d) All of these

(I.T.I. August 2019 Electrician)

70. What is the tool used to bend conduits ?
(a) Coupler (b) Hickey
(c) Pipe vice (d) Bench vice

(I.T.I. August 2019 Electrician)

71. Cold bending could be done by using a.....
(a) Spring (b) Hacksaw frame
(c) Hand (d) Knee

(I.T.I. August 2019 Electrician)

72. What is the name of P.P.E. (Personal Protective Equipment) ?

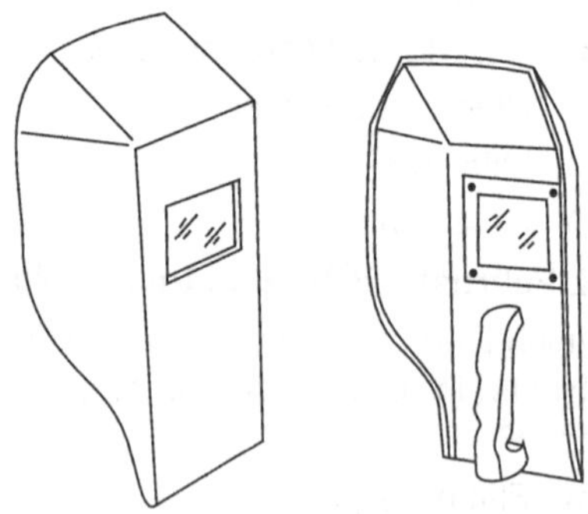

(a) Nose mask (b) Hand screen
(c) Face shield (d) Head shield

(I.T.I. August 2019 Electrician, Electroplator, Wireman, Lift & Escalator Mechanic)

73. What is starving in extinguishing of fire?
(a) Adding fuel to the fire
(b) Removing fuel element from the fire
(c) Using water to cool the fire
(d) Preventing oxygen supply to the fire

(I.T.I. August 2019 Electrician, Electroplator, Wireman, Lift & Escalating Mechanic)

74. What type of fire extinguisher would you use on a fire, that is categorised as a class "F" fire?

(a) Carbon dioxide fire extinguishers
(b) Water fire extinguishers
(c) Wet chemical fire extinguishers
(d) Dry powder fire extinguishers

(I.T.I. August 2019 Electrician)

75. You should use a dry powder fire extinguisher in order to fight in an electrical fire, only in the event of :

(a) A carbon dioxide fire extinguisher is unavailable
(b) A foam fire extinguisher is unavailable
(c) You should never use a dry powder fire extinguisher on an electrical fire
(d) A water fire extinguisher is unavailable

(I.T.I. August 2019 Electrician)

76. How will you diagnose the victim is suffering under cardiac arrest ?

(a) Gets pain in spinal cord
(b) Mouth will be closed tightly
(c) Appears blue colour around his lips
(d) Heavy swelling on his stomach

(I.T.I. August 2019 Electrician, Eletroplator, Wireman, Lift & Escalator Mechanic)

77. Which artificial respiration method is to be performed to the victim with injuries on the chest and belly ?

(a) Schafer's method
(b) Mouth to mouth method
(c) Mouth to nose method
(d) Nelson's arm lift back pressure method

(I.T.I. August 2019 Electrician, Electroplator, Wireman, Lift & Escalator Mechanic)

78. Which type of occupational health hazard is caused for infection ?

(a) Electrical hazard
(b) Biological hazard
(c) Physiological hazard
(d) Psychological hazard

(I.T.I. August 2019 Electrician, Electroplator, Wireman, Lift & Escalator Mechanic)

79. Which disposal method of waste saves lot of energy?

(a) Land fill (b) Incineration
(c) Recycling (d) Composting

(I.T.I. August 2019 Electrician, Electroplator, Wireman, Lift & Escalator Mechanic)

KEY TO MULTIPLE CHOICE QUESTIONS

1. (d)	**2.** (a)	**3.** (c)	**4.** (b)	**5.** (b)	**6.** (b)	**7.** (a)	**8.** (b)	**9.** (c)	**10.** (d)
11. (a)	**12.** (b)	**13.** (c)	**14.** (a)	**15.** (d)	**16.** (a)	**17.** (c)	**18.** (a)	**19.** (a)	**20.** (c)
21. (b)	**22.** (d)	**23.** (b)	**24.** (c)	**25.** (c)	**26.** (d)	**27.** (b)	**28.** (d)	**29.** (c)	**30.** (c)
31. (c)	**32.** (c)	**33.** (a)	**34.** (c)	**35.** (b)	**36.** (b)	**37.** (b)	**38.** (b)	**39.** (c)	**40.** (d)
41. (b)	**42.** (c)	**43.** (c)	**44.** (a)	**45.** (b)	**46.** (d)	**47.** (b)	**48.** (a)	**49.** (a)	**50.** (c)
51. (a)	**52.** (a)	**53.** (c)	**54.** (b)	**55.** (b)	**56.** (d)	**57.** (b)	**58.** (d)	**59.** (a)	**60.** (b)
61. (d)	**62.** (d)	**63.** (d)	**64.** (b)	**65.** (a)	**66.** (d)	**67.** (d)	**68.** (b)	**69.** (b)	**70.** (b)
71. (c)	**72.** (b)	**73.** (b)	**74.** (c)	**75.** (a)	**76.** (c)	**77.** (b)	**78.** (b)	**79.** (c)	

Module 2
Basic Workshop Practice (Allied Trades)

Multiple Choice Questions

Cutting Tools and Fasteners

1. The taper shank drills should be removed from the machine spindle by means of
   (a) hammer (b) tang of file
   (c) punch (d) drift
2. Drill chucks are held on the machine spindle by means of
   (a) arbor (b) drift
   (c) draw-in bar (d) chuck nut
3. The lip clearance angle of twist drill for general purpose should be
   (a) 3 to 5° (b) 4 to 8°
   (c) 8 to 12° (d) 12 to 15°
4. The suitable cutting fluid for drilling M.S. plate is
   (a) soluble oil (b) water
   (c) compressed air (d) dry
5. Taps are re-sharpened by grinding
   (a) flutes (b) threads
   (c) diameter (d) relief
6. The tapping hole should be
   (a) larger than the tap size
   (b) smaller than the tap size
   (c) equal to the tap size
   (d) equal to the core (minor) diameter of the tap
7. The shape of the frame of thread snap gauge is in the form of
   (a) Round (b) Square
   (c) 'Y' type (d) "C" type
8. The shape of the thread in the 'Not Go' end of a thread snap gauge is
   (a) Beveled (b) Truncated
   (c) Taper (d) Sharp
9. The handle of thread plug gauge is made from ebonite material for
   (a) reducing the cost
   (b) reducing the weight
   (c) easy manufacturing
   (d) easy handling
10. A rivet is specified by the diameter of its
    (a) Head
    (b) Tail
    (c) Shank
    (d) None of the above
11. Which one of the following rivets is used for girders and heavy constructional work ?
    (a) Pan head rivet
    (b) Snap head rivet
    (c) Flat head rivet
    (d) Countersunk head rivet
12. Which one of the following rivets is used for general structural work ?
    (a) Pan head rivet
    (b) Snap head rivet
    (c) Flat head rivet
    (d) Countersunk head rivet

13. For riveting aluminium sheets, the material of rivet should be

(a) Aluminium (b) Copper
(c) Brass (d) Wrought iron

14. Identify the type of riveted joint shown in Figure.

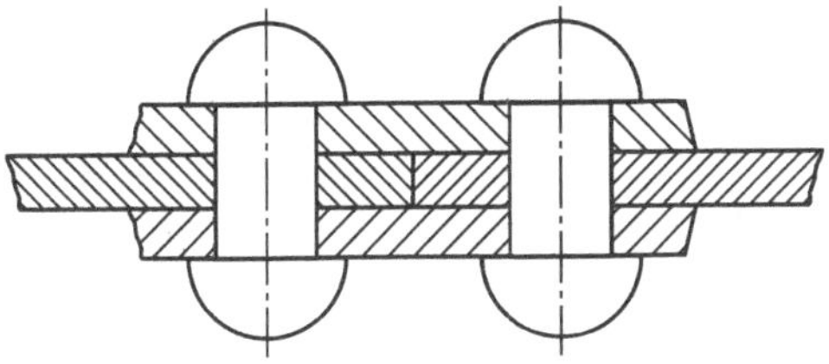

(a) Lap joint
(b) Single strap butt joint
(c) Double strap butt joint
(d) Double riveted lap joint

15. The height of the snap head rivet is standardized to

(a) 0.5 × diameter of rivet
(b) 0.7 × diameter of rivet
(c) 0.9 × diameter of rivet
(d) 1.1 × diameter of rivet

16. In a riveted joint the edges of plates are simply laid over each other and riveted. This joint is called

(a) Lap joint (b) Butt joint
(c) Edge joint (d) Corner joint

17. In a riveted joint the plates are placed end-to-end and jointed through cover plates. This joint is called

(a) Lap joint (b) Butt joint
(c) Edge joint (d) Corner joint

18. In a riveted joint pitch is the distance between two adjacent rivets in the same row, measured parallel to the seam. It is usually equal to

(a) Diameter of the rivet hole
(b) 2 × Diameter of the rivet hole
(c) 3 × Diameter of the rivet hole
(d) 4 × Diameter of the rivet hole

19. In a riveted joint the minimum distance from a rivet hole centre to the nearest edge of the plate is called

(a) Back pitch (b) Allowance
(c) Tolerance (d) Margin

20. Which one of the following reasons is responsible for crushing of metal in the riveted joint ?

(a) Thickness of plates is more
(b) Thickness of plates is less
(c) Diameter of rivet is too large
(d) Diameter of rivet is too small

21. Which one of the following reasons is responsible for shearing of rivet in a riveted joint ?

(a) Diameter of rivet is too small
(b) Diameter of rivet is too large
(c) Thickness of plates vary
(d) Pitch of the hole is too small

22. Which one of the following reasons is responsible for splitting of the metal in the riveted joint ?

(a) Pitch of the rivets is too large
(b) Holes of the rivets are drilled too near from the edge of the plate
(c) Pitch of the rivets is too small
(d) Holes for rivets are drilled too far from the edge of the plate

23. Taper shank drills are held on machine by means of:

(a) Chucks (b) Drift
(c) Sleeve (d) Vice

24. Which tool will you use for removing taper shank drill from the socket?

(a) The slot (b) The sleeve
(c) The tank (d) The drift

(I.T.I. February 2015, Electrician)

25. A coolant is used for cooling the:

(a) Machine (b) Job only
(c) Cutting tool (d) Hot chips

26. Which tap of a set is sufficient to finish the thread of a through hole?

(a) Taper tap
(b) Intermediate tap
(c) Plug tap
(d) Bottoming tap

(I.T.I. February 2015, Electrician)

27. Countersinking is done on drilled holes to accommodate countersunk head of wood screw. What is the general bit size available?

(a) 6 to 15 mm (b) 8 to 20 mm
(c) 10 to 15 mm (d) 10 to 25 mm

28. If a drill bit does not cut the metal easily, the reason is this that there is:

(a) less or no lip clearance angle
(b) more clearance angle
(c) no body clearance angle
(d) high cutting angle

29. You find a V-groove at the bottom of the base of a universal surface gauge. What is the purpose of this V-groove?

(a) For setting the spindle in any position
(b) To reduce the weight of the base
(c) To reduce surface contact for easy movement
(d) For setting the gauge on cylindrical surface

30. In order to check the clearance between two mating surfaces a .... gauge is used.

(a) Ring (b) Thickness
(c) Feeler
(d) Plug

(I.T.I. Februray 2017, Electrician, Electroplator, Wireman)

31. The fastener used to hold a screw in case of a porous or frittle wall is ...

(a) Rawl plug (b) Bakelite holder
(c) Allen key (d) All of these

(I.T.I. February 2017, Wireman)

32. Which type of job motion is there in drilling operation?

(a) Rotary (b) Translating
(c) Fixed (d) None of these

(I.T.I. August 2019 Electrician)

33. Which of the following is not a main part of rivet ?

(a) Head (b) Shank
(c) Point (d) Thread

(I.T.I. August 2019 Electrician)

34. Which one is cheaper: "punching or drilling"

(a) Punching (b) Drilling
(c) Both (d) None of these

(I.T.I. August 2019 Electrician)

Carpentry

1. The size of the hammer is specified by its ...

(a) length of the peem
(b) Diameter of the face
(c) Weight of the hammer
(d) Height of the head

(I.T.I. February 2019, Electrician, Electroplator, Wireman)

2. Which type of file you would select for filing a wooden plank?

(a) Rough file (b) Rasp cut file
(c) Bastard file (d) Single cut file

(I.T.I. February 2015, Electrician)

3. Which tool is used for cutting square holes in mica sheets used on wooden boards in electric wiring?

(a) Tenon saw (b) Key hole saw
(c) Hack saw (d) Carpenter saw

4. Which type of chisal is used for making ractangular holes in wood?

(a) Mortise chisel
(b) Paring chisel
(c) Firmer chisel
(d) Bevel firmer chisel

(I.T.I. February 2015, Electrician)

5. The cutting blades of jack plane is made up of ......

(a) High speed steel
(b) High carbon steel
(c) Tungsten steel
(d) Alloy steel

(I.T.I. February 2019, Electrician, Electroplator, Wireman)

6. Which type of bench plane is suitable to plane a small wooden piece?

(a) Jack plane

(b) Smoothing plane
(c) Coarse plane
(d) Rebate plane

(I.T.I. February 2015, Electrician)

7. What is the type of chisel you would select for squaring the corners?
(a) Cross cut chisel
(b) Half round nose chisel
(c) Diamond point chisel
(d) Web chisel

(I.T.I. February 2015, Electrician)

8. There are several application of a try square what cannot be done by using a try square?
(a) Check parallel of workpiece surface
(b) Check squarences of field surface
(c) Check flatness of surface
(d) Mark lines at 90° to the edge of a workpiece

(I.T.I. February 2015, Electrician)

9. The pitch of the wood screw is the distance ......
(a) From the tip to the end of threaded portion
(b) Between the adjacent threads
(c) From the head to the unthreaded portion
(d) From the head to the tip

(I.T.I. February 2015, Electrician)

10. Which type of half lap joint is shown in figure?

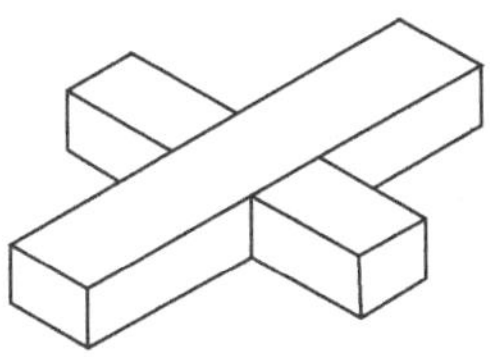

(a) End lap joint (b) Cross lap joint
(c) Middle lap joint (d) Corner lap joint

(I.T.I. February 2015, Electrician)

11. Middle lap type, half lap joint is used where ......
(a) Two parts of a frame cross each other
(b) One part of a job meets another part
(c) Two parts of a job cross each other at the ends
(d) Two parts of a job meet mean the ends

(I.T.I. February 2015, Electrician)

12. Which type of half lap joint is used where one part of a job meets another part at some distance from the ends?
(a) End-lap joint (b) Middle-lap joint
(c) Cross-lap joint (d) Corner-lap joint

(I.T.I. August 2015, Electrician)

13. Which type of chisel is used to clean sharp corners of the wood ?
(a) Firmer chisel
(b) Bevel-edge firmer chisel
(c) Paring chisel
(d) Mortise chisel

(I.T.I. August 2015, Electrician, Wireman)

14. Which type of chisel is used for finishing the wood joints ?
(a) Firmer chisel
(b) Bevel-edge firmer chisel
(c) Paring chisel
(d) Mortise chisel

(I.T.I. August 2015, Electrician)

15. Which type of hammer is used for extracting the nails out of the wood ?
(a) Ball pein hammer
(b) Claw pein hammer
(c) Cross pein hammer
(d) Straight pein hammer

(I.T.I. August 2015, Electrician, Wireman)

16. Which type of hammer is used for extracting the nails out of the wood ?
(a) Ball pein hammer
(b) Cross pein hammer
(c) Straight pein hammer
(d) Claw hammer

(I.T.I. February 2016, Electrician, Wireman)]

17. The type of file used for wood work is ...
(a) Single cut file (b) Rasp cut file
(c) Double cut file (d) All of these

(I.T.I. February 2017, Electrician, Electroplator, Wireman)

18. The caliper meant for measuring width of the slot is ...

(a) Inside caliper (b) Jenny caliper
(c) Odd leg caliper (d) Outside caliper

(I.T.I. February 2017, Electrician, Electroplator, Wireman)

19. A rip saw ...

(a) is used for cutting along the grains
(b) has a narrow blade with two wooden handles
(c) is a two man's saw
(d) all of these

(I.T.I. February 2017, Electrician, Electroplator, Wireman)

20. What is the use of this tool?

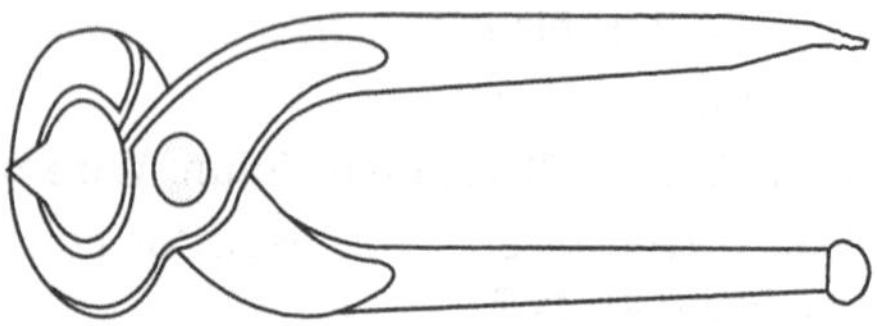

(a) Holding the hot substances
(b) Extracting nails from the wood
(c) Cutting and twisting the wires
(d) Loosening and tightening the bolts and nuts

(I.T.I. August 2019 Electrician, Electroplator, Wireman, Lift & Escalator Mechanic)

21. What is the use of firmer chisel?

(a) Paring and finishing joints
(b) Making rectangular holes in wood
(c) General chiselling work
(d) Light chiselling and to clean sharp corner

(I.T.I. August 2019 Electrician, Electroplator, Wireman, Lift & Escalator Mechanic)

22. Which defect in the timber is caused to reduce its strength?

(a) Twisting (b) Irregularity
(c) Cracking (d) Cupping

(I.T.I. August 2019 Electrician, Electroplator, Wireman, Lift & Escalator Mechanic)

23. Which type of half lap joint is used if one part of a job meets another point at some distance from the ends?

(a) End lap joint
(b) Cross lap joint
(c) Corner half lap joint
(d) Middle lap joint

(I.T.I. August 2019 Electrician, Electroplator, Wireman, Lift & Escalator Mechanic)

Sheet Metal Work

1. The thickness of sheet metal is indicated by a series of numbers, which is called as

(a) Number size
(b) Gauge
(c) Standard size
(d) None of the above

2. Bent snip is used for

(a) Removing the burr
(b) Making holes on the sheet
(c) Bending the sheet
(d) Cutting the sheet along curved lines

3. Which one of the following is the purpose for using hand groover in sheet metal worker ?

(a) To lock the grooved seam
(b) To unlock the grooved seam
(c) To strike the sheet
(d) To hold the sheet in position

4. Mallet is made of

(a) Lead (b) Brass
(c) Hard wood (d) Cast iron

5. Copper and brass sheets are

(a) Hard and brittle
(b) Malleable and ductile
(c) Malleable and brittle
(d) Hard and ductile

6. Which among the following joints is suitable for roofing work?

(a) Lap joint
(b) Butt joint
(c) Hinged joint
(d) Double grooved seam joint

7. Which among the following tools is used for flattening the metal around punched hole ?
(a) Ball pane hammer
(b) Riveting hammer
(c) Setting hammer
(d) Sledge hammer

8. Which among the following types of joints is used in which the end of sheet is placed over the end of another sheet and joined together ?
(a) Lap joint (b) Butt joint
(c) Knocked up joint
(d) Grooved seam joint

9. Sheet metal work is carried out only on metal sheets, which are
(a) Forged (b) Cast
(c) Rolled
(d) None of the above

10. Which one of the following is the common reason for using aluminium sheets ?
(a) Lightness (b) Brightness
(c) Dullness (d) Darkness

11. Which one of the following notches is used when making a job with a 90o bend ?
(a) Square notch (b) V – notch
(c) Wired notch (d) Slant notch

12. Which one of the following stakes is used when shaping and seaming funnels and tapered articles?
(a) Hatchet stake
(b) Half-moon stake
(c) Funnel stake
(d) Creasing stake

13. Which among the following stakes provides a double-ended support ?
(a) Hatchet stake
(b) Half-moon stake
(c) Creasing stake
(d) Horse stake

14. The least bend radius varies depending on the
(a) Material and thickness of plate
(b) Direction of plate
(c) Working temperature
(d) All the above

15. For making various types of hems and seams from the sheet metal of thickness less than 0.4 mm, the allowance should be
(a) Twice the thickness of sheet.
(b) Three times the thickness of sheet.
(c) Four times the thickness of sheet.
(d) No allowance.

16. Which one of the following is a type of hem as shown in Fig.

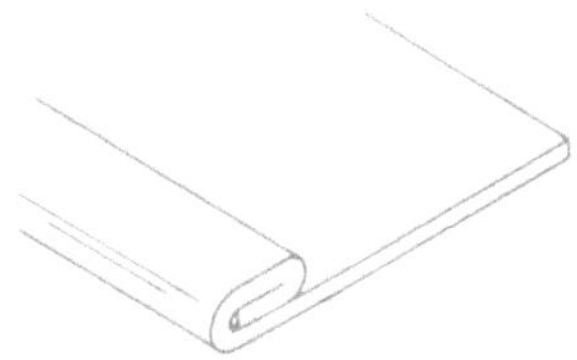

(a) Single hem (b) Double hem
(c) Wired edge (d) Locked hem

17. Which among the following is the calculated length of 'G' of the grooved joint as shown in Fig.
(a) 11.5 mm (b) 12.5 mm
(c) 13.5 mm (d) 14.0 mm

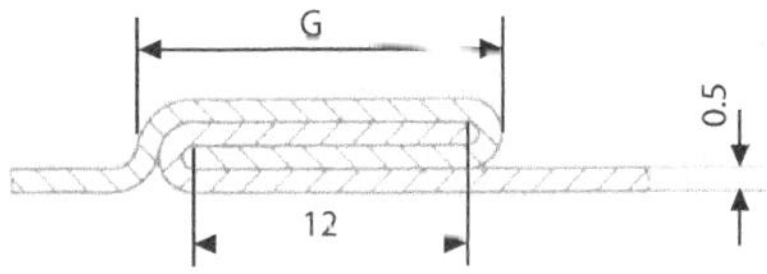

18. Which one of the following sheets is used for making highly corrosive acid tanks ?
(a) Black iron sheets
(b) Galvanised iron sheets
(c) Stainless steel sheets
(d) Lead sheets

19. Which one of the following is the name of a special nut as shown in Fig.
(a) Check nut (b) Castle nut
(c) Speed nut (d) Grooved nut

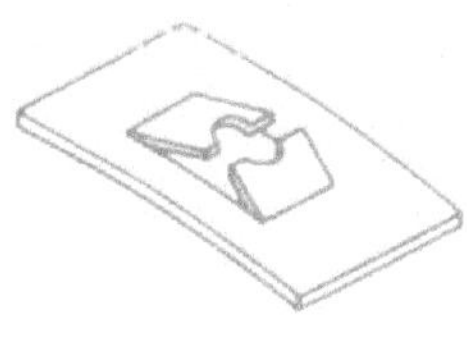

20. Which type of notch is shown in Fig.

(a) Straight notch (b) V-notch
(c) Square notch (d) Slant notch

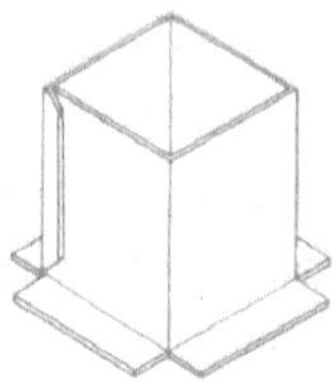

21. Which type of metal sheet is used for furnace?

(a) Sheet steel
(b) Galvanized iron steel
(c) Copper sheet
(d) Aluminium sheet

(I.T.I. February 2015, Electrician)

22. Which type of sheet is used for making house hold utensils?

(a) Sheet steel
(b) Aluminium
(c) Copper sheets
(d) Galvanised iron sheet

(I.T.I. February 2015, Electrician)

23. What type of sheet metal is suitable for making food containers?

(a) Aluminium sheet
(b) Sheet steel
(c) Tin sheet
(d) Galvanized iron sheet

(I.T.I. February 2015, Electrician)

24. The thickness of the metal used in sheet metal works above 5 mm is called as ......

(a) Sheets (b) Plates
(c) Strips (d) Foils

(I.T.I. February 2015, Electrician)

25. What is the name of tool shown in figure?

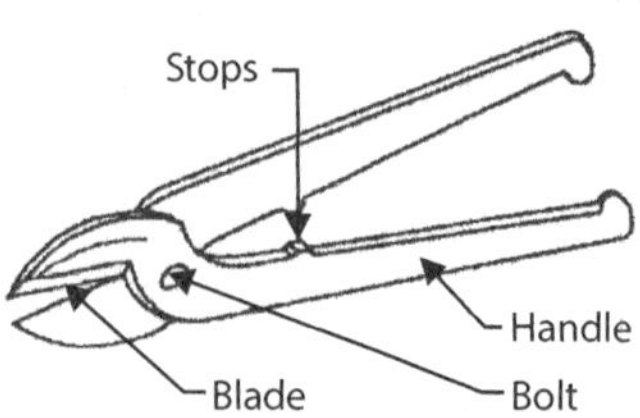

(a) Side cutting plier
(b) Combination plier
(c) Bent snip
(d) Straight snip

(I.T.I. February 2014, Electrician)

26. Which tool is used for cutting thin sheets of metal ?

(a) Cutting plier
(b) Straight snip
(c) Flat nose plier
(d) chisel

(I.T.I. February 2016, Electrician, Wireman)

27. Which type of sheet metal is mostly used for manufacturing house hold utensils ?

(a) Tin plater
(b) G.I. sheet
(c) Copper sheet
(d) Aluminium sheet

(I.T.I. February 2016, Electrician, Wireman)

28. The operation of bending a sheet of metal along the circular axis is known as ...

(a) Plunging (b) Notching
(c) Slitting (d) Forming

(I.T.I. February 2017, Electrician, Electroplator, Wireman)

29. What is the purpose of hem folding in sheet metal work?

(a) Prevents the sheet from damage
(b) Helps for bending sheet metal easily
(c) Prevents burrs forming after cutting
(d) Helps for forming a square box

(I.T.I. August 2019 Electrician, Electroplator, Wireman, Lift & Escalator Mechanic)

30. Which type of notch is used for making a metal tray with 90° bend and an inside flange?

(a) Square notch
(b) Slant notch
(c) "V" notch
(d) Straight notch

(I.T.I. August 2019 Electrician, Elactroplator, Wireman, Lift & Escalator Mechanic)

KEY TO MULTIPLE CHOICE QUESTIONS

Cutting Tools and Fasteners

1. (d)	**2.** (a)	**3.** (d)	**4.** (a)	**5.** (a)	**6.** (d)	**7.** (d)	**8.** (b)	**9.** (b)	**10.** (c)
11. (a)	**12.** (b)	**13.** (a)	**14.** (c)	**15.** (b)	**16.** (a)	**17.** (b)	**18.** (c)	**19.** (d)	**20.** (c)
21. (a)	**22.** (b)	**23.** (c)	**24.** (d)	**25.** (c)	**26.** (b)	**27.** (d)	**28.** (a)	**29.** (d)	**30.** (c)
31. (a)	**32.** (a)	**33.** (d)	**34.** (b)						

Carpentry

1. (c)	**2.** (b)	**3.** (b)	**4.** (a)	**5.** (c)	**6.** (b)	**7.** (c)	**8.** (a)	**9.** (b)	**10.** (b)
11. (d)	**12.** (b)	**13.** (b)	**14.** (c)	**15.** (b)	**16.** (d)	**17.** (b)	**18.** (a)	**19.** (a)	**20.** (b)
21. (c)	**22.** (b)	**23.** (d)							

Sheet Metal Work

1. (b)	**2.** (d)	**3.** (a)	**4.** (c)	**5.** (b)	**6.** (d)	**7.** (a)	**8.** (a)	**9.** (c)	**10.** (a)
11. (b)	**12.** (c)	**13.** (d)	**14.** (d)	**15.** (d)	**16.** (b)	**17.** (c)	**18.** (d)	**19.** (c)	**20.** (a)
21. (b)	**22.** (b)	**23.** (c)	**24.** (b)	**25.** (d)	**26.** (b)	**27.** (d)	**28.** (d)	**29.** (a)	**30.** (c)

Module 3

Wires, Joints and Soldering

MULTIPLE CHOICE QUESTIONS

1. A solid substance or insulated conductor subjected tensile stress with or without screen is called as ....
   (a) cable (b) core
   (c) conductor (d) wire
   (I.T.I. February 2015, Electrician)
2. A conductor should have ....
   (a) low specific resistance
   (b) medium specific resistance
   (c) high specific resistance
   (d) any specific resistance
3. Which one of the following conductors is used in heating elements?
   (a) Tungsten (b) Carbon
   (c) Copper (d) Nichrome
4. When compared to copper, the aluminium has .... of conductivity.
   (a) 60% (b) 60.2%
   (c) 60.4% (d) 60.6%
   (I.T.I. February 2015, Wireman)
5. Which one of the following is NOT a conductor?
   (a) Silver (b) Copper
   (c) Aluminium (d) PVC
   (I.T.I. February 2015, Wireman)
6. The material that contains free electron is ....
   (a) Silver (b) PVC
   (c) Mica (d) Rubber
   (I.T.I. February 2015, Wireman)
7. Which among the following materials is used for making split rings?
   (a) Wrought iron (b) Copper
   (c) Cast iron (d) Aluminium
8. Which material is NOT a good conductor?
   (a) Aluminium (b) Copper
   (c) Porcelain (d) Steel
   (I.T.I. August 2015, Electrician)
   (I.T.I. February 2014, Wireman)
9. Which of the following is a good conductor?
   (a) China clay (b) Porcelain
   (c) Graphite (d) Glass
10. The standard wire gauge (swg) can measure sizes from ....
    (a) 0 to 32 (b) 0 to 34
    (c) 0. to 35 (d) 0. to 36
    (I.T.I. February 2015, Wireman)
11. What is the advantage of stranded conductors over solid conductors?
    (a) more flexible
    (b) more area of cross section
    (c) more resistance
    (d) more weight
    (I.T.I. August 2014, Electrician)
12. What is the use of Eureka wire?
    (a) Rheostat
    (b) Heating element
    (c) Earth wire
    (d) Winding wire

13. The good conductor of heat and electricity ....
(a) stainless steel wire
(b) iron wire
(c) copper wire
(d) brass wire(I.T.I. February 2014, Fitter)

14. Normally metals are good conductor of electricity. Which is the good conductor in terms of conductivity?
(a) Copper (b) Aluminium
(c) Silver (d) Brass
(I.T.I. Mock Test 2013, Electrician)

15. We find, for a required current carrying capacity, aluminium conductor is larger in cross sectional area than copper. The reason for this is ......
(a) aluminium gets heated up more than copper
(b) resistivity of aluminium is greater than copper
(c) aluminium is mechanically weaker than copper
(d) aluminium is cheaper than copper
(I.T.I. January 2014, Wireman)

16. Which of the following is NOT an insulator?
(a) Zinc (b) Glass
(c) Rubber (d) Plastic
(I.T.I. February 2015, Wireman)

17. What is the insulating material used for heater base?
(a) Asbestos (b) Mica
(c) Porcelain (d) Bakelite
(I.T.I. Mock Test 2013, Electrician)

18. Insulating materials are classified according to their withstanding temperature. Which class of material can withstand temperature upto 225°C?
(a) "Y" class (b) "A" class
(c) "B" class (d) "C" class
(I.T.I. January 2014, Wireman)

19. Which insulating material is having high dielectric strength?
(a) Porcelain (b) Fibre
(c) Mica (d) Glass
(I.T.I. January 2014, Wireman)

20. Which material is NOT an insulating material?
(a) Wood (b) Carbon
(c) Asbestos (d) Bakelite
(I.T.I. February 2015, Wireman)

21. The potential difference that causes the break down of insulation is called the ......
(a) breakdown voltage
(b) dielectric strength
(c) insulation resistance
(d) conductivity

22. Which is the high temperature withstanding insulating material?
(a) Bakelite (b) Wood
(c) Asbestos (d) Mica
(I.T.I. February 2014, Wireman)

23. Which type of insulating tape is having high dielectric strength?
(a) Rubber tape
(b) Friction tape
(c) PVC tape
(d) Varnished cambric tape
(I.T.I. February 2015, Wireman)

24. Why soldering is done on joints?
(a) To improve tensile strength
(b) To improve resistivity
(c) To improve ductility
(d) To improve conductivity
(I.T.I. February 2014, Wireman)

25. What purpose is served by flux during soldering?
(a) It cleans the surface to be soldered
(b) It lowers the melting point of the solder
(c) It makes the flow of solder in the joint easy
(d) It does all the above

26. Name the chemical which is used as soldering flux for the electricians solder?
(a) Zinc chloride
(b) Hydrochloric acid
(c) Sal ammonia rosin
(d) Rosin
(I.T.I. August 2014, Electrician)

27. A substance used to dissolve oxides on the surface of conductor using the soldering process is called as ....

(a) solder (b) soldering bit
(c) bonding agent (d) soldering flux

(I.T.I. February 2015, Electrician)

28. The solder wire used in electronic soldering melts at ......

(a) 200°C (b) 400°C
(c) 600°C (d) 800°C

29. For soldering aluminium conductors, Alca-P solder is used. What are the material compositions of Alca-P solder?

(a) Lead 50%, Tin 30%, Zinc 10% and Cadmium 10%
(b) Lead 30%, Tin 50% and Zinc 20%
(c) Lead 51%, Tin 31%, Zinc 9% and Cadmium 9%
(d) Lead 63%, Tin 35% and Zinc 2%

30. Which is the application of the tinman solder?

(a) Heavy duty solder
(b) Tinning
(c) Fine soldering
(d) Electrical joint soldering

31. The solders used for soft soldering is .....

(a) Tinman's solder
(b) Electrician's solder
(c) Fine solder
(d) Plumber's solder

(I.T.I. February 2015, Wireman)

32. A Britannia joint is used in:

(a) O.H. lines
(b) underground cables
(c) conduit wiring
(d) power wiring

33. Which type of soldering flux is suitable for soldering zinc metals?

(a) Zinc chloride
(b) Hydrochloric acid
(c) Rosin
(d) Tallow

(I.T.I. August 2015, Electrician)

34. Which types of flux is used for coarse solder?

(a) Zinc chloride
(b) Hydrochloric acid (Dilute)
(c) Resin
(d) Sal ammonia resin

(I.T.I. August 2015, Electrician, Wireman)

35. What is the suitable wattage of soldering iron to solder resistors on a P.C.B.?

(a) 240 W (b) 150 W
(c) 35 W (d) 25 W

(I.T.I. February 2016, Electronic Mechanic)

36. Which materials is the good conductor of current from the following?

(a) Rubber (b) Copper
(c) Ebonite (d) Bakelite

(I.T.I. February 2016, Electrician, Wireman)

37. If the cable contains one or more cores each formed as a group of wires, having small diameter is known as ....

(a) cable
(b) wire
(c) flexible cable
(d) armoured cable

(I.T.I. August 2015, Wireman)

38. The main property of insulating material is ....

(a) quick absorption of moisture
(b) high specific resistance
(c) high dielectric strength
(d) high temperature coefficient of expansion

(I.T.I. February 2016, Electronics Mech., Electrician)

39. The insulating material used in between commutator segment is ....

(a) ebonite (b) mica
(c) empire (d) glass

(I.T.I. February 2015, Electrician)

40. Which one of the following insulating material belongs to class "E" insulation?

(a) Cotton
(b) Silicon elastomer

(c) Leatheroid paper
(d) Glass fibre

(I.T.I. February 2016, Electrician)

41. The full form of VIR is ....
(a) Vulcanized Indian Rubber
(b) Vulcanized Insulation Rubber
(c) Vulcanized Insulation Reinforced
(d) Vulcanized Insulator Reinforced

(I.T.I. February 2016, Wireman)

42. The main draw back of using paper as an insulating material is...
(a) Poor dielectric strength
(b) Hygroscopic
(c) Low insulation resistivity
(d) High capacitance

(I.T.I. February 2017, Electrician)

43. Which material has extremely large resistance?
(a) Conductor
(b) Intrinsic semi conductor
(c) Insulator
(d) Extrinsic semi conductor

(I.T.I. February 2017, Electrician, Electroplator, Wireman)

44. The thickness of insulation provided on the conductor in cable depends upon which of the following factors?
(a) Operating voltage
(b) Resistivity of conductor
(c) Current to be carried
(d) Power factor

(I.T.I. February, 2017 Electrician)

45. The insulating material for cables should...
(a) Be acid proof
(b) Be non-inflammable
(c) Be non-hygroscopic
(d) have all above properties

(I.T.I. August 2016, Electrician)

46. Which of the following insulation is used in cables?
(a) Varnished cambric
(b) Rubber
(c) Paper
(d) All of these

(I.T.I. August 2016, Electrician)

47. Vulcanised India rubber has dielectric strength of...
(a) 20 kV/mm (b) 17 kV/mm
(c) 15 kV/mm (d) 11 kV/mm

(I.T.I. February 2017, Wireman)

48. Soldering filler material eutectic alloy (63/37) melts at ...
(a) 100°C to 183°C
(b) 183°C
(c) 183°C C to 190°C
(d) 185°C to 215°C

(I.T.I. February 2017, Wireman)

49. In soldering the melting point of filler metal is ...
(a) $< 420°$ F (b) $> 420°$ F
(c) $< 520°$ F (d) $> 520°$ F

(I.T.I. August 2019, Electrician)

50. Fluxes for soft solder are available as ...
(a) Water soluble fluxes
(b) No clean fluxes
(c) Rosin fluxes
(d) All of these

(I.T.I. February 2017, Wireman)

51. The composition of Electrician solder is ...
(a) Lead 37%; Tin 63%
(b) Lead 50%; Tin 50%
(c) Lead 63%; Tin 37%
(d) Lead 70%; Tin 30%

(I.T.I. February 2017, Electrician, Electroplator, Wireman)

52. PVC stands for ...
(a) Polyvinyl chloride
(b) Post varnish conductor
(c) Pressed and varnished cloth
(d) Positive voltage conductor

(I.T.I. August 2016, Electrician)

53. One way to short a device is:
(a) With a cold-solder joint
(b) With a solder bridge

(c) By disconnecting it

(d) By opening it

(I.T.I. February 2017, Electrician)

54. Which type of wire joint is found in the junction box?

(a) Aerial tap joint (b) Plain tap joint

(c) Married joint (d) Rat tail joint

(I.T.I. August 2019 Electrician, Electroplator, Wireman, Lift & Escalator Mechanic)

55. Which type of the wire joint is used to extend the length of conductor in OH line?

(a) Married joint

(b) Scarfed joint

(c) Western union joint

(d) Britannia "T" joint

(I.T.I. August 2019 Electrician, Electroplator, Wireman, Lift & Escalator Mechanic)

56. What is the use of Britannia "T" joint?

(a) Extending the length of the lines

(b) Inside and outside wiring installation

(c) Tapping the service connection from overhead lines

(d) Mechanical stress not required on conductor

(I.T.I. August 2019 Electrician, Electroplator, Wireman, Lift & Escalator Mechanic)

57. What is the name of the joint?

(a) Britannia straight joint

(b) Scarfed joint

(c) Western union joint

(d) Married joint

(I.T.I. August 2019 Electrician, Electroplator, Wireman, Lift & Escalator Mechanic)

58. Which insulating material is used as hot pouring compound for making joints in underground cable?

(a) Polyamine hardner

(b) Cast resin compound

(c) Epoxy cast resin compound

(d) Bituminous compound

(I.T.I. August 2019 Electrician, Electroplator, Wireman, Lift & Escalator Mechanic)

59. Empire tape is ....

(a) varnished cambric

(b) vulcanised rubber

(c) impregnated paper

(d) None of these

(I.T.I. August 2019 Electrician)

60. Materials that allow electrons to flow through them are known as ......

(a) Insulators (b) Conductors

(c) Electrolytes (d) Bases

(I.T.I. August 2019 Electrician)

61. Which of the following is bad conductor of electricity?

(a) Distilled water (b) Alcohol

(c) Salt water (d) Tap water

(I.T.I. August 2019 Electrician)

62. Copper as conductance for cable is used as .....

(a) Annealed

(b) Hardness and Tempering

(c) Hard drawn

(d) Alloy with chromium

(I.T.I. August 2019 Electrician)

63. The surge resistance of cable is ....

(a) 5 ohms (b) 20 ohms

(c) 50 ohms (d) 100 ohms

(I.T.I. August 2019 Electrician)

64. What does the number 1.40 represent if a stranded conductor is designated as 10 sq. mm cable of size 7/1.40?

(a) Diameter of each conductor

(b) Radius of one conductor

(c) Diameter of all conductors

(d) Area of cross section

(I.T.I. August 2019 Electrician, Electroplator, Wireman, Lift & Escalator Mechanic)

65. What is the value of electrical conductivity of Aluminium conductor?

(a) 35 mho/m (b) 56 mho/m
(c) 61 mho/m (d) 28 mho/m

(I.T.I. August 2019 Electrician, Electroplator, Wireman, Lift & Escalator Mechanic)

66. Why the soldering iron must be kept into a stand that is not in use while soldering?

(a) It prevents burns and fire
(b) To control the excessive heat
(c) To save the time of soldering process
(d) To save the operator from electric shock

(I.T.I. August 2019 Electrician, Electroplator, Wireman, Lift & Escalator Mechanic)

67. Which type of soldering method is used for servicing and repairing work?

(a) Dip soldering
(b) Soldering with soldering gun
(c) Soldering with a flame
(d) Soldering with a soldering iron

(I.T.I. August 2019 Electrician, Electroplator, Wireman, Lift & Escalator Mechanic)

68. What is the use of dip soldering method?

(a) Soldering miniature components on P.C.B.
(b) Piping and cable soldering work
(c) Soft soldering
(d) Soldering sensitive electric components

(I.T.I. August 2019 Electrician, Electroplator, Wireman, Lift & Escalator Mechanic)

KEY TO MULTIPLE CHOICE QUESTIONS

1. (d)	**2.** (a)	**3.** (d)	**4.** (d)	**5.** (d)	**6.** (a)	**7.** (b)	**8.** (c)	**9.** (c)	**10.** (d)
11. (a)	**12.** (a)	**13.** (c)	**14.** (c)	**15.** (b)	**16.** (a)	**17.** (c)	**18.** (d)	**19.** (c)	**20.** (b)
21. (a)	**22.** (c)	**23.** (c)	**24.** (d)	**25.** (d)	**26.** (d)	**27.** (d)	**28.** (a)	**29.** (c)	**30.** (a)
31. (b)	**32.** (a)	**33.** (b)	**34.** (d)	**35.** (d)	**36.** (b)	**37.** (c)	**38.** (b)	**39.** (b)	**40.** (c)
41. (a)	**42.** (b)	**43.** (c)	**44.** (a)	**45.** (d)	**46.** (d)	**47.** (c)	**48.** (b)	**49.** (b)	**50.** (d)
51. (a)	**52.** (a)	**53.** (b)	**54.** (d)	**55.** (c)	**56.** (c)	**57.** (d)	**58.** (d)	**59.** (a)	**60.** (b)
61. (b)	**62.** (a)	**63.** (c)	**64.** (a)	**65.** (a)	**66.** (a)	**67.** (b)	**68.** (a)		

Module 4

Basic Electrical Practice

Multiple Choice Questions

Fundamentals of Electricity

Choose the correct answer :

1. How many numbers of electrons in a copper atom?
   (a) 29 (b) 39
   (c) 49 (d) 59
   (I.T.I. February 2019, Electrician, Elactroplator, Wireman)
2. One joule per second is ....
   (a) one volt (b) one ampere
   (c) one watt (d) one ohm
   (I.T.I. February 2015, Electrician)
3. What is the unit of power?
   (a) kilo volt (b) watt hour
   (c) kilo watt (d) kilo watt hour
   (I.T.I. February 2019, Electrician, Electroplator, Wireman)
4. What is the value of resistivity of copper conductor?
   (a) 1.72 micro ohm metre
   (b) 1.73 ohm/cm^3
   (c) 1.72 micro ohm/m^3
   (d) 1.72×10^{-8} micro ohm/cm^3
   (I.T.I. February 2015, Electrician)
5. What is the S.I. unit of specific resistance?
   (a) ohm/cm (b) ohm metre
   (c) micro ohm/cm (d) ohm/metre
   (I.T.I. February 2014, Wireman)
6. The increase in resistance per ohm per degree centrigrade rise in temperature is ....
   (a) positive temperature coefficient
   (b) difference in temperature
   (c) linear coefficient
   (d) negative temperature coefficient
   (I.T.I. February 2019, Electrician, Electroplator, Wireman)
7. Which is the example for negative temperature coefficient substance?
   (a) Copper (b) Nichrome
   (c) Aluminium (d) Mica
   (I.T.I. February 2015, Electrician)
8. As per Ohm's law statement in the circuit, current is directly proportional to ....
   (a) voltage (b) resistance
   (c) temperature
   (d) specific resistance
   (I.T.I. February 2019, Electrician, Electroplator, Wireman)
9. Which law states that, in each closed circuit, the sum of all voltage drops are equal to zero?
   (a) Ohm's law
   (b) Kirchhoff's first law
   (c) Kirchhoff's second law
   (d) Coulomb's law
   (I.T.I. August 2019, Electrician, Electroplator, Wireman)

10. The work or energy in an electrical circuit is calculated by ....

(a) $W = V \,.\, C$ joules

(b) $W = I \,.\, R \,.\, C$ joules

(c) $W = V \,.\, I \,.\, E$ joules

(d) $W = I^2 \,.\, R \,.\, T$ joules

(I.T.I. February 2015, Wireman)

11. An appliance has a resistance of 100 ohms, takes 2 amperes. What is the power consumed by the appliance?

(a) 400 watts (b) 300 watts

(c) 450 watts (d) 350 watts

(I.T.I. February 2019, Electrician, Electroplator, Wireman)

12. The given Wheatstone bridge is balanced when P = 100 W; Q = 1 kW and S = 120 W. Calculate the value of the unknown resistance.

(a) 1.1 kW (b) 1.2 kW

(c) 1.3 kW (d) 1.5 kW

(I.T.I. February 2015, Electrician)

13. The voltage 'V' is determined through 'R' and 'I' with reference from the circuit shown is ....

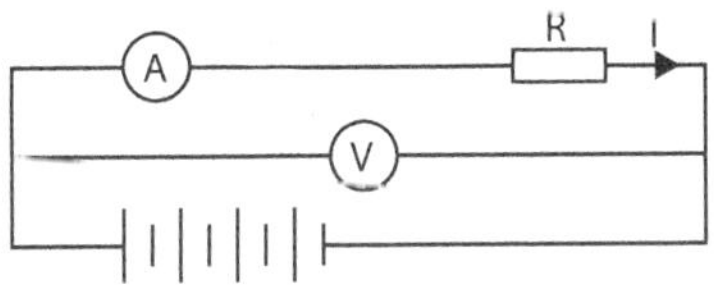

(a) $V = I \times R$ (b) $V = I/R$

(c) $V = I^2R$ (d) $V = R/I$

(I.T.I. February 2015, Electrician)

14. From the given figure calculate the resistance measured between x and y.

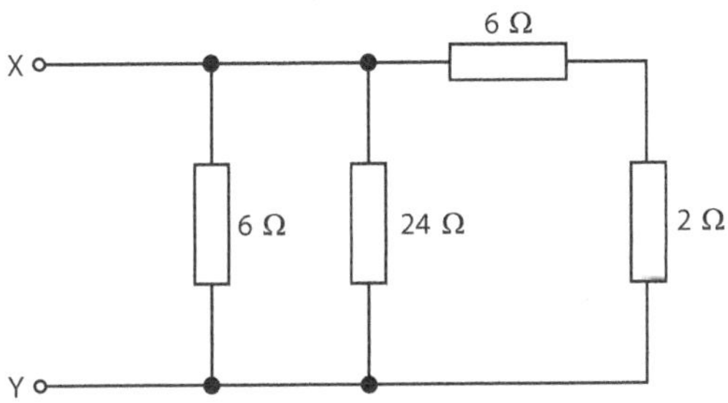

(a) 2 W (b) 3 W

(c) 8 W (d) 38 W

(I.T.I. February 2015, Electrician)

15. Calculate the hot resistance value of the lamp 200 watts/250 volts?

(a) 6250 ohms (b) 625 ohms

(c) 3125 ohms (d) 312.5 ohms

(I.T.I. February 2014, Electrician)

16. Two lamps rated 200 W/240 V are connected in series and combined set is connected across 240 V, AC supply mains. The total power consumed in the circuited is ....

(a) 400 W (b) 300 W

(c) 200 W (d) 100 W

(I.T.I. February 2015, Electrician)

17. What will happen to the indication of the voltmeter when the switch is closed?

(a) Increases (b) Decreases

(c) Zero (d) No change

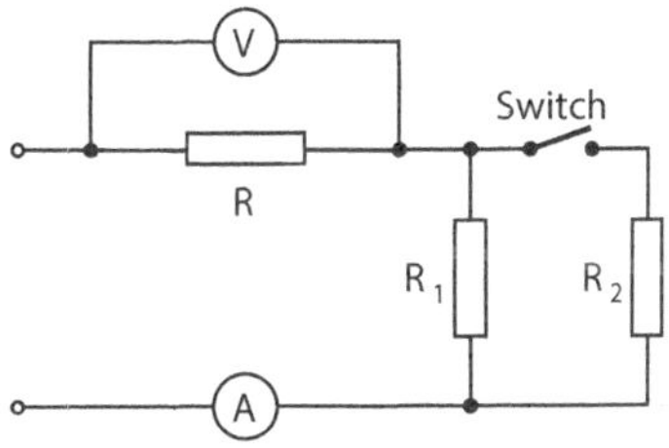

(I.T.I. February 2014, Electrician)

18. What is the value of medium resistance based on its ohmic value?

(a) One ohm to 1000 ohms

(b) Above one ohm upto 10 kilo ohms

(c) Above one ohm upto 100 kilo ohms

(d) Above 100 kilo ohms

(I.T.I. February 2015, Electrician)

19. A 12 ohm resistor connected across a 60 V, DC source. Calculate the current flows through the resistor.

(a) 4 amperes (b) 5 amperes

(c) 10 amperes (d) 12 amperes

(I.T.I. February 2014, Electrician)

20. What is the commercial unit of electrical energy?

(a) watt hour (b) kilo volt ampere
(c) kilo watt hour (d) volt ampere

(I.T.I. February 2019, Electrician, Electroplator, Wireman)

21. The specific resistance of a substance depends on ....

(a) its length
(b) its cross sectional area
(c) its kind of material
(d) all the above stated factors

22. The amount of current flowing through a resistance depends on

(a) applied voltage only
(b) ohmic value of the resistor only
(c) applied voltage and ohmic value of the resistor
(d) none of the above

23. What will be the voltage drop across a resistor of 60 ohms if the current flowing through it is 0.1 ampere —

(a) 60 V (b) 6 V
(c) 1 V (d) 600 V

24. What will be the total resistance of the three resistances R_1, R_2 and R_3 which are connected as shown in the given figure —

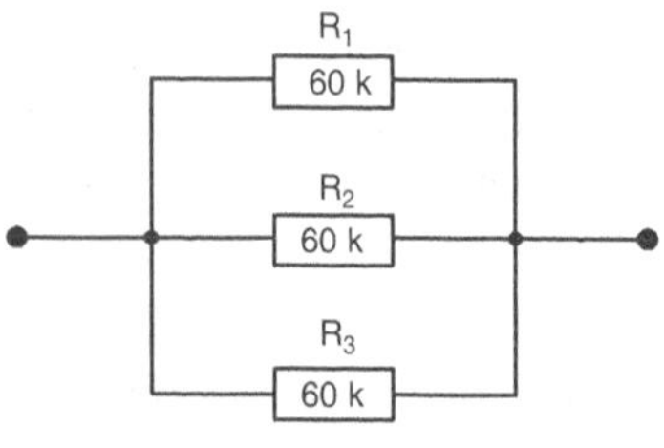

(a) 20 kW (b) 30 kW
(c) 60 kW (d) 180 kW

25. If three resistances of 100 W , 200 W and 300 W are connected in parallel and the applied voltage is V then—

(a) the current flowing through the resistors is the same, whereas, voltages developed across them will be different.
(b) the current flowing through the resistors and the voltages developed across them will be the same
(c) the voltages across the resistors will be the same, whereas, currents flowing through them will be different
(d) the current flowing through the resistors and the voltages developed across them will be different

26. What will be the amount of curret (I) flowing through the given circuit —

(a) 6 A (b) 3 A
(c) 1 A (d) 0 A

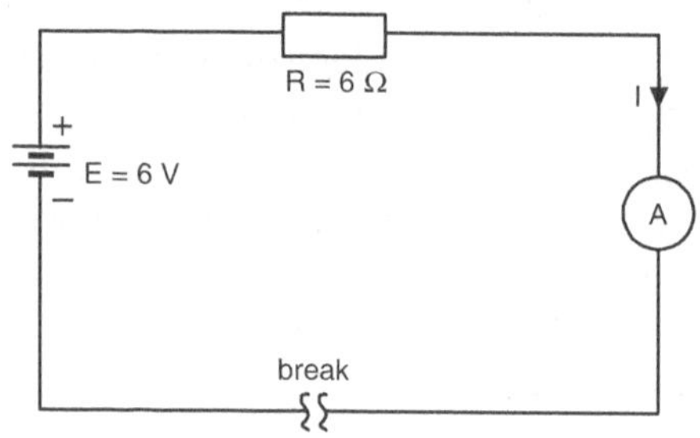

27. If in the given circuit the value of R3 is doubled then for a given voltage Vs ......

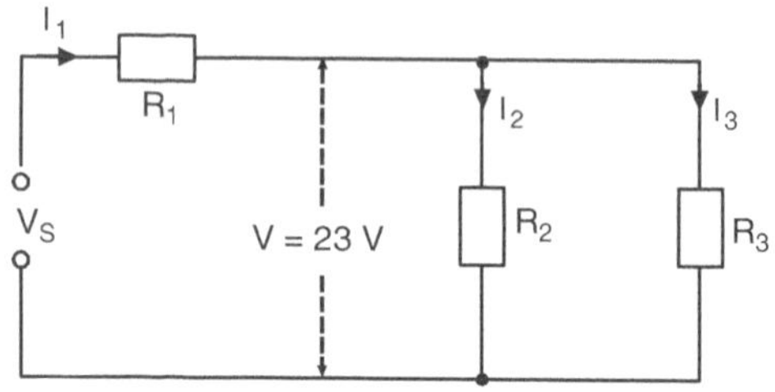

(a) I_3 would be halved.
(b) I_2 and I_3 would be halved.
(c) I_2 would be unaffected.
(d) All currents would be affected.

28. It was experimentally found by James Precott Joule that the heat produced in a current carrying conductor is directly proportional to :

(a) the square of current
(b) the square of resistance
(c) the current
(d) the square of the applied voltage.

29. If a 200 watts lamp is connected across a 250 volt source then what will be the circuit current ?

(a) 4 A (b) 2 A
(c) 0.8 A (d) 1 A

30. The monthly power consumption of 10 lamps of 100 watts each used 8 hours a day will be :

(a) 240 units (b) 245 units
(c) 225 units (d) 250 units

31. What is the magnitude of current in branch OX of the given figure ?

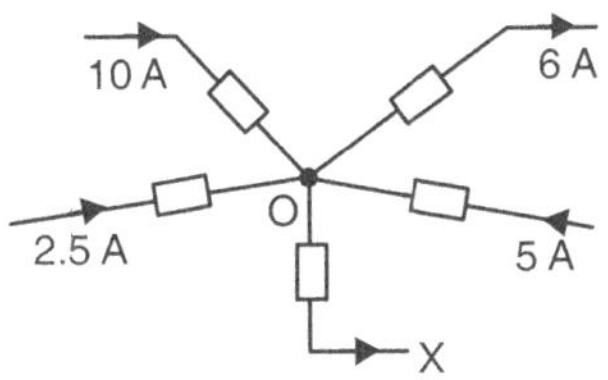

(a) 1.5 A (b) 6.5 A
(c) 10.5 A (d) 11.5 A

(I.T.I. February 2015, Electrician)

32. How many electrons are in the outer most orbit of a semi conductor substance?

(a) 5 only (b) 4 only
(c) 3 only (d) 1 only

(I.T.I. February 2015, Electrician)

33. The number of electrons in a silicon atom is ....

(a) 7 (b) 14
(c) 29 (d) 32

(I.T.I. August 2015, Electrician, Wireman)

34. How many electrons are in aluminium atom?

(a) 13 (b) 18
(c) 29 (d) 43

(I.T.I. February 2016, Electrician, Wireman)

35. How many protons in germanium atom?

(a) 14 (b) 29
(c) 32 (d) 34

(I.T.I. February 2016, Electrician)

36. The unit of resistance is ....

(a) mho (b) volt
(c) ohm (d) apmpere

(I.T.I. February 2016, Electrician, Wireman)

37. The potential difference is measured by ....

(a) ammeter (b) ohmmeter
(c) voltmeter (d) wattmeter

(I.T.I. February 2016 Electrician, Wireman)

38. The unit of the quantity of electricity is ....

(a) ampere/second (b) coulomb
(c) mho (d) watt-second

(I.T.I. February 2016, Electrician, Wireman)

39. The formula to find the specific resistance (ρ) where length (l), cross sectional area (a), and Resistance (R) is ....

(a) $\rho = \frac{l}{Ra}$ ohm/cm^3

(b) $\rho = \frac{Ra}{l}$ ohm/cm^3

(c) $\rho = \frac{R}{al}$ ohm/cm^3

(d) $\rho = \frac{al}{R}$ ohm/cm^3

(I.T.I. February 2016, Electrician, Wireman)

40. If the diameter of a metal wire of a given length is doubled, then its resistance will ....

(a) be doubled (b) be halved
(c) remain the same (d) be 1/4th time

(I.T.I. August 2015, Electrician)

41. Which one of the following has negative temperature coefficient?

(a) Brass (b) Mercury
(c) Electrolytes (d) Silver

(I.T.I. February 2015, Mechanic Consumer Electronics)

42. The formula to find the temperature co-efficient of resistance (R_t) of a material is ....

(a) $R_t = R_o - \alpha t$ (b) $R_t = R_o (1 + \alpha t)$

(c) $R_t = \frac{R_o + a}{t}$ (d) $R_t = (1 + Ro)\,\alpha t$

(I.T.I. August 2015, Electrician)

43. The heat generated is calculated using Joules law is ....

(a) $H = 0.24\ IRt$ calories
(b) $H = 0.24\ I^2Rt$ calories
(c) $H = 0.24\ I^2R$ calories
(d) $H = 0.24\ I^2t$ calories

(I.T.I. February 2016, Wireman)

44. In any electric network, the sum of the incoming current at each junction is equal to the sum of outgoing current. What is the name of the law?
(a) Ohm's law
(b) Kirchhoff's second law
(c) Laws of resistance
(d) Kirchhoff's first law

(I.T.I. August 2015, Electrician, Wireman)

45. In the Wheatstone bridge network, ABCD is balanced where P = 100 ohm, Q = 1000 ohm and S = 130 ohms. Determine the value of R_x.

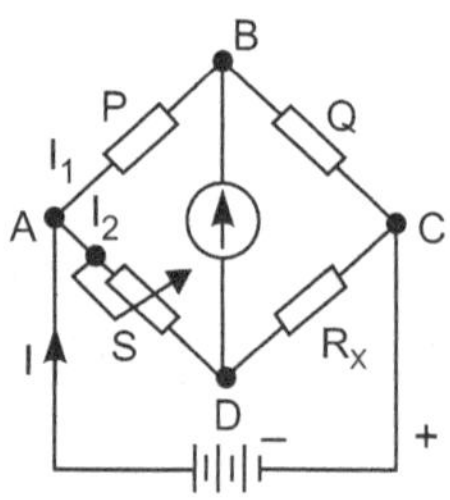

(a) 1300 ohm (b) 10000 ohm
(c) 13000 ohm (d) 15000 ohm

(I.T.I. February 2016, Electrician, Wireman)

46. Under balanced condition, the current flowing through the detector is equal to ...
(a) One ampere
(b) Zero ampere
(c) Sum of the currents flowing in the adjacent arms
(d) Difference between the currents flowing in the adjacent

(I.T.I. February 2019, Electrician)

47. The accuracy in a bridge measurement depends on ...
(a) Sensitivity of detector
(b) Applied voltage
(c) Accuracy of indicator
(d) Sensitivity of detector or Applied Voltage

(I.T.I. February 2019, Electrician)

48. A null type of bridge with D.C. excitation is commonly known as ...
(a) Anderson bridge
(b) Wheatstone bridge
(c) Wein bridge
(d) Schering bridge

(I.T.I. February 2019, Electrician)

49. The condition for the validity under ohm's law is that the ..
(a) Temperature should remain constant
(b) Current should be proportional to Voltage
(c) Resistance should be wire wound type
(d) All of these

(I.T.I. February 2019, Wireman)

50. The relation between voltage, current and resistance in a closed circuit is given as ...
(a) $I = V \times R$ (b) $I = \frac{V}{R}$
(c) $R = \frac{I^2}{V}$ (d) None of these

(I.T.I. February 2019, Wireman)

51. The ratio of voltage and electric current in closed circuit ...
(a) Remains constant
(b) Varies
(c) Increases
(d) Falls

(I.T.I. February 2019, Wireman)

52. If in a closed circuit, the voltage is reduced to half and resistance is doubled, the current becomes ...
(a) Four times (b) double
(c) A quarter (d) half

(I.T.I. February 2019, Wireman)

53. The curve representing ohm's law is ...
(a) Sine function (b) Linear
(c) A parabola (d) A hyperbola

(I.T.I. February 2019, Wireman)

54. A linear resistor are one's which obey...
(a) Ampere's law (b) Lenz's law
(c) Ohm's law (d) Kirchhoff's law

(I.T.I. February 2017, Electrician, Electroplator, Wireman)

55. Which is the lightest particle in an atom?
(a) Proton
(b) Neutron
(c) Electron
(d) All are of equal weight

56. Flow of how many electrons per second will produce current of one ampere?
(a) 6.25 × 1018 (b) 6.25 × 1019
(c) 1.602 × 1019 (d) 6.25 × 108
(I.T.I. February 2019, Wireman)

57. In gases, the flow of current is due to ...
(a) Electrons only
(b) Positive and negative ions
(c) Electrons and positive ions
(d) Electrons, positive ions and negative ions
(I.T.I. February 2019, Wireman)

58. Silicon has ...... valence electrons.
(a) 4 (b) 3
(c) 5 (d) 6
(I.T.I. February 2019, Electrician, Electroplator, Wireman)

59. When a normal atom loses an electron, the atom ...
(a) Becomes a positive ion
(b) Becomes a negative ion
(c) Becomes electrically neutral
(d) is then forced to move about
(I.T.I. February 2019, Electrician, Electroplator, Wireman)

60. The flow of electric current in a conductor is due to flow of ...
(a) Electrons
(b) Protons
(c) Electrons and ions
(d) Charged particles
(I.T.I. February 2019, Wireman)

61. An electric current is the ......
(a) Random movement of electrons in a conductor
(b) Movement of free electrons predominantly in one direction
(c) Pressure difference between two poles
(d) The power that causes drift of electrons
(I.T.I. February 2019, Wireman)

62. Which of the following is an active element in a circuit?
(a) Current source
(b) Resistance
(c) Inductance, capacitance
(d) Capacitance
(I.T.I. February 2019, Wireman)

63. An ideal voltage source should have ...
(a) Zero source resistance
(b) Infinite source resistance
(c) Terminal voltage in proportion to current
(d) Terminal voltage in proportion to load
(I.T.I. February 2019, Wireman)

64. For a voltage source ...
(a) Terminal voltage is equal to source e.m.f
(b) Terminal voltage cannot exceed source e.m.f
(c) Terminal voltage is always lower than source e.m.f
(d) Terminal voltage is higher than the source e.m.f
(I.T.I. February 2019, Wireman)

65. The intornal resistance of practical voltage source is considered to be connected in ...
(a) Series
(b) Parallel
(c) Either in series or parallel
(d) None of these
(I.T.I. February 2017, Electrician, Elatroplator, Wireman)

66. If three resistances each of R ohms are connected in parallel, then value of effective resistance in ohms will be ...
(a) 3 R (b) R
(c) $\frac{R}{3}$ (d) R^3
(I.T.I. August 2016, Wireman)

67. Which quantity consists of a unit 1 kWh?
(a) Energy (b) Power
(c) Time (d) Charge
(I.T.I. February 2017, Electrician, Electroplator, Wireman)

68. Intrinsic semiconductor material is characterised by a valence shell of .... electrons.
(a) 1 (b) 2
(c) 4 (d) 6
(I.T.I. August 2016, Electrician, Elatroplator, Wireman)

69. One eV is equal to ...
(a) 6.02×10^{23} joules
(b) 1.6×10^{-19} joules
(c) 6.25×10^{18} joules
(d) 1.66×10^{-24} joules
(I.T.I. August 2016, Electrician, Electroplator, Wireman)

70. Resistance of human body is about ...
(a) 10 Ω (b) 1 kΩ
(c) 100 Ω (d) 10 kΩ
(I.T.I. February 2017, Electrician, Electroplator, Wireman)

71. The effect of connecting an additional parallel load to an electrical supply source is to increase the ...
(a) Resistance of the load
(b) Voltage of the source
(c) Potential difference across the load
(d) Current taken from the source
(I.T.I. February 2017, Electrician, Electroplator, Wireman)

72. The resistance of the material depends on ...
(a) Length
(b) Area of cross section
(c) Material
(d) All of these
(I.T.I. August 2016, Electrician, Electroplator, Wireman)

73. Which of the following lamps has least resistance?
(a) 200 W, 220 V (b) 100 W, 220 V
(c) 60 W, 220 V (d) 15 W, 220 V
(I.T.I. February 2017, Electrician, Electroplator, Wireman)

74. Which resistor is physically largest in size?
(a) 100 Ω, 10 W (b) 10 Ω, 50 W
(c) 1 MΩ, $\frac{1}{2}$ W (d) 1 kΩ, 1 W
(I.T.I. February 2017, Electrician, Electroplator, Wireman)

Hint: Physical size depends upon power rating.

75. For the same voltage, the ratio of resistance of 25 W and 100 W lamp will be ....
(a) 1 : 2 (b) 2 : 1
(c) 4 : 1 (d) 1 : 4
(I.T.I. Februray 2017, Electrician, Electroplator, Wireman)

76. Which of the following materials has least resistivity?
(a) Lead (b) Copper
(c) Zinc (d) Mercury
(I.T.I. Februray 2017, Electrician, Electroplator, Wireman)

77. Which of the following materials possesses the least specific resistance?
(a) Aluminium (b) Copper
(c) Silver (d) Iron
(I.T.I. February 2019, Wireman)

78. The resistance of wire varies inversely as ...
(a) Area of cross section
(b) Length
(c) Resistivity
(d) Temperature
(I.T.I. February 2019, Wireman)

79. Kirchhoff's law are valid for ...
(a) Linear circuits only
(b) Passive time invariant circuits
(c) Non linear circuits only
(d) Both linear and non linear circuits
(I.T.I. February 2017, Electrician, Electroplator, Wireman)

80. Kirchhoff's voltage law is based on law of conservation of ...
(a) Charge (b) Energy
(c) Momentum (d) Mass
(I.T.I. August 2016, Electrician, Electroplator, Wireman)

81. "Sum of all the currents meeting at a point is zero" is stated by ...
(a) Kirchhoff's current law
(b) Kirchhoff's voltage law
(c) Ohm's law
(d) Gauss's law
(I.T.I. August 2016, Electrician, Electroplator, Wireman)

82. According to Kirchhoff's voltage law, the algebraic sum of all I.R. drops and E.M.Fs in any closed loop of a network is always......
(a) Zero
(b) Positive
(c) Negative
(d) Greater than unity
(I.T.I. August 2019 Electrician)

83. Nodal analysis is generally used to determine.....
(a) Voltage (b) Current
(c) Resistance (d) Power
(I.T.I. August 2019 Electrician)

84. Resistances $R_1 = 4\ \Omega$; $R_2 = 6\ \Omega$ and unknown resistance R_3 are connected in parallel across a supply of 24 volts. Find the value of resistor R_3 if current in the circuit is 12 amperes.
(a) 4 Ω (b) 12 Ω
(c) 8 Ω (d) 6 Ω
(I.T.I. August 2019 Electrician, Electroplator, Wireman, Lift & Escalator Mechanic)

85. A 10 ohms resistor is powered by a 5 volt battery. Current flowing through the source is.....
(a) 10 amperes (b) 50 amperes
(c) 2 amperes (d) 0.5 amperer
(I.T.I. August 2019 Electrician)

86. A 100 Ω resistance is to be used in a circuit carrying a current of 0.5 ampere. Power rating should be ...... watts.
(a) 50 (b) 20
(c) 200 (d) 500
(I.T.I. August 2019 Electrician)

87. In 5 hours, a 200 W, 250 V lamp will consume ......
(a) 10 kWh (b) 5 kWh
(c) 1 kWh (d) none of these
(I.T.I. August 2019 Electrician)

88. When electric current is passed through bulb, the bulb gives light because of ......
(a) Electric effect of current
(b) Heating effect of current
(c) Glowing effect of current
(d) Light effect of current
(I.T.I. August 2019 Electrician)

89. What is the effect on opened resistor in series circuit?
(a) No effect in opened resistor
(b) Total supply voltage will appear across the opened resistor
(c) Full circuit current will flow in opened resistor
(d) No voltage will appear across the opened resistor
(I.T.I. August 2019 Electrician, Electroplator, Wireman, Lift & Escalator Mechanic)

90. The resistivity of a semiconductor is ...... conductors and insulators.
(a) More than that of
(b) Lies between that of
(c) Less than that of
(d) None of these
(I.T.I. August 2019 Electrician)

91. What is the effect of the circuit, if " a b " points are shorted?

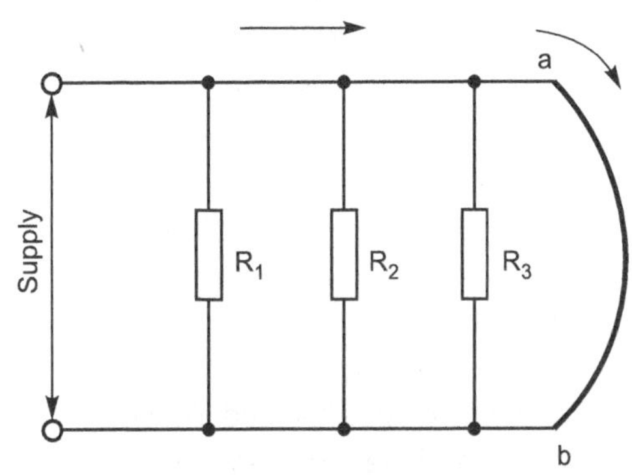

(a) Same current will flow in all branches
(b) Circuit resistance will be zero
(c) Supply voltage will exist in each branch
(d) Total circuit current is equal to each branch circuit current

(I.T.I. August 2019 Electrician, Electroplator, Wireman, Lift & Escalator Mechanic)

Resistors

1. The fundamental unit of measurements of the resistance is —
   (a) megaohm (b) millohm
   (c) ohm (d) kiloohm
2. A resistor is used in an electric circuit for —
   (a) blocking the flow of current
   (b) increasing the flow of current
   (c) reducing the flow of current
   (d) none of the above

   (I.T.I. February 2019, Electrician, Electroplator, Wireman)
3. For measuring an unknown resistance, choose ... range initially — in a multimeter.
   (a) the lowest (b) the highest
   (c) any (d) $R \times 10$

   (I.T.I. February 2019, Electrician, Electroplator, Wireman)
4. For measuring insulation resistance, the best choice would be —
   (a) digital multimeter
   (b) megger
   (c) multimeter
   (d) Wheat Stone Bridge

   (I.T.I. February 2019, Electrician, Electroplator, Wireman)
5. Potentiometer is used to provide —

   (I.T.I. February 2019, Electrician, Electroplator, Wireman)
   (a) a fixed resistance value
   (b) a variable resistance value
   (c) any two resistance values
   (d) a resistance value in steps
6. Carbon resistors are made in —
   (a) composition type only
   (b) film type only
   (c) both composition and film types
   (d) none of the above stated types
7. In an open circuit ....
   (a) resistance and current both are zero
   (b) resistance and current both are infinity
   (c) resistance is zero and current is infinity
   (d) resistance is infinity and current is zero
8. To compensate the lower value of cold resistance of the filament which one of the following is connected in series with the serial set?
   (a) NTC thermistor (b) PTC thermistor
   (c) Voltage dependent resistor
   (d) Flasher

   (I.T.I. August 2014, Electrician)
9. Which type of resistors are called as thermistors?
   (a) PTC resistors
   (b) Varistors (VDR)
   (c) Light dependent resistor (LDR)
   (d) NTC resistors

   (I.T.I. February 2015, Electrician)
10. Which type of resistor is depending upon the voltage?
    (a) Sensistors (b) Varistors
    (c) LDR (d) Thermistor

    (I.T.I. February 2014, Wireman)
11. VDR stands for ....
    (a) Voltage Depended Resistor
    (b) Voltage Drop Resistance
    (c) Voltage Depended Resistance
    (d) Voltage Depressing Resistor

    (I.T.I. February 2016, Wireman)
12. Thermistors are ....
    (a) less sensistive than RTD
    (b) equal sensitive as RTD
    (c) more sensitive than RTD
    (d) less accurate than RTD

    (I.T.I. February 2016, Electronic Mechanic)

13. The material to be used in manufacture of a standard resistor should be of ...

(a) Low resistivity
(b) High resistivity and low temperature coefficient
(c) High temperature coefficient
(d) Low resistivity and high temperature coefficient

(I.T.I. February 2017, Electrician, Electroplator, Wireman)

14. What is the name of the resistor, if its resistance value increase with increase in temperature?

(a) Varistors
(b) Light Dependent Resistors (L.D.R.)
(c) Thermistors
(d) Sensistors

(I.T.I. August 2019 Electrician, Electroplator, Wireman, Lift & Escalator Mechanic)

Resistor Colour Codes

Choose the correct answer:

1. In a colour code resistor, the fourth band has no colour which indicates the tolerance ....

(a) ± 5% (b) ± 10%
(c) ± 15% (d) ± 20%

(I.T.I. August 2014, Electronic Mechanic)

2. Colour codes used in 22 ohms ± 10% resistor will be ....

(a) red, red, brown, silver
(b) red, red, black, silver
(c) red, red, black, gold
(d) red, red, brown, gold

(I.T.I. February 2016, Electronic Mech. Trade Theory)

3. What is the value of resistance given in the figure by colour coding?

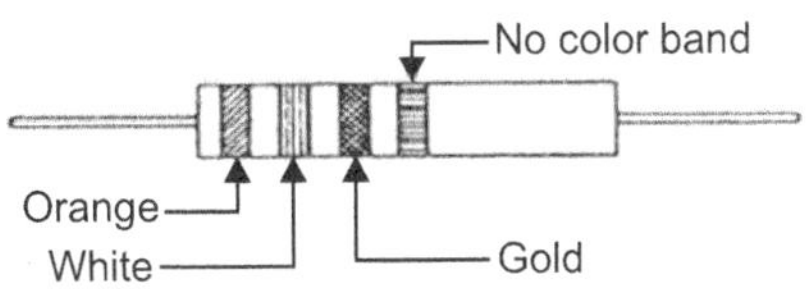

(a) 3.9 Ω (b) 39 Ω
(c) 390 Ω (d) 3900 Ω

(I.T.I. February 2014, Electrician)

4. The colour bands on a fixed carbon resistor are brown, red and black. Its value is :

(a) 12 Ω (b) 100 Ω
(c) 21 Ω (d) 120 Ω

(I.T.I. February 2019, Electrician, Electroplator, Wireman)

5. We need a resistor of value 47 kΩ with ± 5% tolerance. The sequence of the colour band on this resistor should be ......

(a) Yellow, violet, yellow and gold
(b) Yellow, violet, orange and gold
(c) Yellow, violet, yellow and silver
(d) Yellow, violet, brown and gold

(I.T.I. February 2019, Electrician, Electroplator, Wireman)

KEY TO MULTIPLE CHOICE QUESTIONS

Fundamentals of Electricity

1. (a)	**2.** (c)	**3.** (c)	**4.** (d)	**5.** (b)	**6.** (a)	**7.** (d)	**8.** (a)	**9.** (c)	**10.** (d)
11. (a)	**12.** (b)	**13.** (a)	**14.** (b)	**15.** (d)	**16.** (d)	**17.** (a)	**18.** (c)	**19.** (b)	**20.** (c)
21. (c)	**22.** (c)	**23.** (b)	**24.** (a)	**25.** (c)	**26.** (d)	**27.** (d)	**28.** (a)	**29.** (c)	**30.** (a)
31. (d)	**32.** (b)	**33.** (b)	**34.** (a)	**35.** (c)	**36.** (c)	**37.** (c)	**38.** (b)	**39.** (b)	**40.** (d)
41. (c)	**42.** (b)	**43.** (b)	**44.** (d)	**45.** (a)	**46.** (b)	**47.** (d)	**48.** (b)	**49.** (a)	**50.** (b)
51. (a)	**52.** (c)	**53.** (b)	**54.** (c)	**55.** (c)	**56.** (a)	**57.** (d)	**58.** (a)	**59.** (a)	**60.** (a)
61. (b)	**62.** (a)	**63.** (a)	**64.** (b)	**65.** (a)	**66.** (c)	**67.** (a)	**68.** (c)	**69.** (b)	**70.** (b)
71. (d)	**72.** (d)	**73.** (a)	**74.** (b)	**75.** (c)	**76.** (b)	**77.** (c)	**78.** (a)	**79.** (d)	**80.** (b)
81. (a)	**82.** (a)	**83.** (a)	**84.** (b)	**85.** (d)	**86.** (b)	**87.** (c)	**88.** (b)	**89.** (b)	**90.** (b)
91. (b)									

Resistors

1. (c)	**2.** (c)	**3.** (b)	**4.** (b)	**5.** (b)	**6.** (c)	**7.** (d)	**8.** (a)	**9.** (d)	**10.** (b)
11. (a)	**12.** (b)	**13.** (b)	**14.** (d)						

Resistor Colour Codes

1. (d)	**2.** (b)	**3.** (a)	**4.** (a)	**5.** (b)

Module 5

Capacitors, Magnetism and Electromagnetism

Multiple Choice Questions

Capacitors

1. A capacitor consists of two
   (a) insulation separated by a dielectric
   (b) conductors separated by an insulator
   (c) ceramic plates and one mica disc
   (d) silver coated insulators
2. The capacitance of a capacitor is NOT influenced by
   (a) plate thickness
   (b) plate area
   (c) plate separation
   (d) nature of the dielectric
3. A capacitor that stores a charge of 0.5 coulomb at 10 volts has a capacitance of .... farads
   (a) 5 (b) 20
   (c) 10 (d) 0.05
4. Find the total effective capacitance of three condensers 5μF, 7μF and 12 μF connected in parallel.
   (a) 2.4 μF (b) 0.24 μF
   (c) 240 μF (d) 24 μF
   (I.T.I. February 2019, Electrician, Electroplator, Wireman)
5. A capacitor of capacitance 50 micro farads is charged to a potential difference of 200 volts. Find out energy stored in the condenser.
   (a) 1 joule (b) 100 milli joules
   (c) 10 joules (d) 100 joules
6. By increasing the charge Q of a capacitor :
   (a) its capacity increases
   (b) its capacity decreases
   (c) its capacity remains unchanged
   (d) its potential difference remains unchanged
   (I.T.I. February 2019, Electrician, Electroplator, Wireman)
7. By bringing a charged conductor closed to an earthed conductor :
   (a) the voltage of the conductor will increase
   (b) the capacity of the conductor will increase
   (c) the capacity of the conductor will decrease
   (d) both the voltage and capacity of the conductor will decrease.
8. The capacity of a parallel plate capacitor is 4 μF and the distance between the two plates is 4 cm. If the distance is reduced to 2 cm then the capacity will become :
   (a) 8 μF (b) 12 μF
   (c) 16 μF (d) 2 μF
9. The relation between the capacity (C), P.D. (V) and charge (Q) for a capacitor is given by :

(a) $C = \frac{Q}{V}$ (b) $C = \frac{V}{Q}$

(c) $C = \frac{Q}{2V}$ (d) $C = \frac{Q.V}{2}$

10. The unit of capacitance is :

(a) coulomb (b) volt

(c) farad (d) ampere

11. The total capacitance of a group of capacitors is increased when they are connected in :

(a) series (b) parallel

(c) series-parallel (d) parallel-series.

12. Opening the switch of an inductive circuit sets an arc across the switch contacts due to large voltage induced and stored energy in magnetic field. The arc is suppressed in such circuits by connecting ......

(a) a capacitor across the supply

(b) a capacitor across the switch contacts

(c) arcing horns across the switch contacts

(d) another small inductor across the switch contacts in reverse direction.

(I.T.I. July 2008, Electrician)

13. The capacitance to the capacitor is directly proportional to the ......

(a) material of the plate

(b) area of the plate

(c) voltage across the plate

(d) polarity of the plates

(I.T.I. February 2015, Electrician)

14. When testing a capacitor with a multimeter, the needle is showing zero position right from the beginning. What does it indicate?

(a) The capacitor is open circuited

(b) The capacitor is short circuited

(c) The capacitor is charged in capacity

(d) None of the above

15. The type of capacitor used in capacitor start motor is ....

(a) electrolytic capacitor

(b) ceramic capacitor

(c) paper capacitor

(d) mica capacitor

(I.T.I. January 2013, Electrician)

16. What is the effect on the capacitance value if the dielectric constant of the material of the capacitor is increased?

(a) Capacitance value increases

(b) Capacitance value decreases

(c) Capacitance value remains same

(d) Capacitance value becomes zero

(I.T.I. February 2014, Wireman)

17. One pico farad (pf) = ...

(a) 10^{-12} μF (b) 10^{-9} μF

(c) 10^{-6} μF (d) 10^{-3} μF

(I.T.I. February 2016, Electrician, Wireman)

18. Capacitor does not allow sudden change in ...

(a) Voltage (b) Current

(c) Power (d) None of these

(I.T.I. February 2017, Electrician, Electroplator, Wireman)

19. Rating of capacitor is 1 μF, 15 V, Also a +ve sign is printed near one of its terminals. The capacitor is ...

(a) Mica capacitor

(b) Ceramic capacitor

(c) Electrolytic capacitor

(d) Paper capacitor

(I.T.I. February 2017, Electrician, Electroplator, Wireman)

20. Capacitor is ...

(a) Active and bilateral component

(b) Active and unilateral component

(c) Passive and bilateral component

(d) Passive and unilateral component

(I.T.I. February 2019, Wireman)

21. Capacitors for power factor correction are rated in ...

(a) kW (b) kVA

(c) kV (d) kVAR

(I.T.I. February 2017, Electrician, Electroplator, Wireman)

22. The filter action in a D.C. power supply can be improved by:
(a) Adding capacitance
(b) Subtracting capacitance
(c) Removing capacitance
(d) Using very low value of capacitance

23. Energy stored in a capacitor is given by the expression-
(a) $W = CV$ (b) $W = \frac{1}{2} CV^2$
(c) $W = \frac{1}{2} C^2V$ (d) $W = \frac{1}{2} C^2V^2$

(I.T.I. February 2019, Electrician, Electroplator, Wireman)

24. Three capacitors 10 μF, 10 μF and 20 μF are connected in series across a supply of 24 volts. Calculate total value of capacitance in micro farads.
(a) 40 (b) 10
(c) 15 (d) 4

(I.T.I. August 2019 Electrician, Electroplator, Wireman, Lift & Escalator Mechanic)

25. What precaution is to be taken before connecting the different voltage rating capacitors in series?
(a) All the capacitors must be of same manufacturer
(b) Break down voltage of each capacitor must be same
(c) Total capacitors value must be less than the lowest value of capacitor
(d) Each capacitors voltage drop must be less than its voltage rating

(I.T.I. August 2019 Electrician, Electroplator, Wireman, Lift & Escalator Mechanic)

Magnetism and Electromagnetism

Choose the correct answer :

1. A natural magnet is called :
(a) steel (b) load stone
(c) magnetism (d) soft iron

2. Externally, the magnetic lines of force travel:
(a) north to south
(b) south to north
(c) negative to positive
(d) in both the directions

(I.T.I. February 2019, Electrician, Electroplator, Wireman)

3. The magnetism that remains in a magnetic material after the withdrawal of magnetising force is called:
(a) permeability
(b) induction
(c) residual magnetism
(d) saturation

(I.T.I. February 2019, Electrician, Electroplator, Wireman)

4. The force acting between two magnetic poles is :
(a) directly proportional to the distance
(b) inversely proportional to the distance
(c) directly proportional to the square of the distance
(d) inversely proportional to the square of the distance

(I.T.I. February 2019, Electrician, Electroplator, Wireman)

5. Which of the following material is used for making permanent magnets ?
(a) Carbon steel (b) Platinum cobalt
(c) Alnico-V (d) All of these

6. In order to induce more magnetic flux, a magnetic circuit should have minimum :
(a) reluctance (b) resistance
(c) reactance (d) conductance

7. The simplest way of varying the flux produced by an electromagnet smoothly in an electrical device is by :
(a) increasing and decreasing the number of turns of the coil
(b) using an exciting coil with number of tappings

(c) varying the current through the exciting coil

(d) varying the core position of the exciting coil

(I.T.I. February 2015, Electrician)

8. Which law/rule states that the magnitude of the induced e.m.f. is directly proportional to the rate of change of flux linkage?

(a) Lenz's law

(b) Faraday's first law of electromagnetic induction

(c) Faraday's second law of electromagnetic induction

(d) Fleming's right hand rule

(I.T.I. February 2015, Electrician)

9. What will be the direction of the magnetic field shown in the figure, when the two wires carrying current in opposite direction?

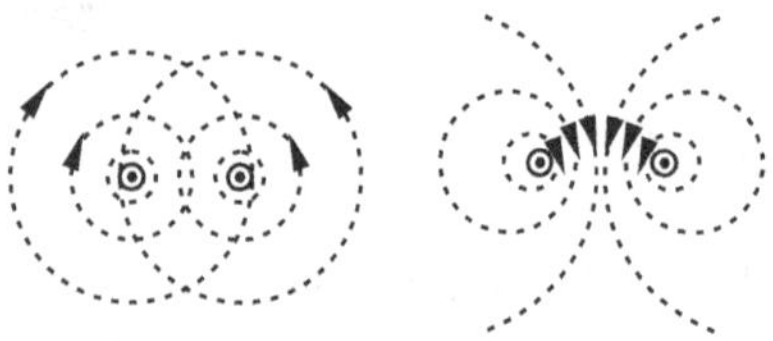

(a) Will be in the same direction

(b) Will form loop around the wires

(c) Will be in opposite direction

(d) Will aid one to another

(I.T.I. August 2014, Electrician)

10. When the distance between the turns of wire in a magnetic coil is increased then the inductance of the coil will ....

(a) increase (b) decrease

(c) become zero (d) remain same

(I.T.I. February 2015, Wireman)

11. The resistance in electrical circuit is analogous to magnetic circuit is ....

(a) retentivity (b) reluctance

(c) reactance (d) permeability

(I.T.I. February 2015, Electrician)

12. Which magnetic term is the analogous (similarity) to the electrical term of current?

(a) Flux density (b) Permeability

(c) Flux (d) Reluctance

(I.T.I. February 2015, Electrician)

13. What is the unit of permeability?

(a) ampere turns

(b) weber/m^2

(c) ampere turns/weber

(d) no unit

(I.T.I. February 2015, Electrician)

14. What is the unit of magneto motive force (mmf)?

(a) tesla (b) ampere turns

(c) ampere/turns (d) weber/$metre^2$

(I.T.I. August 2014, Electrician)

15. Which of the following materials is a paramagnetic substance?

(a) Iron (b) Glass

(c) Aluminium (d) Nickel

(I.T.I. July 2006, Electrician)

16. Which substance is the example of diamagnetic material?

(a) Manganese (b) Nickle

(c) Glass (d) Cobalt

(I.T.I. February 2015, Electrician)

17. In which class of magnetic substance, water does belong to?

(a) Diamagnetic substances

(b) Paramagnetic substances

(c) Non-magnetic substances

(d) Ferromagnetic substance

18. Which of the following materials is a ferromagnetic substance?

(a) Platinum (b) Aluminium

(c) Steel (d) Copper

(I.T.I. July 2014, Electrician)

19. The example of diamagnetic substance is ....

(a) aluminium (b) bismuth

(c) manganese (d) copper

(I.T.I. February 2016, Electrician, Wireman)

20. Which one is the paramagnetic material ?

(a) Bismuth (b) Wood

(c) Air (d) Glass

(I.T.I. August 2015, Electrician, Wireman)

21. The reciprocal of permeability is ....

(a) magneto motive force

(b) field intensity
(c) flux density
(d) reluctivity

(I.T.I. February 2016, Electrician, Electronic Mech.)

22. Which one of the following is the similarity magnetic term to the electrical term of conductance ?

(a) Reluctance (b) Flux density
(c) Permeability (d) Permeance

(I.T.I. February 2016, Electrician, Wireman)

23. What is the property which enables material to conduct the magnetic lines?

(a) Retentivity (b) Permeability
(c) Reluctance (d) Susceptance

(I.T.I. August 2015, Electrician)

24. The unit of flux density (B) is ....

(a) ampere-turns
(b) weber
(c) web/m^2
(d) ampere turns/web

(I.T.I. February 2016, Electrician, Wireman)

25. The .... is the unit for measuring the m.m.f. of a current carrying coil.

(a) A.T. (b) current
(c) emf (d) none of these

(I.T.I. August 2015, I.T. and E.S.M.)

26. What is the unit of permeability?

(a) ampere - turns (b) weber/m^2
(c) no unit (d) tesla

(I.T.I. February 2016, Electrician, Wireman)

27. An electromagnet of length 10 cm with magnetizing force of 1500 A.T./mt works in magnetizing circuits.

Calculate its magneto-motive force (mmf):

(a) 15000 A.T. (b) 1500 A.T.
(c) 150 A.T. (d) 15 A.T.

(I.T.I. August 2015, Electrician)

28. The formula to find the field intensity (H) of the magnetic field is ....

(a) $H = \frac{\text{flux}}{\text{Area of the core}}$

(b) $H = \frac{\text{M.M.F.}}{\text{length of coil in meter}}$

(c) $H = \frac{\text{MMF}}{\text{reluctance}}$

(d) $H = \frac{\text{length of coil in meter}}{\text{M.M.F.}}$

(I.T.I. February 2016, Electrician, Wireman)

29. The magnitude of the induced emf at zero degree angle is given by the formula E_0 = BLV sin θ.

In this, V' represents ....

(a) induced voltage
(b) velocity of the loop rotation
(c) value of the magnetic flux density
(d) velocity of the cutting force by magnetic lines

(I.T.I. February 2019, Electrician)

30. The e.m.f. induced in a coil depends upon ...

(a) Number of its turns
(b) The change of flux linked with it
(c) The time taken to change the flux
(d) All of these

(I.T.I. February 2019, Wireman)

31. Whenever a magnet is quickly put in an open circulted stationary coil ...

(a) A current is induced in it
(b) e.m.f is induced in it
(c) Power is spent
(d) None of these

(I.T.I. February 2019, Wireman)

32. The direction of the induced e.m.f. in a coil is found by ...

(a) Faraday's Law (b) Lenz's law
(c) Fleming's left hand rule
(d) Fleming's right hand rule

(I.T.I. February 2019, Wireman)

33. When the coil moves at right angle to the magnetic flux, the induced e.m.f. will be

(a) oscillating (b) minimum
(c) zero (d) maximum

(I.T.I. February 2018, Electrician)

34. Magnetic flux has unit of ...

(a) newton (b) weber

(c) ampere turns (d) tesla

(I.T.I. February 2017, Electrician, Electroplator, Wireman)

35. The magnetic flux density is measured in ...

(a) tesla (b) webers

(c) ampere turns (d) weber per metre

(I.T.I. August 2016, Electrician, Electroplator, Wireman)

36. The unit for reluctance is ...

(a) tesla

(b) ampere turns/weber

(c) ampere turns/metre

(d) weber

(I.T.I. February 2017, Electrician, Electroplator, Wireman)

37. Which part of magnetic path requires largest m.m.f.?

(a) Air gap (b) Coil

(c) Inductance (d) Core

(I.T.I. February 2017, Electrician, Electroplator, Wireman)

38. Which of the following statements is true about the force between two charges?

(a) Unlike charges repel each other

(b) Like charges repel each other

(c) Like charges attract each other

(d) No force exists between two charges.

(I.T.I. February 2019, Electrician, Electroplator, Wireman)

39. To reduce eddy current losses, the core of magnetic material is/has ...

(a) Laminated (b) Solid

(c) High resistivity (d) Both (a) and (b)

(I.T.I. February 2019, Wireman)

40. The A.C. Bridge used for the measurement of inductance is/are ...

(a) Maxwell's inductance bridge

(b) Hay's bridge

(c) Both (a) and (b)

(d) None of these

(I.T.I. February 2019, Electrician)

41. Which of the following circuit element stores energy in the form of electromagnetic field?

(a) Inductor (b) Condenser

(c) Variable resistor (d) Capacitor

(I.T.I. February 2017, Electrician, Electroplator, Wireman)

42. The e.m.f. induced is 2 volts when current is changing at the rate of 0.5 ampere per second. The self inductance of coil is ...

(a) 4 henries (b) 2 henries

(c) 1 henry (d) 3 henries

(I.T.I. February 2019, Wireman)

43. A 100 turns coil has an inductance of 6 milli henries. If the number of turns is increased to 200 and other quantities remain the same, the inductance will be ...

(a) 1.5 mH (b) 3 mH

(c) 12 mH (d) 24 mH

(I.T.I. February 2019, Wireman)

HINT:

The inductance of a coil is proportional to the square of the number of turns, when all other parameters remain same.

44. Instantaneous power in inductor is proportional to ...

(a) Product of instantaneous current and rate of change of current

(b) Square of instantaneous current

(c) Square of rate of change of current

(d) Temperature of the inductor

(I.T.I. February 2017, Electrician, Electroplator, Wireman)

45. What is the unit of permeance?

(a) ampere turns

(b) weber/ampere turns

(c) ampere turns/weber

(d) weber/square metre

(I.T.I. August 2019 Electrician, Electroplator, Wireman, Lift & Escalator Mechanic)

46. Which rule is applied to find the direction of magnetic fields in a solenoid coil?

(a) Right hand palm rule

(b) Cork screw rule

(c) Fleming's left hand rule
(d) Fleming's right hand rule

(I.T.I. August 2019 Electrician, Electroplator, Wireman, Lift & Escalator Mechanic)

47. The magnetic material follow ....
(a) Faraday's law
(b) Ampere's law
(c) Lenz's law
(d) Curie Weiss's law

(I.T.I. August 2019 Electrician)

KEY TO MULTIPLE CHOICE QUESTIONS

Capacitors

1. (b)	**2.** (a)	**3.** (d)	**4.** (d)	**5.** (a)	**6.** (c)	**7.** (a)	**8.** (a)	**9.** (a)	**10.** (c)
11. (b)	**12.** (b)	**13.** (b)	**14.** (b)	**15.** (a)	**16.** (a)	**17.** (c)	**18.** (a)	**19.** (c)	**20.** (c)
21. (d)	**22.** (a)	**23.** (b)	**24.** (d)	**25.** (d)					

Magnetism and Electromagnetism

1. (b)	**2.** (a)	**3.** (c)	**4.** (d)	**5.** (d)	**6.** (a)	**7.** (c)	**8.** (c)	**9.** (c)	**10.** (b)
11. (b)	**12.** (c)	**13.** (d)	**14.** (b)	**15.** (c)	**16.** (c)	**17.** (a)	**18.** (c)	**19.** (b)	**20.** (c)
21. (d)	**22.** (d)	**23.** (b)	**24.** (c)	**25.** (a)	**26.** (c)	**27.** (c)	**28.** (b)	**29.** (b)	**30.** (d)
31. (b)	**32.** (b)	**33.** (d)	**34.** (b)	**35.** (a)	**36.** (b)	**37.** (a)	**38.** (b)	**39.** (a)	**40.** (c)
41. (a)	**42.** (a)	**43.** (d)	**44.** (a)	**45.** (b)	**46.** (a)	**47.** (d)			

Module 6

A.C. Circuits

MULTIPLE CHOICE QUESTIONS

1. Power factor of an A.C. circuit is equal to :
   (a) cosine of angle between voltage and current
   (b) tangent of angle between voltage and current
   (c) impedance/resistance
   (d) apparent power/real power.
   (I.T.I. February 2019, Electrician, Electroplator, Wireman)
2. The resistance of an open circuit is equal to:
   (a) zero
   (b) lesser than one ohm
   (c) infinite
   (d) lesser than before.
   (I.T.I. February 2019, Electrician, Electroplator, Wireman)
3. The unit of inductance is :
   (a) farad (b) ohm
   (c) mho (d) henry.
   (I.T.I. February 2019, Electrician, Electroplator, Wireman)
4. The power dissipation of a pure inductor is:
   (a) very low (b) zero
   (c) very high (d) lesser than unity.
5. The magnitude of inductance .... with an increase in the frequency.
   (a) decreases
   (b) remains unchanged
   (c) increases
   (d) becomes zero
6. The difference between peak positive and peak negative values of an a.c. voltage is called :
   (a) maximum value (b) average value
   (c) effective value (d) peak-to-peak value.
   (I.T.I. February 2019, Electrician, Electroplator, Wireman)
7. The r.m.s. value of A.C. is equal to :
   (a) 0.637 of max. value
   (b) 0.707 of max. value
   (c) 1.414 of max. value
   (d) the maximum value.
   (I.T.I. February 2019, Electrician, Electroplator, Wireman)
8. The value of form factor for a pure sine-wave is :
   (a) 1.414 (b) 0.707
   (c) 0.637 (d) 1.11.
   (I.T.I. February 2019, Electrician, Electroplator, Wireman)
9. If the voltage and current have a phase difference of 90° in an a.c. circuit, the power will be
   (a) maximum (b) minimum
   (c) $1.1 \times V \times I$ (d) zero.
10. Which of the following frequencies has the longest time period ?
    (a) 1 Hz. (b) 10 Hz.
    (c) 1 kHz. (d) 100 kHz.
    (I.T.I. February 2019, Electrician, Electroplator, Wireman)

11. The total voltage in a series R – L circuit leads the current by an angle :
    (a) of 90°
    (b) between 0° and 90°
    (c) between 90° and 180°
    (d) between 0° and 360°.

(I.T.I. February 2019, Electrician, Electroplator, Wireman)

12. The frequency at which $X_L = X_C$ is called :
    (a) natural frequency
    (b) resonance frequency
    (c) neutral frequency
    (d) super frequency.

13. In a series R - L - C circuit the impedance is equal to _____ and current equal to ________ at resonance.
    (a) minimum, maximum
    (b) zero, zero
    (c) infinite, zero
    (d) zero, infinite.

14. What is the power factor of a pure resistive circuit?
    (a) 1.0 (b) 0.95
    (c) 0.9 (d) 0.85

(I.T.I. February 2015, Wireman)

15. The power factor is defined as the ratio ....
    (a) $\cos\theta = \dfrac{\text{True power}}{\text{Apparent power}}$
    (b) $\cos\theta = \dfrac{\text{Apparent power}}{\text{True power}}$
    (c) $\cos\theta = \dfrac{\text{True power}}{\text{Reactive power}}$
    (d) $\cos\theta = \dfrac{\text{Apparent power}}{\text{Reactive power}}$

(I.T.I. February 2015, Wireman)

16. The average value of alternating current for a pure sine wave is equal to ....
    (a) 0.536 × maximum value of current
    (b) 0.636 × maximum value of current
    (c) 0.707 × maximum value of current
    (d) 0.737 × maximum value of current

(I.T.I. February 2015, Electrician)

17. The formula to calculate the capacitive reactance "X_C" of the AC circuit is ......
    (a) $X_C = \sqrt{2} \cdot pfC$ ohm
    (b) $X_C = 2\pi fC$ ohm
    (c) $X_C = \dfrac{1}{2pfC}$ ohm
    (d) $X_C = \dfrac{1}{\sqrt{2}pfC}$ ohm

(I.T.I. February 2015, Wireman, Electrician)

18. What is the time period for the frequency of 50 Hz?
    (a) 200 milli seconds
    (b) 20 milli seconds
    (c) 2 milli seconds
    (d) 0.2 milli second

(I.T.I. February 2015, Electrician)

19. If the supply voltage V = 10 volts and the voltage across the coil V_L = 8 volts, the amount of voltage read by the voltmeter V_R in the given figure is :

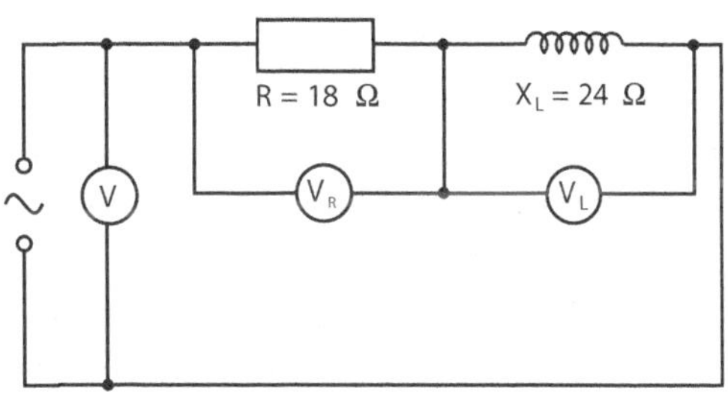

    (a) 2 volts (b) 6 volts
    (c) 10 volts (d) 18 volts

20. The capacitive reactance (X_C) of a capacitor varies :
    (a) inversely with the frequency
    (b) directly with the frequency
    (c) directly with the applied voltage
    (d) inversely with the applied voltage.

21. Which one of the following relationships is true between true power (P_T), reactive power (P_R) and apparent power (P_A) ?
    (a) $P_A = P_T + P_R$ (b) $P_A = P_T - P_R$
    (c) $P_A = \sqrt{P_T^2 + P_R^2}$ (d) $P_A = P_T^2 + P_R^2$

22. What is the power consumed in the circuit given below at a power factor of 0.6 ?

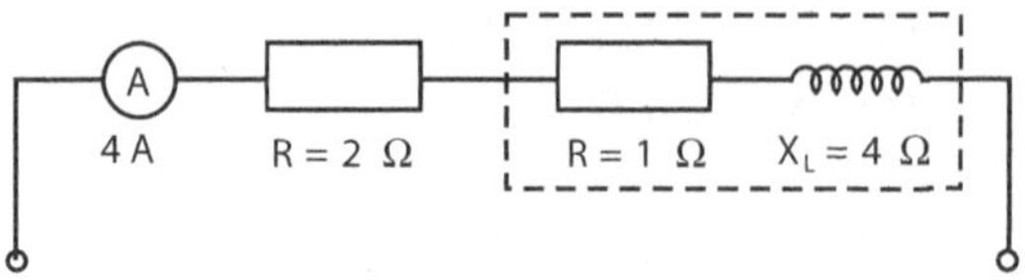

(a) 32 W (b) 48 W
(c) 60 W (d) 112 W

23. In a 3-phase generator 220 volts a. c. is induced in each coil. Which of the following values is indicated by the voltmeters ?

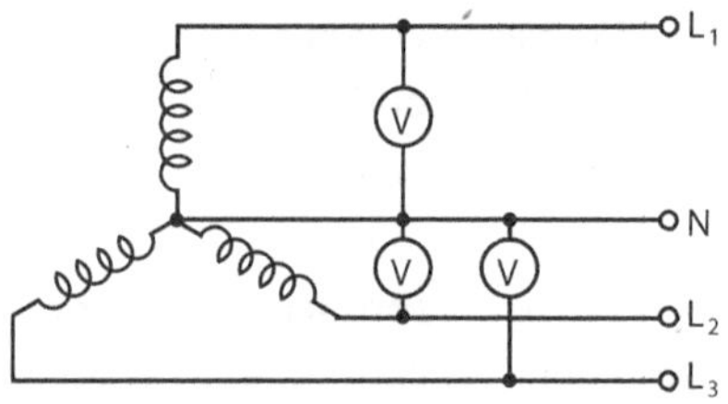

(a) 220 V × $\sqrt{3}$ = 380 V
(b) 220 V
(c) 220 V ÷ $\sqrt{3}$ = 127 V
(d) 220 V × $\sqrt{2}$ = 310 V

24. The positive and negative halves of a complex wave are symmetrical when
(a) it contains even harmonics
(b) phase difference between even harmonics and fundamental is 0 to π
(c) it contains odd harmonics
(d) phase difference between even harmonics and fundamental is either $\frac{p}{2}$ or $\frac{3p}{2}$

25. In a 3-phase system .... the harmonic has negative phase sequence of RBY.
(a) 9 (b) 13
(c) 5 (d) 15

26. A complex voltage wave is applied across a pure capacitor. As compared to the fundamental voltage, the reactance offered by the capacitor to the third harmonic voltage would be
(a) nine times (b) three times
(c) one third (d) one ninth

27. When a pure inductive coil is fed by a complex voltage wave, its current wave
(a) has larger harmonic content
(b) is more distorted
(c) is identical with voltage wave
(d) shows less distortion

28. Which of the following harmonic voltage components in a 3-phase system would be in phase with each other?
(a) 3rd, 9th, 15th etc.
(b) 7th, 13th, 19th etc.
(c) 5th, 11th, 17th etc.
(d) 2nd, 4th, 6th etc.

29. In the figure given below you see the wave forms of voltage and current in a AC circuit.

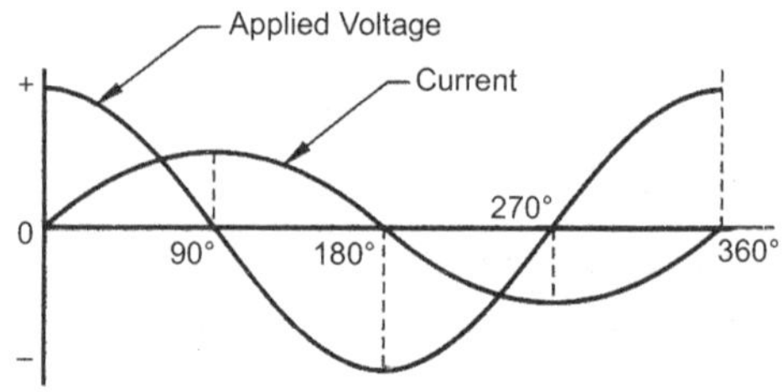

Which electrical components does this AC circuit contain?
(a) *R* and *L* in series
(b) *R* and *C* in series
(c) *C* only (d) *L* only

30. What is the reading of Ammeter in the coil shown below?

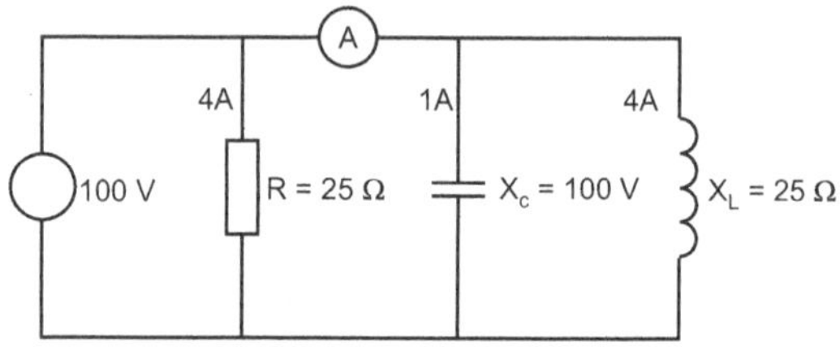

(a) 9 A (b) 5 A
(c) 4 A (d) 3 A

(I.T.I. August 2014, Electrician)

31. When several impedances say Z_1, Z_2 and Z_3 are connected in parallel, their combined impedance *Z* is ....

(a) $Z = \dfrac{1}{\dfrac{1}{Z_1} + \dfrac{1}{Z_2} + \dfrac{1}{Z_3}}$

(b) $Z = \dfrac{1}{Z_1 + Z_2 + Z_3}$

(c) $Z = Z_1 + Z_2 + Z_3$

(d) $Z = \dfrac{1}{Z_1} + \dfrac{1}{Z_2} + \dfrac{1}{Z_3}$

(I.T.I. February 2015, Wireman)

32. In a 3 phase system, three phased voltages are displaced by an angle of ....

(a) 90° (b) 120°
(c) 180° (d) 270°

(I.T.I. February 2015, Wireman)

33. The power consumed for a balanced load in star or delta connection is ....

(a) $3\ V_L \cdot I_L \cdot \cos\theta$
(b) $\sqrt{3}\ V_L \cdot I_L \cdot \cos\theta$
(c) $3 \cdot V_P \cdot I_P \cdot \cos\theta$
(d) $\sqrt{3} \cdot V_L \cdot I_L \cdot \sin\theta$

(I.T.I. February 2015, Electrician)

34. The power taken by a AC single phase circuit is ....

(a) $P = V \cdot I$ (b) $P = V \cdot I \cdot \cos\theta$
(c) $P = V \cdot I \cdot \sin\theta$ (d) $P = \dfrac{V \cdot I}{\cos q}$

(I.T.I. February 2015, Wireman)

35. You have to find true power, apparent power, reactive power and power factor of a 3 phase circuit. From which one of the following values can be determined?

(a) By 2 element type 3 phase wattmeter
(b) By 2 wattmeter method
(c) By 3 element 4 wire 3 phase wattmeter
(d) By 3 element 3 wire 3 phase wattmeter

(I.T.I. July 2009, Electrician)

36. The wattmeter, 'W_2' indicates the reading in reverse direction in the figure given below:

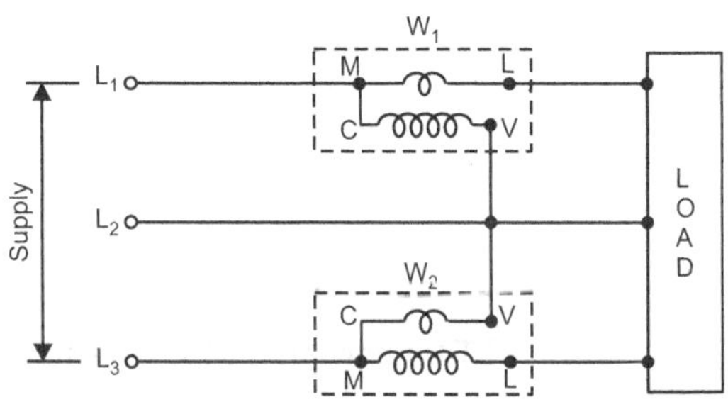

How will you get the forward reading?

(a) Interchange the connections to the meters W_1 and W_2 (i.e.) L_1 and L_3
(b) Reverse the potential coil of the connection of the meter W_1
(c) Reverse the potential coil of the connection of the meter W_2
(d) Reverse the connections of both the current and pressure coil of the W_2

(I.T.I. February 2015, Electrician)

37. The formula to find 'tan θ', when the 3 phase power is measured by two wattmeter method is ....

(a) $\dfrac{W_1 + W_2}{W_1 + W_2}$ (b) $\dfrac{\sqrt{3}(W_1 - W_2)}{W_1 + W_2}$

(c) $\dfrac{\sqrt{3}(W_1 + W_2)}{W_1 - W_2}$ (d) $\dfrac{W_1 + W_2}{W_1 - W_2}$

(I.T.I. February 2016, Wireman)

38. As per BIS 1293-1988, the A.C. Single phase voltage rating is ....

(a) 200 V (b) 220 V
(c) 230 V (d) 240 V

(I.T.I. February 2015, Wireman)

39. The reciprocal of the frequency is known as ....

(a) time period (b) amplitude
(c) time constant (d) resonance

(I.T.I. February 2016, Electrician, Wireman)

40. What will be the effect of frequency, when the value of capacitance is increased in series resonance circuit?

(a) Decreases (b) Increases
(c) No change (d) Becomes zero

(I.T.I. August 2015, Electrician)

41. If the RMS voltage (V_{rms}) is 240 V, the maximum voltage (V_{max}) in AC single phase circuit is ....

(a) 389.45 volt (b) 376.76 volt
(c) 339.46 volt (d) 252.0 volt

(I.T.I. February 2016, Electrician, Wireman)

42. The main cause of low power factor is ....

(a) the fluctuation of supply voltage of the equipment

(b) the reactive power in the circuit
(c) the incorrect rating of controlling devices
(d) the loose connections terminations in the circuit

43. If the impedance (Z) formula is $Z = \sqrt{R^2 + X_L^2}$. What is the power factor formula for this circuit?

(a) $\frac{R}{X_L}$ (b) $\frac{R}{Z}$
(c) $\frac{Z}{R}$ (d) $\frac{X_L}{R}$

(I.T.I. February 2016, Electrician, Wireman)

44. What is the value of inductive reactance of 0.2 henry inductor connected to 50 Hz supply?

(a) 6.28 ohm (b) 62.8 ohm
(c) 628 ohm (d) 6280 ohm

(I.T.I. February 2016, Wireman)

45. In RLC series circuit the resosance will occur when ....

(a) $X_L < X_C$ (b) $X_L = X_C$
(c) $X_L > X_C$ (d) $X_L < R$

(I.T.I. August 2015, Electrician)

46. The line voltage (E_L) in a star connected system is equal to ....

(a) $\sqrt{3}\, E_P$ (b) $\frac{E_P}{\sqrt{3}}$
(c) $3\, E_P$ (d) $\frac{\sqrt{3}}{E_P}$

(I.T.I. February 2016, Wireman)

47. The formula to find the reactive power (Q) in 3 phase star connection?

(a) $Q = \sqrt{3}\, V_L I_L \cos\theta$
(b) $Q = 3\, V_P L_P \cos\theta$
(c) $Q = \sqrt{3}\, V_L I_L \sin\theta$
(d) $Q = 3\, V_P I_P \sin\theta$

(I.T.I. August 2015, Electrician, Wireman)

48. What is the formula to find true power (P_T) in 3 phase delta connected circuit?

(a) $P_T = 3\, V_L I_L \cos\theta$
(b) $P_T = \sqrt{3}\, V_{ph} I_{ph} \cos\theta$
(c) $P_T = \sqrt{3}\, V_L I_L \cos\theta$
(d) $P_T = 3\, V_{ph} I_{ph} \sin\theta$

(I.T.I. February 2016, Electrician, Wireman)

49. Three phase power can be measured by two wattmeter method with the asumption that ...

(a) Load is balanced
(b) Load is purely resistive
(c) Load is purely capacitive
(d) Load is purely inductive

(I.T.I. February 2019, Wireman)

50. Which is apparent power in A.C. operation?

(a) Product of voltage & current
(b) Product of RMS voltage & RMS current
(c) Product of voltage and current at no load
(d) Product of voltage & current at full load

(I.T.I. February 2017, Electrician, Elatroplator, Wireman)

51. Power consumed in a purely inductive circuit over a complete cycle is ...

(a) V.I. $\cos\theta$ (b) V.I. $\sin\theta$
(c) Zero (d) V.I.

(I.T.I. August 2016, Electrician, Electroplator, Wireman)

52. Power dissipated in ideal inductor is ...

(a) $\frac{1}{2}\, L \,.\, i^2_{max}$
(b) $\frac{1}{2}\, 7L \,.\, i^2_{min}$
(c) Zero
(d) cannot be determined

(I.T.I. February 2017, Electrician, Electroplator, Wireman)

53. In three phase delta connection, if line voltage is 400 volts, phase voltage will be ...

(a) 400 volts (b) $400\sqrt{2}$ volts
(c) $400\sqrt{3}$ volts (d) None of these

(I.T.I. February 2019, Wireman)

54. In star connection system ...
(a) Line current is equal to phase current
(b) Line voltage is equal to phase voltage
(c) Line current is equal to $\sqrt{3}$ times the phase current
(d) None of these

(I.T.I. February 2019, Wireman)

55. In three phase system, phase voltages differ by ...
(a) 180° (b) 120°
(c) 90° (d) 360°

(I.T.I. February 2019, Wireman)

56. If the frequency of supply is 50 Hz, the time period of one cycle will be ...
(a) 0.02 second (b) 0.1667 second
(c) 16.67 seconds (d) 0.2 second

(I.T.I. February 2019, Wireman)

57. The equation of instantaneous value of e.m.f. is ...
(a) $e = E_{r.m.s} \sin \omega t$ (b) $e = E_{max}$
(c) $e = E_{max} \cdot \sin \omega t$ (d) $e = E_{r.m.s}$

(I.T.I. February 2019, Wireman)

58. A sinusoidal current has an R.M.S. value of 14 mA. The peak to peak value is ...
(a) 45.12 mA (b) 16 mA
(c) 39.6 mA (d) 22.6 mA

(I.T.I. August 2016, Electrician, Electroplator, Wireman)

59. For a sinusoidal wave, peak factor is given by ...
(a) 1.212 (b) 1.11
(c) 1.414 (d) 1.155

(I.T.I. February 2019, Wireman)

60. The sine wave has a frequency of 50 Hz. Its angular velocity is .... radians per second.
(a) $\frac{50}{p}$ (b) $\frac{50}{2p}$
(c) 50π (d) 100π

(I.T.I. February 2019, Wireman)

61. One cycle of sinusoidal wave is equal to ...
(a) $\frac{p}{2}$ radians (b) $\frac{3p}{2}$ radians
(c) π radians (d) 2π radians

(I.T.I. February 2019, Wireman)

62. Two sinusoidal quantities are said to be in phase quadrature when their differeence is ...
(a) 0° (b) 30°
(c) 45° (d) 90°

(I.T.I. Februray 2017, Electrician, Electroplator, Wireman)

63. A.C. main supply in India has the frequency of ....
(a) 100 Hz (b) 60 Hz
(c) 50 Hz (d) 40 Hz

(I.T.I. Februray 2019, Electrician, Elatroplator, Wireman)

64. The frequency of an alternating voltage is given by ...
(a) $f = \frac{w}{2p}$ (b) $f = \frac{w}{2p \sin q}$
(c) $f = \frac{w}{2pf}$ (d) None of these

(I.T.I. Februray 2019, Wireman)

65. According to I.E. rules, the permissible variation of frequency is ...
(a) 3% (b) 5%
(c) 8% (d) 10%

(I.T.I. February 2019, Electrician)

66. The unit of impendance is ...
(a) henry (b) ohm
(c) hertz (d) mho

(I.T.I. August 2016, Electrician, Electroplator, Wireman)

67. The unit of admittance is ...
(a) siemens (b) ohms
(c) hernry (d) volts

(I.T.I. August 2016, Electrician, Electroplator, Wireman)

68. Peak factor is the ratio of ...
(a) Peak value to R.M.S. value
(b) Peak value to average value
(c) R.M.S. value to average value
(d) Average value to R.M.S. value

(I.T.I. August 2016, Electrician, Electroplator, Wireman)

69. Form factor = ...

(a) $\frac{\text{Average value}}{\text{R.M.S. value}}$ (b) $\frac{\text{Average Value}}{\text{Peak value}}$

(c) $\frac{\text{Instantaneous value}}{\text{Aerage value}}$

(d) $\frac{\text{R.M.S. value}}{\text{Average value}}$

(I.T.I. February 2017, Electrician, Electroplator, Wireman)

70. In a pure inductive A.C. circuit ...

(a) Voltage leads the current vector by 90°
(b) Voltage lags the current vector by 90°
(c) Current leads the voltage vector by 90°
(d) Current is in phase with the voltage vector

(I.T.I. February 2018, Electrician, Electroplator, Wireman)

71. R.M.S value depends on which of the following?

(a) Heating effect (b) Charge transfer
(c) Current (d) Voltage

(I.T.I. February 2017, Electrician, Electroplator, Wireman)

72. A heater is rated as 230 volts, 10 kW and A.C. The value 230 volts refers to...

(a) Average voltage (b) R.M.S. voltage
(c) Peak voltage (d) None of these

(I.T.I. August 2016, Electrician, Electroplator, Wireman)

73. The R.M.S. of sinusoidal voltage is given by the relation:

(a) $E_{RMS} = \frac{E_{max}}{2}$ (b) $E_{RMS} = \frac{E_{max}}{\sqrt{2}}$

(c) $E_{RMS} = \sqrt{2}\ E_{max}$ (d) $E_{RMS} = \frac{E_{max}}{2\sqrt{2}}$

(I.T.I. February 2019, Wireman)

74. Inductive reactance X_L = ...

(a) $2\pi fL$ (b) $\frac{1}{2\text{p}fL}$

(c) πfL (d) $\frac{1}{\text{p}fL}$

(I.T.I. August 2016, Electrician, Electroplator, Wireman)

75. For a symmetrical wave form, the average value of one full cycle is ...

(a) 1 (b) 1.11
(c) 2.22 (d) zero

(I.T.I. August 2016, Electrician, Electroplator, Wireman)

76. KVAR is also called as ...

(a) Reactive power (b) Apparent power
(c) Active power (d) Capacitive power

(I.T.I. August 2016, Electrician, Electroplator, Wireman)

77. A resistor and a capacitor are in series across a 20 volts A.C. source, circuit impedance is 4.33 kilo ohms, current flow in the circuit is ...

(a) 9.2 mA (b) 92 mA
(c) 4.6 mA (d) 460 mA

(I.T.I. August 2016, Electrician, Electroplator, Wireman)

78. The direction of current in an A.C. circuit ...

(a) is from +ve to –ve
(b) is always in one direction
(c) varies from instant to instant
(d) cannot be determined

(I.T.I. February 2017, Electrician, Electroplator, Wireman)

79. Power factor of A.C. circuit is equal to ....

(a) tan of phase angle
(b) cosine of phase angle
(c) sine of phase angle
(d) None of these

(I.T.I. August 2016, Electrician, Elatroplator, Wireman)

80. Power factor can be improved by ...

(a) Static capacitor
(b) Synchronous condenser
(c) Phase advancers
(d) All of these

(I.T.I. August 2016, Electrician)

81. In purely capacitive load, the power factor can be given as ...

(a) unity power factor
(b) 0.8 power factor leading

(c) ZPF leading
(d) ZPF lagging

(I.T.I. February 2017, Wireman)

HINT:

ZPF means zero power factor. In pure capacitive load, the phase angle is 90° leading.

∴ Power factor = cosine of angle
= cos 90° = 0

Hence we can say ZPF leading.

82. A resistance of 20 kilo ohms has a conductance of ...

(a) 20 siemen (b) 0.05 siemen
(c) 0.05 milli siemen (d) 20 kilo siemens

(I.T.I. February 2017, Electrician, Electroplator, Wireman)

83. Three elements having conductance G_1, G_2 and G_3 are connected in parallel. Their combined conductance will be ...

(a) $(G_1 + G_2 + G_3)^{-1}$
(b) $G_1 + G_2 + G_3$
(c) $1/G_1 + 1/G_2 + 1/G_3$
(d) $(1/G_1 + 1/G_2 + 1/G_3)^{-1}$

(I.T.I. February 2018, Electrician, Electroplator, Wireman)

84. How the low power factor can be improved in A.C. circuits?

(a) By connecting resistors in series
(b) By connecting capacitors in series
(c) By connecting inductors in series
(d) By connecting capacitors in parallel

(I.T.I. August 2019 Electrician, Electroplator, Wireman, Lift & Escalator Mechanic)

85. Calculate the power factor of R.L.C. circuit having resistance (R) = 15 Ω, resultant reactance (X) = 20 Ω connected across 240 volts, 50 Hz supply.

(a) 0.5 (b) 0.8
(c) 0.7 (d) 0.6

(I.T.I. August 2019 Electrician, Electroplator, Wireman, Lift & Escalator Mechanic)

86. What is the power factor if one of the wattmeters reading is zero and the other reads total power, in two wattmeter method, three phase power measurement?

(a) unity (b) zero
(c) 0.5 (d) Below 0.5

(I.T.I. August 2019 Electrician, Electroplator, Wireman, Lift & Escalator Mechanic)

87. What is the formula to calculate the impedance (Z) of the R.L.C. series circuit, if the inductive reactance (X_L) is less than capacitive reactance (X_C)?

(a) $Z = \sqrt{R^2 + (X_C - X_L)^2}$

(b) $Z = \sqrt{R^2 + (X_L - X_C)^2}$

(c) $Z = \sqrt{R^2 + (X_L + X_C)^2}$

(d) $Z = R^2 + \sqrt{X_L^2 - X_C^2}$

(I.T.I. August 2019 Electrician, Electroplator, Wireman, Lift & Escalator Mechanic)

88. What is the formula to calculate the three phase active power (P) if the line voltage = V_L; Line current = I_L and phase angle is "θ"?

(a) $P = 3\ V_L \cdot I_L \sin\theta$
(b) $P = 3\ V_L \cdot I_L \cdot \cos\theta$
(c) $P = \sqrt{3}\ V_L \cdot I_L \cdot \cos\theta$
(d) $P = \sqrt{3}\ V_L \cdot I_L \cdot \sin\theta$

(I.T.I. August 2019 Electrician, Electroplator, Wireman, Lift & Escalator Mechanic)

89. Calculate the apparent power of a star connected 3 phase load, if it is connected across 3 phase 415 volt, 50 Hz supply at 0.8 power factor and the phase current is 10 amperes.

(a) 12.45 kVA (b) 7.188 kVA
(c) 3.320 kVA (d) 57.50 kVA

(I.T.I. August 2019 Electrician, Electroplator, Wireman, Lift & Escalator Mechanic)

90. What is the relation between the line current I_L and phase current I_P in delta connected system?

(a) $I_L = I_P$ (b) $I_L = \sqrt{3}\ I_P$

(c) $I_L = 3\ I_P$ (d) $I_L = \frac{I_P}{\sqrt{3}}$

(I.T.I. August 2019 Electrician, Electroplator, Wireman, Lift & Escalator Mechanic)

91. Three similar inductive coils of resistance 20 ohms and reactance of 30 ohms are connected in mesh across a 3 phase, 400 V supply. What will be phase current?

(a) 11.1 amperes (b) 1.11 amperes

(c) 111 amperes (d) 0.111 ampere

(I.T.I. August 2019 Electrician)

92. What is the formula to calculate line current (I_L) of this single phase R.C. parallel circuit?

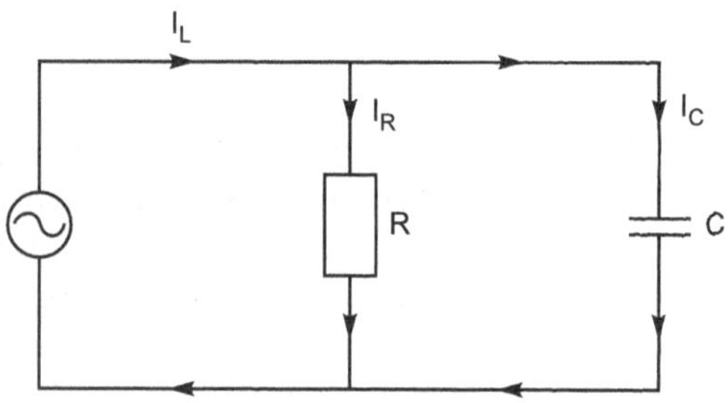

(a) $I_L = I_R - I_C$ (b) $I_L = I_R + I_C$

(c) $I_L = I_R^2 + I_C^2$ (d) $I_L = \sqrt{I_R^2 + I_C^2}$

(I.T.I. August 2019 Electrician, Electroplator, Wireman, Lift & Escalator Mechanic)

93. Calculate the value of admittance (Y) of the R.L.C. parallel circuit connected across 240 volts, 50 Hz A.C. supply and 8 amperes current is passed through it.

(a) 0.033 mho (b) 0.33 mho

(c) 3.33 mho (d) 0.003 mho

(I.T.I. August 2019 Electrician, Electroplator, Wireman, Lift & Escalator Mechanic)

94. The frequency of a tuned circuit at resonance (F_T) is ....

(a) $\frac{1}{\sqrt{2pLC}}$ (b) $\frac{1}{2p\sqrt{LC}}$

(c) $\frac{1}{2pLC}$ (d) $\frac{2p}{\sqrt{LC}}$

95. Calculate the frequency oscillation of a tank if value of inductance L = 100 mH and value of capacitance C = 490 μF.

(a) 22.7 Hz (b) 49.2 Hz

(c) 100 Hz (d) 250 Hz

KEY TO MULTIPLE CHOICE QUESTIONS

1. (a)	**2.** (c)	**3.** (d)	**4.** (b)	**5.** (b)	**6.** (d)	**7.** (b)	**8.** (d)	**9.** (d)	**10.** (a)
11. (b)	**12.** (b)	**13.** (a)	**14.** (a)	**15.** (a)	**16.** (b)	**17.** (c)	**18.** (b)	**19.** (b)	**20.** (a)
21. (c)	**22.** (b)	**23.** (b)	**24.** (c)	**25.** (c)	**26.** (c)	**27.** (d)	**28.** (d)	**29.** (d)	**30.** (d)
31. (a)	**32.** (b)	**33.** (c)	**34.** (b)	**35.** (b)	**36.** (c)	**37.** (b)	**38.** (d)	**39.** (a)	**40.** (a)
41. (c)	**42.** (b)	**43.** (b)	**44.** (b)	**45.** (b)	**46.** (a)	**47.** (c)	**48.** (c)	**49.** (a)	**50.** (b)
51. (c)	**52.** (c)	**53.** (a)	**54.** (a)	**55.** (b)	**56.** (a)	**57.** (c)	**58.** (c)	**59.** (c)	**60.** (d)
61. (d)	**62.** (d)	**63.** (c)	**64.** (a)	**65.** (a)	**66.** (b)	**67.** (a)	**68.** (a)	**69.** (d)	**70.** (a)
71. (a)	**72.** (b)	**73.** (b)	**74.** (a)	**75.** (d)	**76.** (a)	**77.** (c)	**78.** (c)	**79.** (b)	**80.** (d)
81. (c)	**82.** (c)	**83.** (b)	**84.** (d)	**85.** (d)	**86.** (c)	**87.** (a)	**88.** (c)	**89.** (b)	**90.** (b)
91. (a)	**92.** (d)	**93.** (a)	**94.** (b)	**95.** (a)					

Module 7

Cells and Batteries

MULTIPLE CHOICE QUESTIONS

1. Electro-chemical equivalent is :
   (a) ratio of atomic weight of an element to atomic weight of hydrogen
   (b) mass of the element liberated per unit quantity of hydrogen
   (c) ratio of atomic weight to valency
   (d) none of the above
2. Impurities in an electrolyte can cause an internal short-circuit condition called :
   (a) depolarization (b) electrolysis
   (c) local action (d) polarization
3. Distilled water is used in electrolytes because it :
   (a) prevents or slows down local action
   (b) speeds up electrochemical action
   (c) improves specific gravity
   (d) prevents polarisation
4. In a primary cell, the local action can be rectified by :
   (a) charging the cell
   (b) amalgamating the zinc electrode with mercury
   (c) using a cell for few minutes only
   (d) making a dry cell
5. Polarisation in a primary cell can be removed by :
   (a) coating the electrodes of the cell
   (b) chemical action
   (c) discharging the cell
   (d) disposing the cell
6. An average dry cell gives an approximate e.m.f. of :
   (a) 1.2 V (b) 1.3 V
   (c) 1.4 V (d) 1.5 V
7. Lechlanche cell is most suited for :
   (a) continuous use
   (b) large current requirements
   (c) intermittent use
   (d) large voltage requirements
8. When n cells, each e.m.f. E volts and internal resistance r ohms are connected in series, the current I through the external resistor of R ohms is given by the formula :
   (a) $I = \frac{E}{R + \frac{r}{n}}$ (b) $I = \frac{E}{R + n}$
   (c) $I = \frac{n.E}{R + n.r}$ (d) $I = \frac{E}{r + n.R}$
9. Which one of the following acts as depolariser in a dry cell ?
   (a) Manganese dioxide
   (b) Zinc chloride
   (c) Ammonium chloride
   (d) Carbon powder
10. Gassing occurs in the process of :
   (a) charging an accumulator
   (b) charging a dry cell
   (c) discharging an accumulator
   (d) discharging a dry cell

11. A lead-acid accumulator should be recharged when the specific gravity of its electrolyte is about :

(a) 1.18 (b) 1.15
(c) 1.25 (d) 1.35

12. The internal resistance of a discharged battery:

(a) is high
(b) is low
(c) remains the same
(d) is negative

13. The advantage of nickel-iron cell over lead-acid cell is this that :

(a) it has a higher efficiency
(b) it needs less maintenance
(c) it is much cheaper
(d) it gives a higher e.m.f.

14. A lead-acid cell is rechargeable because :

(a) its electrolyte is sulphuric
(b) it is a wet cell
(c) its chemical action is reversible
(d) its electrolyte has a high specific gravity

15. 6 cells of 1.5 V each are connected in series as shown in the figure. Voltmeter connected across the terminals + and – reads :

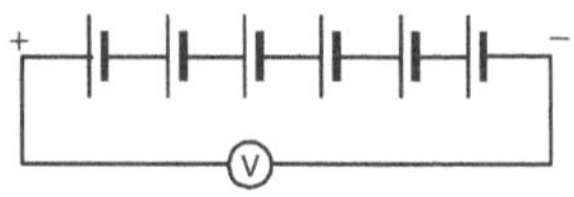

(a) 4.5 V (b) 6.0 V
(c) 7.5 V (d) 9.0 V

16. Which one of the following is not a maintenance free battery ?

(a) Ni - Cd
(b) Ni - Fe
(c) Alkaline magnesium
(d) Lead-acid

17. As compared to lead-acid cell, the efficiency of a Ni - Fe cell is less due to :

(a) lower e.m.f.
(b) smaller quantity of electrolyte
(c) higher internal resistance
(d) compactness of the cell

18. The e.m.f. produced by a zinc-carbon cell is —

(a) 1.2 V (b) 1.5 V
(c) 1.8 V (d) 2.0 V

19. In order to increase the ampere-hour rating of a battery, cells are connected in —

(a) series (b) parallel
(c) series parallel (d) star

20. A lead-acid battery should not be used, if it is discharged beyond the minimum value of —

(a) 1.2 V per cell (b) 1.5 V per cell
(c) 1.7 V per cell (d) 2.0 V per cell

21. A discharged battery should not be tested by using—

(a) high rate discharge tester
(b) voltmeter
(c) hydrometer
(d) any of the above stated instruments

22. In a lead-acid battery, sulphation accurs due to—

(a) keeping the battery idle for a long time
(b) trickle charging
(c) high rate discharging
(d) high rate charging

23. The active materials of Ni-Fe cell are —

(a) nickel hydroxide
(b) powdered iron and ferrous hydroxide
(c) 21% solution of caustic potash
(d) all the above stated materials

24. During charging or discharging of a Ni-Fe cell—

(a) its e.m.f. remains constant
(b) water is neither formed nor absorbed
(c) corrosive fumes are produced
(d) nickel-hydroxide remained unsplit

25. The battery charging method shown in the given figure is —

(a) constant voltage charging method
(b) constant current charging method
(c) rectifier battery charging method
(d) trickle charging method

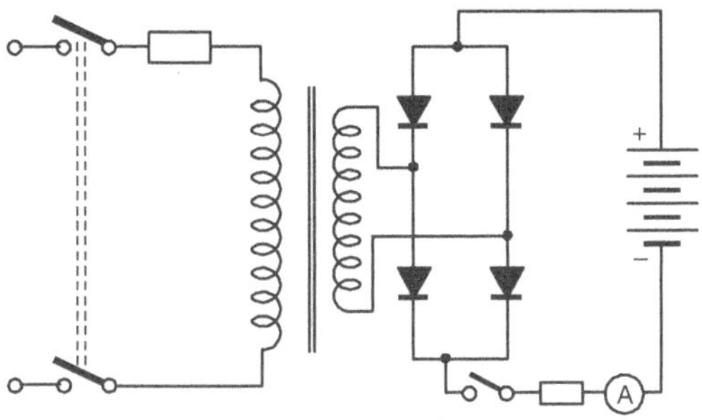

26. In constant voltage battery charging method, the voltage applied across the battery is —

(a) equal to the battery nominal voltage
(b) twice the battery nominal voltage
(c) 2.3 to 2.5 V per cell
(d) half the nominal voltage of the battery

27. Boost charge is given to a —

(a) dead battery
(b) battery in danger of becoming over-discharged during a working shift
(c) fully charged battery
(d) completely discharged battery

28. Which type of battery is generally used in emergency lights —

(a) Primary (b) Alkaline
(c) Lead-acid (d) Edison

29. Which gas is produced during charging a load-acid battery —

(a) Hydrogen (b) Methane
(c) Oxygen (d) Carbon dioxide

30. The nominal e.m.f. of a Ni-Cd cell is —

(a) 1.25 V (b) 1.50 V
(c) 1.00 V (d) 2.00 V

31. Which type of cell is shown in the given figure?

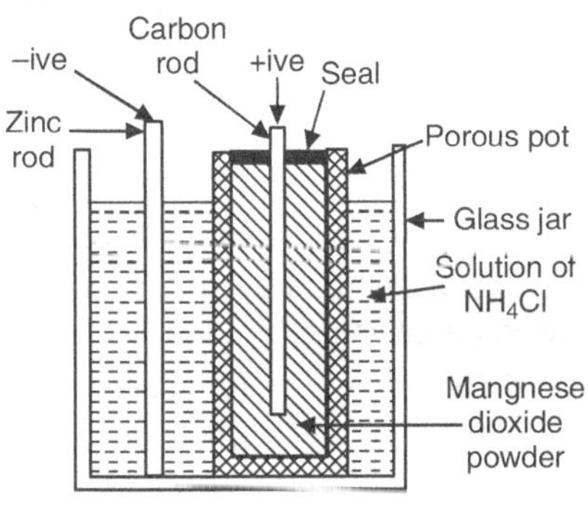

(a) Daniel cell (b) Valtaic cell
(c) Lechlanche cell (d) Lead-acid cell

32. As compared to a lead-acid cell, the efficiency of a nickel-iron cell (Edison cell) is less due to the —

(a) lower e.m.f.
(b) smaller quantity of electrolyte
(c) higher internal resistance
(d) compactness of the cell

33. The colour of positive plate of a lead acid battery is ......

(a) brown (b) grey
(c) white (d) black

34. The colour of negative plate of a lead acid battery is

(a) brown (b) grey
(c) white (d) black

35. The normal voltage of a fully charged lead acid cell is ......

(a) 1.1 V (b) 1.5 V
(c) 2.2 V (d) 3.0 V

36. The relative density of the electrolyte in a lead acid battery is 1.280. This value indicates that the battery is ....

(a) fully discharged
(b) half charged
(c) three quarter charged
(d) fully charged

37. The active material used in positive plates of lead acid cell is ......

(a) Pb (b) PbO_3
(c) PbO_2 (d) $PbSO_4$

(I.T.I. February 2015, Electrician)

38. Which electrolyte is used in lead acid battery?

(a) Potassium Hydroxide (KOH)
(b) Ammonium Chloride (NH_4CL)
(c) Dilute Sulphuric acid (H_2SO_4)
(d) Alkaline solution

(I.T.I. February 2015, Electrician)

39. What defect will occur in the lead acid cell, when the cell being left in a discharged condition for a long time?

(a) Partial short (b) Buckling
(c) Hard sulphation (d) Depolarization

(I.T.I. August 2014, Electrician)

40. Which material is applied to the terminal posts of a lead acid battery to avoid corrosion?

(a) Ammonium sulphate
(b) Silica gel
(c) Petroleum jelly
(d) Sulphur powder

(I.T.I. February 2015, Electrician)

41. What is the reason to keep open the vent plugs of lead acid cell during charging?

(a) To escape the gas freely
(b) To speed up the charging rate
(c) To increase the specific gravity of the electrolyte
(d) To reduce the sulphation defect.

(I.T.I. February 2015, Electrician)

42. How many lead acid cells are to be grouped for 12 V battery and by which connection?

(a) 6 cells in parallel
(b) 6 cells in series
(c) 12 cells in parallel
(d) 12 cells in series

(I.T.I. February 2014, Electrician)

43. A 120 Ah capacity battery can deliver a current of 8 A for a period of ......

(a) 20 hrs (b) 15 hrs
(c) 12 hrs (d) 8 hrs

(I.T.I. February 2015, Electrician)

44. Which instrument is used to check the battery for its charging condition by measuring the voltage?

(a) Hydrometer
(b) Volt meter
(c) High rate discharge tester
(d) Multimeter

(I.T.I. February 2015, Electrician)

45. In electrolysis process the mass of the substances liberated (or) deposited is directly proportional to ....

(a) current and time
(b) voltage and resistance
(c) Internal resistance
(d) specific resistance of the plate

(I.T.I. August 2014, Electrician)

46. Battery electrolyte consists of approximately (by volume) ....

(a) 35 percent sulphuric acid and 65 percent water.
(b) 65 percent sulphuric acid and 35 percent water.
(c) 65 percent nitric acid and 35 percent water.
(d) 35 percent nitric acid and 65 percent water.

47. The number of positive plates in a battery cell is ....

(a) one more than the negative plates.
(b) one less than the negative plates.
(c) two less than the negative plates.
(d) two more than the negative plates.

(I.T.I. February 2014, Mech. Motor Vehicle)

48. Figure shows an instrument being used to check the specific gravity of the electrolyte of the battery. This instrument is called—

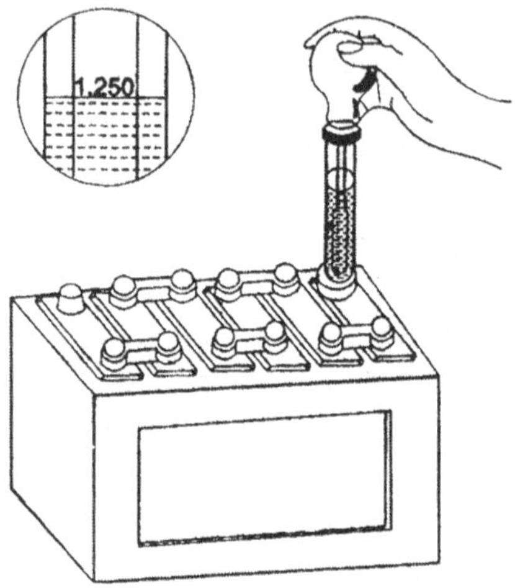

(a) Hydrometer (b) Hygrometer
(c) Hydroscope (d) none of these

49. Each solar cell has a open circuit voltage having equal value to ....

(a) 0.55 V (b) 0.45 V
(c) 0.35 V (d) 0.25 V

50. The amount of voltage that can be obtained from a single cell of a storage battery is about ....

(a) 2 V (b) 4 V
(c) 6 V (d) 8 V

51. What is the output voltage of the battery shown in figure?

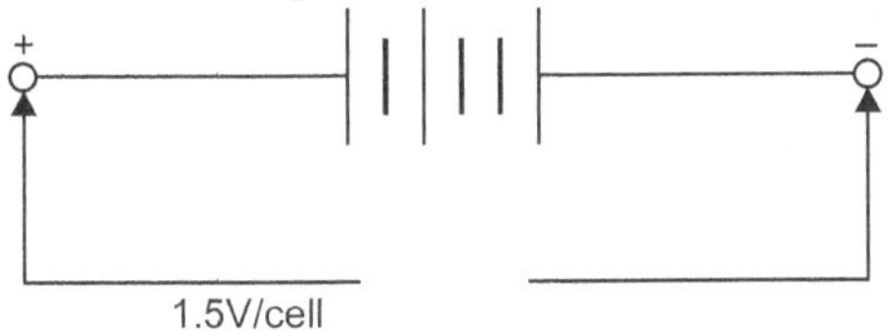

(a) 6 V (b) 4.5 V
(c) 3 V (d) 1.5 V

(I.T.I. February 2014, Electrician)

52. What is the effect on internal resistance of the cell, if the cell, it the cell is kept idle for long time?

(a) Remain the same
(b) Increased
(c) Decreased
(d) Reduced to zero

(I.T.I. February 2014, Electrician)

53. What is the name of the defect occured in simple voltaic cell, due to the presence of impurities on the zinc plate?

(a) Polarisation (b) Local action
(c) Sulphation (d) Buckling

(I.T.I. February 2015, Electrician)

54. The mass of a substance liberated from an electrolyte by one coulomb of electricity is termed as ....

(a) quantity of electricity
(b) electro chemical equivalent
(c) electric charge
(d) electro chemical expression

(I.T.I. February 2015, Electrician)

55. Four number of cells are connected in parallel and the rating of each cell is 1.5 V and 16 Ah. What will be the "Ah" rating of this battery?

(a) 16 Ah (b) 32 Ah
(c) 48 Ah (d) 64 Ah

(I.T.I. February 2015, Electrician)

56. At the time of charging a lead acid battery which one of the harmful gas evolute?

(a) Oxygen (b) Nitrogen
(c) Hydrogen (d) Carbon dioxide

(I.T.I. February 2014, Mech. Motor Vehicle)

57. Identify the fault in the battery charging circuit.

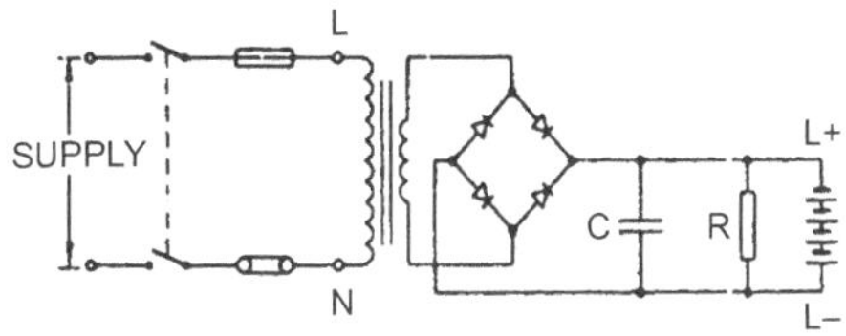

(a) Line and Neutral terminal of transformer are reversed with respect to incoming supply.
(b) Polarity of battery connection to rectifier is reversed.
(c) Capacitor should be in series with Battery.
(d) One of the diode is reversed.

(I.T.I. January 2014, Electrician)

58. What type method of charge can be done on a battery with fully discharged and kept for a long time?

(a) Trickle charge
(b) Constant current
(c) Constant voltage
(d) Rectifier charging

(I.T.I. August 2015, Electrician, Wireman)

59. When the current is passed through an elecrtrolyte solution, it decomposes into ions. What is the type of effect of electric current of the above process ?

(a) Heating effect
(b) Chemical effect
(c) Electrostatic effect
(d) 'X'-ray effect

(I.T.I. August 2015, Electrician)

60. The formula to find the mass liberated from an electrode based on Faraday's first law of electrolysis is ....

(a) $m = \frac{z}{i.t}$ (b) $m = z.i.t$

(c) $m = \frac{2.i}{t}$ (d) $m = \frac{i.t}{z}$

(I.T.I. August 2015, Electrician)

61. The name of the electrolyte used in the form of moist paste in dry cell is called ....
(a) potassium hydroxide
(b) diluted sulphuric acid
(c) ammonium chloride
(d) alkaline solution
(I.T.I. August 2015, Electrician)

62. The cause for occurring buckling defect in lead acid battery is due to ......
(a) keeping the cell in discharged condition for long period
(b) the short circuiting between the plates
(c) over charging and discharging
(d) the over load to the battery
(I.T.I. February 2016, Electrician, Wireman)

63. When the battery is charged at a very low rate 2 to 3% of the normal rate for a long period it is said to be ....
(a) constant current charge
(b) trickle charge
(c) constant voltage change
(d) rectifier change
(I.T.I. February 2016, Electrician, Wireman)

64. The capacity of the battery is expressed in ....
(a) watt - hour (b) ampere/sec
(c) ampere - hour (d) kilo watt
(I.T.I.February 2016, Electrician, Wireman)

65. The material used as positive electrode in lithium cell is ...
(a) lithium
(b) carbon
(c) zinc
(d) manganese di-oxide
(I.T.I. February 2016, Electrician, Wireman)

66. The primary cell which has the longest life is .....
(a) Lithium cell
(b) Alkaline cell
(c) Mercury cell
(d) Carbon - Zinc cell
(I.T.I. August 2015, Electrician, Wireman)

67. An example of active photo semiconductor is ....
(a) Photo resistor
(b) Photo diode
(c) Photo transistor
(d) Photo voltaic cells
(I.T.I. February 2016, Electronic Mechanic)

68. The material used in solar cell as positive output terminal is ....
(a) nickel plated ring
(b) chromium plated ring
(c) Silver coated ring
(d) zinc plated ring
(I.T.I. August 2015, Electrician)

69. The material used as negative output terminal of solar cell is ...
(a) nickel plated ring
(b) chromium plating
(c) button plating
(d) zinc plating
(I.T.I. February 2016, Electrician)

70. In a online UPS system, the switch over time from mains to battery is ....
(a) 10 mill seconds (b) 9 mill seconds
(c) 8 mill seconds (d) zero seconds
(I.T.I. August 2015, Electrician)

71. An electronic device, that converts Direct Current (DC) to Alternating Current (AC), is called as .....
(a) converter (b) rectifier
(c) inverter (d) diode
(I.T.I. February 2016, Electrician)

72. Five cells of 2 V each having R_{in} = 0.2 Ω are connected in series to a load of resistance 14 Ω. The current flowing in the circuit is .....
(a) 10 A (b) 1.4 A
(c) 1.5 A (d) $\frac{2}{3}$A
(I.T.I. February 2017, Electrician, Electroplator, Wireman)

73. When an electric current flows through an electrolyte and produces some the chemical action, the process is called ....

(a) Photo synthesis
(b) Electrolysis
(c) Electromagnetism
(d) Electro chemical

(I.T.I. February 2016, Electrician, Electroplator, Wireman)

74. A battery consists of a ....

(a) cell (b) circuit
(c) generator (d) number of cells

(I.T.I. February 2017, Electrician, Electroplator, Wireman)

75. The open circuit voltage of a 12 V battery is .....

(a) 12 volts
(b) More than 12 volts
(c) Less than 12 volts
(d) None of these

(I.T.I. August 2016, Electrician, Electroplator, Wireman)

76. ampere second could be the unit of ....

(a) Power (b) Conductance
(c) Energy (d) Charge

(I.T.I. August 2016, Electrician, Electroplator, Wireman)

77. Electroplating takes place on ....

(a) The cathode only
(b) The anode only
(c) Both the anode and the cathode
(d) Either the anode or the cathode

(I.T.I. August 2016, Electrician, Electroplator, Wireman)

78. The positive pole of dry battery is made up of ....

(a) carbon (b) copper
(c) zinc (d) mercury

(I.T.I. February 2017, Electrician, Electroplator, Wireman)

79. When main supply fails, alternate supply is obtained immediately and automatically by ....

(a) Diesel generator
(b) Gas turbine generator
(c) U.P.S.
(d) Battery source

(I.T.I. August 2016, Wireman)

80. The storage battery generally used in Electric power station is.........

(a) Nickel cadmium battery
(b) Zinc carbon battery
(c) Lead acid battery
(d) None of these

(I.T.I. August 2016, Electrician, Electroplator, Wireman)

81. The life of a lead acid battery is expected to be....

(a) 6 months (b) 1 years
(c) 2 to 5 years (d) 10 to 15 years

(I.T.I. August 2016, Electrician, Electroplator, Wireman)

82. The energy of secondary cells is usually renewed by ....

(a) Passing current through it
(b) Heating it
(c) It cannot be renewed at all
(d) Renewing its chemicals

(I.T.I. February 2017, Electrician, Electroplator, Wireman)

83. Cells are connected in series in order to increase the ....

(a) current capacity
(b) Life of the cells
(c) Voltage rating
(d) Terminal voltage

(I.T.I. February 2019, Wireman)

84. Electric battery is a device that:

(a) Generates e.m.f. by chemical action
(b) Converts heat energy into electrical energy
(c) Converts mechanical energy into electrical energy
(d) Converts fuel energy into elecrical energy

(I.T.I. February 2019, Wireman)

85. Which effect causes by-passing electric current in liquids?

(a) Heating (b) Lighting
(c) Chemical (d) Magnetic

(I.T.I. August 2019 Electrician)

86. Which material is used to make negative plates in lead acid battery?

(a) Lead dioxide (b) Lead sulphate
(c) Lead peroxide (d) Sponge lead

(I.T.I. August 2019 Electrician)

87. Which technique is used to control the corrosion of a metal surface?

(a) Cathodic protection
(b) Anodic protection
(c) Electrolytic protection
(d) Electrostatic protection

(I.T.I. August 2019 Electrician)

88. According to Faraday's second law, the weights of different substances evolved by the passage of same quantity of electricity are proportional of their ......

(a) chemical equivalent weight
(b) current supply
(c) chemical equivalent density
(d) volume

(I.T.I. August 2019 Electrician)

89. Easy to shape materials like ceramics are used as ......

(a) Conductors (b) Insulators
(c) Electrodes (d) Catalysts

(I.T.I. August 2019 Electrician)

90. What is the effect if one cell is connected with reverse polarity in a parallel combination circuit?

(a) Voltage becomes zero
(b) Becomes open circuit
(c) No effect will function normally
(d) Will get short circuited

(I.T.I. August 2019 Electrician)

91. The process of producing chemical decomposition of a compound by passing electricity through the compound is called .....

(a) Electrolysis
(b) Electroplating
(c) Electrolyte
(d) None of the these

(I.T.I. August 2019 Electrician)

92. What is the function of fine selector switch in battery charger?

(a) Selection of voltage range
(b) Selection of charging time
(c) Selection of current rating
(d) Selection of charging method

(I.T.I. August 2019 Electrician)

93. For what purpose the hydrometer is used during charging of battery?

(a) Determine the specific gravity of electrolyte
(b) Assess the battery voltage level
(c) Assess the discharge level of battery
(d) Determine the ampere hour capacity

(I.T.I. August 2019 Electrician)

94. Which cell is most often used in digital watches?

(a) Voltaic (b) Lithium
(c) Silver oxide (d) Mercury

(I.T.I. August 2019 Electrician)

95. What is the effect on output power with respect to temperature in solar cells?

(a) Decreases with decrease in temperature
(b) Increases with increase in temperature
(c) Decreases with increase in temperature
(d) No effect on change in temperature

(I.T.I. August 2019 Electrician)

KEY TO MULTIPLE CHOICE QUESTIONS

1. (b)	**2.** (c)	**3.** (a)	**4.** (b)	**5.** (b)	**6.** (d)	**7.** (c)	**8.** (c)	**9.** (a)	**10.** (a)
11. (a)	**12.** (a)	**13.** (b)	**14.** (c)	**15.** (d)	**16.** (d)	**17.** (c)	**18.** (b)	**19.** (b)	**20.** (c)
21. (a)	**22.** (a)	**23.** (d)	**24.** (b)	**25.** (c)	**26.** (c)	**27.** (b)	**28.** (c)	**29.** (a)	**30.** (a)
31. (c)	**32.** (c)	**33.** (a)	**34.** (b)	**35.** (c)	**36.** (d)	**37.** (c)	**38.** (c)	**39.** (c)	**40.** (c)
41. (a)	**42.** (b)	**43.** (b)	**44.** (c)	**45.** (a)	**46.** (a)	**47.** (b)	**48.** (a)	**49.** (a)	**50.** (a)
51. (d)	**52.** (b)	**53.** (b)	**54.** (b)	**55.** (d)	**56.** (c)	**57.** (b)	**58.** (a)	**59.** (b)	**60.** (b)
61. (c)	**62.** (c)	**63.** (b)	**64.** (c)	**65.** (b)	**66.** (a)	**67.** (d)	**68.** (a)	**69.** (c)	**70.** (d)
71. (c)	**72.** (d)	**73.** (b)	**74.** (d)	**75.** (a)	**76.** (d)	**77.** (a)	**78.** (a)	**79.** (c)	**80.** (c)
81. (c)	**82.** (a)	**83.** (c)	**84.** (a)	**85.** (c)	**86.** (d)	**87.** (a)	**88.** (a)	**89.** (c)	**90.** (d)
91. (a)	**92.** (c)	**93.** (a)	**94.** (d)	**95. (a)**					

Module 8

Basic Wiring Practice

Multiple Choice Questions

Electrical Accessories

1. The fuse current in amperes is related to fuse wire's diameter is:
   (a) $I \propto \frac{1}{D}$ (b) $I \propto D$
   (c) $I \propto D^{3/2}$ (d) $I \propto D^2$
2. Over-load current protection for a motor is provided by:
   (a) cartridge fuse
   (b) kit-kat fuse
   (c) over-load current relay
   (d) all the above
3. In a motor circuit, a fuse provides:
   (a) over-load current protection
   (b) short-circuit protection
   (c) open-circuit protection
   (d) none of the above
4. The quick response type catridge fuse is used for ......
   (a) lighting circuits (b) heating circuits
   (c) motor circuits (d) wiring circuits
   (I.T.I. February 2015, Electrician)
5. Different type of switches are being used in lighting circuits. Which switch is used for stair case wiring?
   (a) One way switch (b) Two way switch
   (c) Bed switch (d) Push switch
   (I.T.I. August 2014, Wireman)
6. How many two way switches are required for the stair case wiring?
   (a) 2 (b) 3
   (c) 4 (d) 5
   (I.T.I. February 2015, Wireman)
7. In which type of accessories, intermediate switch belongs?
   (a) Holding accessories
   (b) Safety accessories
   (c) Controlling accessories
   (d) Outlet accessories
   (I.T.I. August 2014, Wireman)
8. In what classification of accessories, the distribution board will belong?
   (a) Safety (b) Controlling
   (c) General (d) Holding
   (I.T.I. February 2014, Wireman)
9. The safe maximum current that can pass continuously without overheated is called as ....
   (a) fusing current (b) current rating
   (c) over load current (d) safe current
   (I.T.I. February 2015, Electrician)
10. Which type of outlet accessory is suitable for double insulated appliances having P.V.C. or insulated body?
    (a) Three pin socket (b) Two pin socket
    (c) Three pin plug (d) Two pin plug
    (I.T.I. August 2014, Electrician)

11. Which tool is used to measure the size of the conductor?

(a) Bevel gauge
(b) Vernier caliper
(c) Standard wire gauge
(d) Depth gauge

(I.T.I. February 2015, Electrician)

12. What is the advantage of stranded conductors over solid conductors?

(a) More flexible
(b) More area of cross section
(c) More resistance
(d) More weight

(I.T.I. August 2014, Electrician)

13. The outer screwed contacts of Edison screw type lamp holder must always be connected to the:

(a) neutral wire of the circuit
(b) phase wire of the circuit
(c) Switch (half wire) of the circuit
(d) earth wire of the circuit.

(I.T.I. February 2015, Electrician)

14. What type of accessory is shown in the figure?

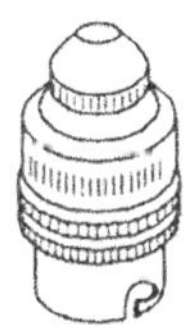

(a) Holding Accessories
(b) Controlling Accessories
(c) Safety Accessories
(d) Outlet Accessories

(I.T.I. February 2014, Electrician)

15. What is the type of metal conduit outlet box shown in the figure?

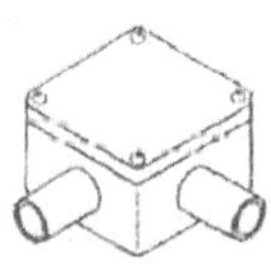

(a) 'T' type outlet box
(b) Angle type outlet box
(c) Through type outlet box
(d) Back type outlet box

(I.T.I. February 2015, Wireman)

16. Which type of junction box is used to connect three directional circuits, shown in the figure.

(a) 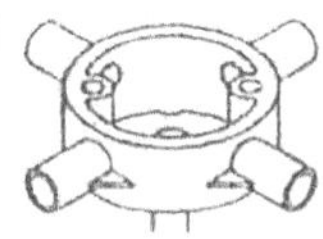
(b) 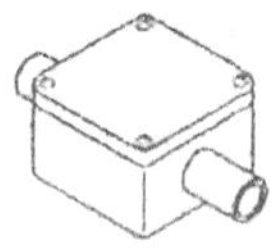
(c) 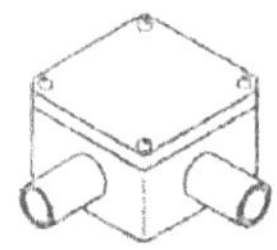
(d) 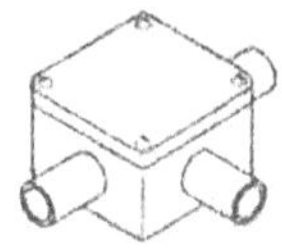

(I.T.I. August 2014, Wireman)

17. Name the part indicated as 'X' in the Distribution board shown in the figure?

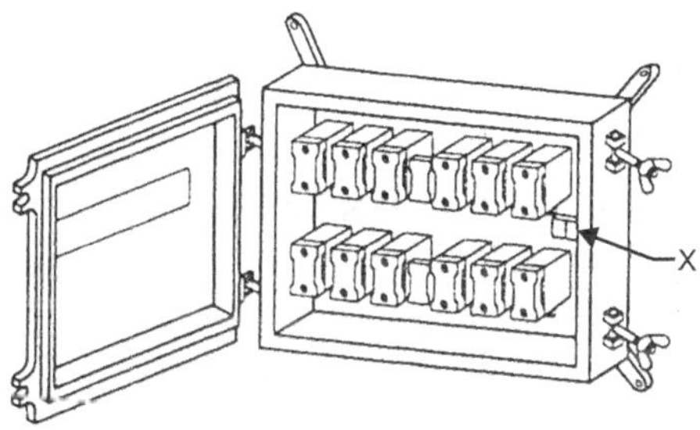

(a) Earth Terminal
(b) Neutral link
(c) Enclosure lock nut
(d) Fixing screw

(I.T.I. Febraury 2015, Wireman)

18. A terminal connector is shown in the figure. What is the specification of this terminal connector?

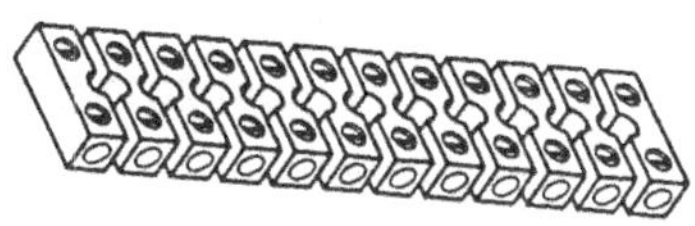

(a) 6 amperes, 12 ways, 250 V
(b) 16 amperes, 12 ways, 250 V
(c) 30 amperes, 12 ways, 500 V
(d) 32 amperes, 12 ways, 500 V

(I.T.I. February 2015, Wireman)

19. Name the electrical accessory shown in the figure?

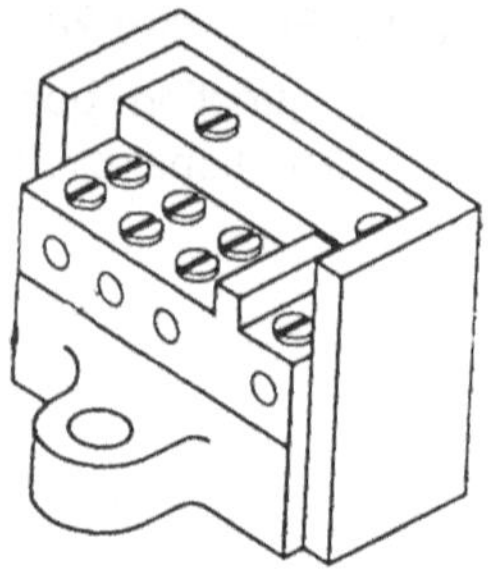

(a) Terminal connector
(b) Fuse cut out
(c) Earth Terminal
(d) Neutral link

(I.T.I. August 2014, Wireman)

20. It is mandatory to use Underwriter's knot in one of the accessories to avoid accidents. Name the accessory?

(a) Ceiling rose (b) Batten holder
(c) Bracket holder (d) 3 pin plug

(I.T.I. August 2015, Electrician Wireman)

21. Which electical accessory comes under the controlling accessory calssification ?

(a) 3 pin wall socket 16A/240V
(b) Two way switch flush type 6A/240V
(c) Batten lamp holder 6A/240V
(d) Rewireable fuse 5A

(I.T.I. February 2016, Electrician, Wireman)

22. The standard current ratings of the socket accessories are.....

(a) 6,16 and 32 amperes
(b) 5,16 and 32 amperes
(c) 5,15 and 32 amperes
(d) 6,15 and 32 amperes

(I.T.I. August 2015, Wireman)

23. What type of conduit accessory shown in the figure ?

(a) Inspection conduit bend
(b) Solid conduit bend
(c) Inspection conduit elbow
(d) Solid conduit elbow

(I.T.I. August 2015, Wireman)

24. What is the name of the accessory shown in the figure ?

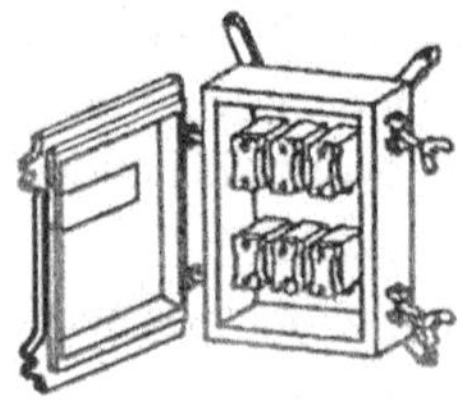

(a) 3 way distribution board
(b) 6 way distribution board
(c) 3 way fuse cut out
(d) Triple pole Iron clad switch

(I.T.I. August 2015, Wireman)

25. In a tunnel wiring, the type of switches used is ....

(a) SPT switch (b) T.W. switch
(c) I.M. switch (d) P.B. switch

(I.T.I. February 2016, Wireman)

26. Name the type of switch shown in the symbol below figure ?

(a) Two way switch
(b) Toggle switch
(c) Bell push switch
(d) Intermediate switch

(I.T.I. February 2016, Electrician)

27. The number of terminals available in an intermediate switch is ....

(a) 2 (b) 3
(c) 4 (d) 5

(I.T.I. February 2016, Wireman)

28. The voltage rating of TPIC switch is ....

(a) 400 V (b) 415 V
(c) 440 V (d) 450 V

(I.T.I. February 2016, Wireman)

29. The current rating of double pole iron clad switch is from ....

(a) 15 A ro 200 A (b) 16 A ro 220 A
(c) 16 A ro 200 A (d) 15 A ro 250 A

(I.T.I. August 2015, Wireman)

30. Edison type lamp holders are used above ....

(a) 200 watt (b) 220 watt
(c) 230 watt (d) 250 watt

(I.T.I. February 2016, Wireman)

31. Which type lamp holder is used for more than 300 watt lamps ?

(a) Bayonet cap lamp holders
(b) Edison screw-type lamp holders
(c) Goliath Edison screw type holders
(d) Swivel lamp-holders

(I.T.I. August 2015, Electrician)

32. Two pin socket and plug is rated as ....

(a) 6 A, 220 V (b) 6 A, 230 V
(c) 6 A, 240 V (d) 6 A, 250 V

(I.T.I. August 2015, Wireman)

33. The fuse blows off by ....

(a) Arcing (b) Burning
(c) Melting (d) Sparking

(I.T.I. February 2016, Electronic Machanic)

34. HRC full form is ....

(a) High Rating Capacity
(b) High Rupturing Capacity
(c) High Rating current
(d) High Rupturing Current

(I.T.I. February 2016, Wireman)

35. Which type of fuse is shown in the figure?

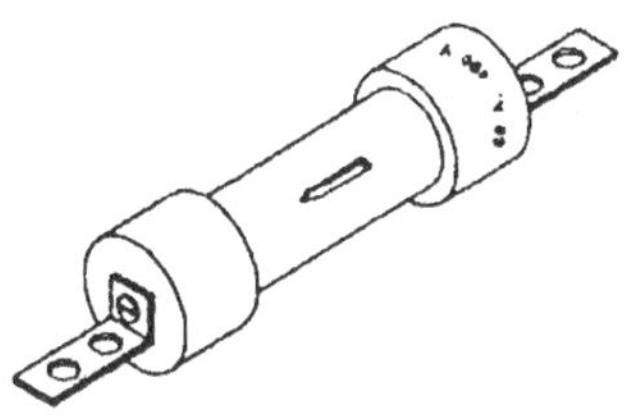

(a) Re-Wirable fuse
(b) High rupturing capacity fuse
(c) Cartridge fuse
(d) Kit-kat type fuse

(I.T.I. August 2015, Electrician)

36. The fusing factor of HRC fuse is ....

(a) 2.0 (b) 1.7
(c) 1.4 (d) 1.1

(I.T.I. August 2015, Wireman)
(I.T.I. February 2016, Electrician)

37. What is the fuse rating that can be selected for a 10 H.P. induction motor?

(a) 10 amperes (b) 15 amperes
(c) 20 amperes (d) 25 amperes

(I.T.I. August 2015, Electrician)

38. Formula for fusing factor is ....

(a) Fusing factor $= \dfrac{\text{Minimum fusing current}}{\text{Rated current}}$

(b) Fusing factor $= \dfrac{\text{Rated current}}{\text{Mimimum fusing current}}$

(c) Fusing factor = Minimum fusing current + Rated current

(d) Fusing factor = Minimum fusing current × Rated current

(I.T.I. August 2015, Wireman)

39. A fuse is ...

(a) Always connected in series with the circuit
(b) Always connected in parallel with the circuit
(c) Normally connected in series with the circuit
(d) Normally connected in parallel with the circuit

(I.T.I. August 2016, Electrician)

40. The material used for the fuse wire should be of ...

(a) Low resistivity and high melting point
(b) High resistivity and high melting point
(c) High resistivity and low melting point
(d) Low resistivity and low melting point

(I.T.I. August 2016, Electrician)

41. For motor circuit fuses are rated as ...

(a) Twice the full load current of the motor
(b) Thrice the full load current of the motor
(c) Starting current of the motor
(d) No load current of the motor

(I.T..August 2016, Wireman)

42. The main function of fuse is to ...
(a) Protect the line
(b) Prevent excessive currents
(c) Protect the appliance
(d) Open the circuit
(I.T.I. February 2017, Electrician)

43. Which among these is/are the main characteristics of a fuse element?
(a) High conductivity
(b) Low conductivity
(c) Maximum deterioration due to oxidation
(d) High melting point
(I.T.I. February 2019, Electrician)

44. The rating of fuse is always expressed in ...
(a) volts (b) amperes
(c) ampere–volts (d) ampere–hours
(I.T.I. August 2016 Electrician)

45. H.R.C. fuse provides best protection against ...
(a) short circuit
(b) Reverse current
(c) overload
(d) open circuits
(I.T.I. February 2017, Electrician)

46. We connect the fuse in ...
(a) Phase line (b) Neutral line
(c) Earth line (d) None of these
(I.T.I. February 2019, Electrician)

47. The lamp holder used for hanging light downward from the ceiling rose with flexible wire is called as ...
(a) Pendant lamp holder
(b) Water tight bracket holder
(c) Angle lamp holder
(d) Swivel lamp holder
(I.T.I. February 2017, Wireman)

48. The longest pin of a three pin top has bigger diameter, so that ...
(a) More phase current can flow in the circuit
(b) More neutral current can flow out of the circuit
(c) A fault current to ground flows with less resistance
(d) More loads can be connected to the same plug point
(I.T.I. February 2017, Wireman)

49. In a 3 phase motor circuit, the type of main switch used is ...
(a) ICDP (b) ICTP
(c) MCB (d) DOL
(I.T.I. Augusts 2016, Wireman)

50. Ceiling roses are available in the following combinations:
(a) 1 plate, 2 plate (b) 1 plate, 3 plate
(c) 1 plate, 4 plate (d) 2 plate, 3 plate
(I.T.I. February 2017, Wireman)

51. Which electrical accessory belongs to general classification of accessories?
(a) Fuse
(b) Ceiling roses
(c) Intermediate switch
(d) Pendent lamp holder
(I.T.I. August 2019 Electrician)

52. Zinc, Copper, Aluminium, or alloys are main composition of ......
(a) M.C.B. (b) Fuse
(c) A.C.B. (d) O.C.B.
(I.T.I. August 2019 Electrician)

53. Selection of fuse is based on ....
(a) Steady load
(b) Fluctuating load
(c) Both (a) and (b)
(d) None of these
(I.T.I. August 2019 Electrician)

54. What is the term for the time taken by a fuse to interrupt the circuit in fault?
(a) Time factor (b) Fusing factor
(c) Fusing current (d) Cut off factor
(I.T.I. August 2019 Electrician)

55. Cord grip and under writer's knot are used in......
(a) Lamp holder's terminals
(b) Ceiling roses
(c) Terminals of pendents
(d) All of these
(I.T.I. August 2019 Electrician)

56. Where is the iron clad double pole (I.C.D.P.) main switch used?
(a) Large industrial installations
(b) Control main or branch circuits
(c) Three phase power circuit installations
(d) Single phase domestic installations
(I.T.I. August 2019 Electrician)

Domestic and Industrial Wiring

Choose the correct answer:

1. The effect of smoke, acid and alkali is negligible on:

(a) T.R.S. wire
(b) V.I.R. wire
(c) Flexible wire
(d) Lead sheathed wire.

2. The minimum cross-sectional area of a copper conductor used in a 'light and fan' sub-circuit should be :

(a) 1 mm^2 (b) 2 mm^2
(c) 2.5 mm^2 (d) 3 mm^2

3. If machanical and fire safety are important factors, the wiring should be :

(a) lead sheathed type
(b) conduit type
(c) (c)T.S. type
(d) casing-caping type.

4. A stair case point is operated with :

(a) two one-way switches
(b) two two-way switches
(c) two intermediate switches
(d) none of these.

5. The rating factor which the normal current carrying capacity should be multiplied to get the permissible current in the cable for the circuits protected by close-excess current protection will be :

(a) 0.81 (b) 0.91
(c) 1.01 (d) 1.23

(I.T.I. July 2010, Electrician)

6.

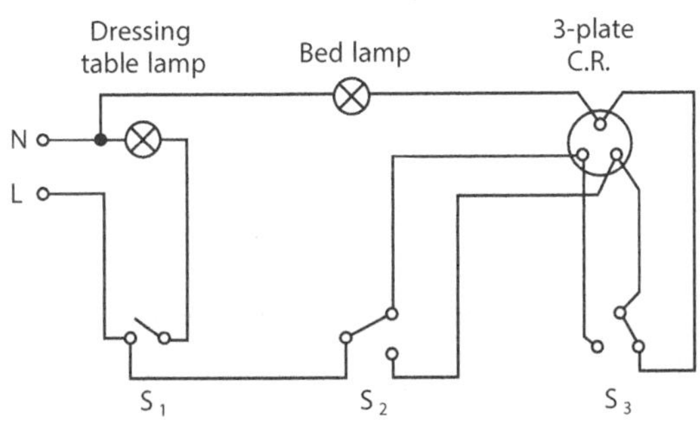

Which switch or switches can control the dressing table lamp ?

(a) Switch nos. 1 and 2
(b) Switch nos. 2 and 3
(c) Switch nos. 2
(d) Switch no. 1

7. At what maximum distance do you fix saddles from a switch box, in a conduit wiring ?

(a) 30 cm (b) 50 cm
(c) 60 cm (d) 70 cm.

8. To estimate the materials and cables, the diagram which is very useful will be :

(a) laying diagram and wiring diagram.
(b) installation plan and circuit diagram.
(c) circuit diagram and installation plan.
(d) wiring diagram and circuit diagram.

(I.T.I. July 2006, Electrician)

9. The current capacities of the cables are specified for certain ambient temperature in BIS regulations. The specified ambient temperature is ....

(a) 20°C (b) 30°C
(c) 40°C (d) 50°C

(I.T.I. February 2015, Wireman)

10. As per standard practice the circuit of different phases should be kept at ....

(a) 3.5 metres (b) 3.05 metres
(c) 2.55 metres (d) 2.05 metres

(I.T.I. February 2015, Electrician)

11. To determine the size of cable, the following formula given below as:

Cross sectional area of the wire

$$= \frac{2 ' K ' L ' I}{\text{Voltage drop}}$$

What does the letter '*L*' represent in the above formula?

(a) *L* = Length of the cable in metres
(b) *L* = Distance between the load centre from the supply ends in metres

(c) L = Distance between the load and earthing point in metres

(d) L = Length of the cable per each load point in metres

(I.T.I. August 2014, Wireman)

12. Polarity test ensures safety to the user and also to the electrician.

Verification of polarity in an installation is to ensure that ....

(a) the fuses are connected in the phase wire and the single pole switches are connected in the neutral wire

(b) the fuses are connected in the neutral wire and the single pole switches are connected to the phase wire

(c) the fuses and the single pole switches are connected in the phase wire

(d) the fuses and the single pole switches are connected in the neutral wire

(I.T.I. August 2014, Electrician)

13. Which statement describes a characteristic feature of routine preventive maintenance?

(a) Maintenance schedule needs to be decided, based on the maintenance reports entered in the maintenance card.

(b) Maintenance could be done either during the working of the machine or during short interval down periods.

(c) Maintenance is performed at irregular frequencies.

(d) Maintenance is only performed, if a machine has a fault or defect.

(I.T.I. July 2014, Electrician)

14. On wiring installation, as per BIS standard the minimum value of Insulation Resistance (R_i) expressed as ....

(a) $R_i = \dfrac{5000}{\text{No. of points + outlets}}$

(b) $R_i = \dfrac{500}{\text{No. of points + outlets}}$

(c) $R_i = \dfrac{250}{\text{No. of points}}$

(d) $R_i = \dfrac{50}{\text{No. of points + outlets}}$

(I.T.I. February 2015, Wireman)

15. During continuity test in an installation, the megger indicated reading is ....

(a) 0 megohm (b) 1 megohm

(c) 500 megohm (d) Infinity megohm

(I.T.I. February 2015, Wireman)

16. How many light points can be allowed in one circuit?

(a) 5 points (b) 10 points

(c) 15 points (d) 20 points

17. How many power points can be allowed in one circuit?

(a) Four (b) Three

(c) Two (d) One

18. The rating factor of the circuit having coarse excess current protection is ....

(a) 1.63 (b) 1.23

(c) 0.81 (d) 0.61

(I.T.I. February 2015, Wireman)

19. The rating factor of the circuit having close excess current protection is ....

(a) 1.81 (b) 1.23

(c) 0.81 (d) 0.61

(I.T.I. August 2014, Wireman)

20. Normal current capacity of a certain cable is 16 amperes. If the circuit is protected by the close excess current protection, then the current capacity of the cable will be ....

(a) 12 amperes (b) 13 amperes

(c) 16 amperes (d) 20 amperes

(I.T.I. August 2014, Wireman)

21. How many number of cables required for connecting the two way switches shown in the layout diagram?

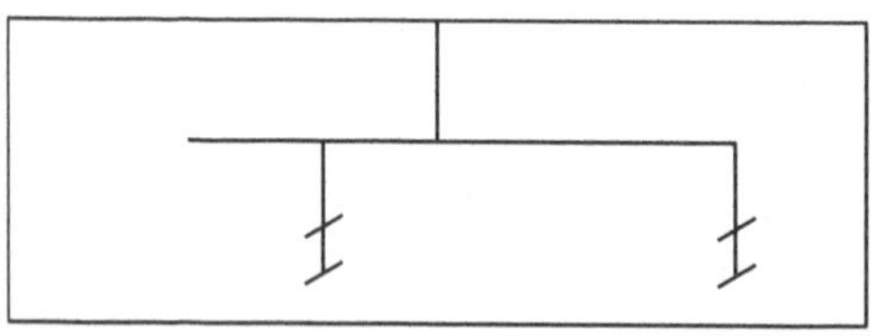

(a) 2 (b) 3
(c) 4 (d) 5

(I.T.I. February 2015, Wireman)

22. Which type of wiring is suitable for multi-storey building?
(a) Tree-system
(b) Ring main system
(c) Distribution board system
(d) Ring main and distribution board system

(I.T.I. February 2015, Electrician)

23. The radius of bends used in conduit wiring should not be less than ....
(a) 3 cm (b) 5 cm
(c) 6 cm (d) 7.5 cm

(I.T.I. February 2015, Electrician)

24. The surface conduit wiring saddles should be fixed at intervals of not more than ....
(a) 0.5 m (b) 0.75 m
(c) 1.0 m (d) 1.5 m

(I.T.I. February 2015, Electrician)

25. What is the wall thickness of 25 mm heavy gauge metal conduit pipe?
(a) 2.24 mm (b) 2.00 mm
(c) 1.80 mm (d) 1.60 mm

(I.T.I. February 2014, Wireman)

26. Which one of the following is the most suitable fish wire to be used in conduit runs having more than a couple of bends?
(a) Copper wire
(b) Mild steel wire
(c) Stainless steel wire
(d) Steel curtain springs

27. What is the main disadvantages of the PVC casing and capping wiring system?
(a) Cost is more than other wiring system
(b) It is inflammable and risk of fire
(c) It is difficult to locate the fault
(d) It requires frequent maintenance

(I.T.I. February 2015, Wireman)

28. The length of the PVC channel for wiring is available ....
(a) 2 metre (b) 2.5 metre
(c) 3.0 metre (d) 4.0 metre

(I.T.I. February 2015, Wireman)

29. The size of PVC pipe for cold bending should not exceed ....
(a) 10 mm (b) 15 mm
(c) 20 mm (d) 25 mm

(I.T.I. February 2014, Wireman)

30. The only wiring which can be taken through shortest route is the roof and required less right angle bends is ....
(a) conduit surface wiring
(b) conduit concealed wiring
(c) casing and capping wiring
(d) batten wiring

(I.T.I. August 2014, Wireman)

31. Laying of concealed wiring in roof starts, soon after completion of ....
(a) pouring concrete
(b) centering
(c) shuttering
(d) load bearing walls

(I.T.I. August 2014, Wireman)

32. Which one of the following is the normal wall thickness of PVC conduit pipes?
(a) 0.8 mm (b) 1.0 mm
(c) 1.2 mm (d) 1.6 mm

33. In which one of the following wirings, three two way switches only are required to operate three lamps?
(a) Staircase wiring (b) Corridor wiring
(c) Tunnel wiring (d) Godown wiring

34. In which type of wiring, two lamps are switched ON at a time?
(a) Tunnel wiring (b) Hostel wiring
(c) Godown wiring (d) Corridor wiring

(I.T.I. August 2014, Wireman)

35. A staircase point can be operated ......
(a) with one one-way and one two-way switches
(b) with two one-way switches
(c) with two two-way switches
(d) with two intermediate switches

36. T.R.S. cables on batten is suitable for ....
(a) high voltage

(b) extra high voltage
(c) low voltage
(d) medium voltage

(I.T.I. August 2014, Wireman)

37. Identity the BIS symbol which represents the wiring in conduit concealed?

(a)  (b) 

(c)  (d) 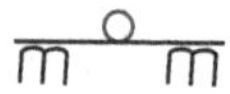

(I.T.I. August 2014, Wireman)

38. Identify the type of diagram/plan from the given diagram.

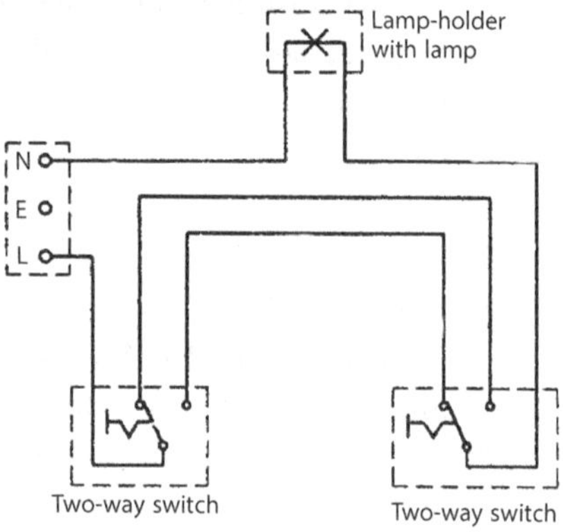

(a) Installation plan (b) Circuit diagram
(c) Wiring diagram (d) Layout diagram

(I.T.I. February 2015, Electrician)

39. What is the name of the diagram, shown in the figure?

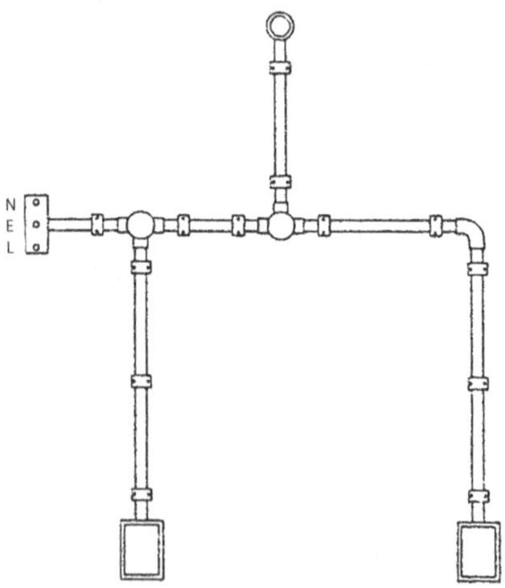

(a) Layout plan
(b) Schcmatic diagram
(c) Installation plan
(d) Wiring diagram

(I.T.I. August 2019, Electrician)

40. Name the type of wiring circuit diagram as shown in the figure.

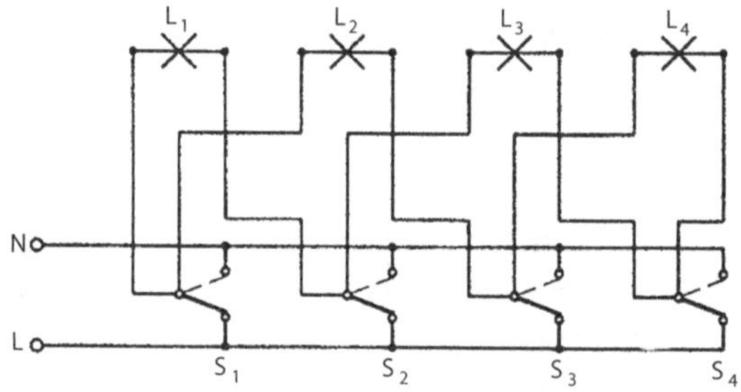

(a) Bank wiring (b) Godown wiring
(c) Corridor wiring (d) Tunnel wiring

(I.T.I. August 2014, Wireman)

41. What is the name of wiring diagram shown in the figure?

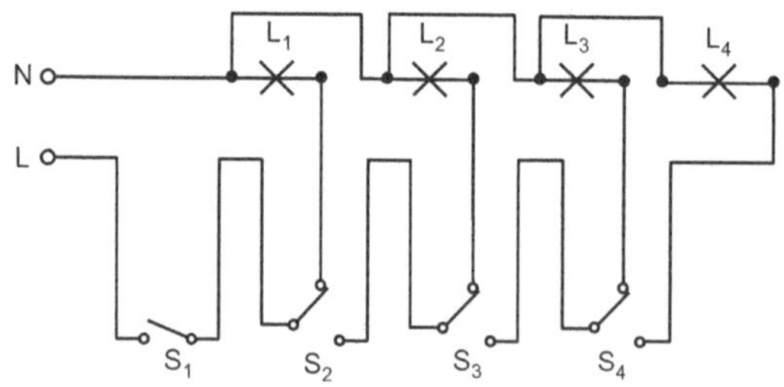

(a) Tunnel wiring
(b) Corridor wiring
(c) Go-down wiring
(d) Stain Case wiring

(I.T.I. August 2019, Electrician)

42. Raising main type distribution system is suitable for which type of multistoried buildings?

(a) Workshop buildings
(b) Substation buildings
(c) Residential flats
(d) Shopping complex buildings

(I.T.I. February 2016, Electrician)

43. PVC channel wiring is suitable for ....

(a) go-down wiring
(b) industrial wiring

(c) out-door wiring
(d) indoor surface wiring

(I.T.I. February 2016, Electrician)

44. Which one is the main disadvantages of P.V.C. Channel surface wiring?
(a) Cost is more than other system
(b) Inflammable and risk of fire
(c) Frequent maintenance required
(d) More difficult to make wiring

(I.T.I. August 2015, Electrician)

45. The number of wires of 4 sq . mm drawn in a 20 mm × 10 mm casing and capping is ....
(a) 1 (b) 2
(c) 3 (d) 4

(I.T.I. August 2015, Wireman)

46. The number of wires of 6 sq. mm drawn in a 25 × 10 mm casing capping is ....
(a) 6 (b) 4
(c) 3 (d) 2

(I.T.I. February 2016, Wireman)

47. Which type of wiring system shall not be used in places exposed to sun and rain?
(a) Cleat wiring
(b) Thermoplastic Heat and Water resistant Nylon coated wiring
(c) P.V.C. casing and capping
(d) Metal conduit wiring

(I.T.I. August 2015, Electrician)

48. The only wiring which can be taken through shortest route in the roof and requires less right angle bends is ....
(a) casing and capping
(b) batten wiring
(c) conduit surface wiring
(d) conduit concealed wiring

(I.T.I. August 2015, Electrician)

49. In case the voltage drop exceeds 3%, then the technician has to choose ....
(a) bigger size of wire
(b) double the size of wire
(c) half the size of wire
(d) next bigger size of wire

(I.T.I. August 2015, Wireman)

50. I.E. rule states the leakage current should not exceed ....
(a) $\frac{1}{5000}$ th part of full load current
(b) $\frac{1}{4000}$ th part of full load current
(c) $\frac{1}{3000}$ th part of full load current
(d) $\frac{1}{2000}$ th part of full load current

(I.T.I. August 2015, Wireman)

51. The full form of BIS is ....
(a) Bureau of Indian Standards
(b) Bureau of Institute Standards
(c) Bureau of Indian Supply
(d) Bureau of Institute Standards

(I.T.I. February 2016, Wireman)

52. As per BIS, number of link clips per box are .... No.
(a) 100 (b) 90
(c) 80 (d) 70

(I.T.I. August 2019, Electrician)

53. Link clips are made out of ....
(a) zinc (b) brass
(c) nickel (d) iron

(I.T.I. February 2016, Wireman)

54. The thickness of link clips of 30 SWG is ....
(a) 0.28 mm (b) 0.30 mm
(c) 0.31 mm (d) 0.32 mm

(I.T.I. February 2016, Wireman)

55. What is the formula to be used to calculate the voltage drop, to select and determine the size of cable for panel wiring installation?
When,
R = Resistance of the conductor / metre
L = Distance of the load centre from the supply ends in metre.
I = Total current in the circuit in ampere
(a) Voltage drop = $2 \times R \times L \times I$
(b) Voltage drop = $\frac{2 \times L}{R \times I}$
(c) Voltage drop = $R \times L \times I$

(d) Voltage drop

$$= \frac{2 \times R \times L \times I}{\text{Cross sectional area in square mm}}$$

(I.T.I. August 2015, Wireman)

56. To calculate the limit of maximum current demand from starting point is ....

(a) Maximum current demand $= \frac{\text{Line current}}{1.5}$

(b) Maximum current demand = Line current × 1.5 times

(c) Maximum current demand = Line current + 1.5 times

(d) Maximum current demand $= \frac{1.5 \times \text{Full load current}}{\text{Line current}}$

(I.T.I. February 2016, Wireman)

57. What is the name of the material used for harnessing cables while making wiring connection of motor?

(a) P.V.C. saddles (b) P.V.C. straps

(c) P.V.C. trays (d) P.V.C. casting

(I.T.I. August 2015, Wireman)

58. Which device is the example for coarse excess current protection?

(a) MCB (b) HRC

(c) Cartridge fuse (d) Rewirable fuse

(I.T.I. August 2015, Wireman)

59. Normal current capacity of a certain cable is 16 amperes. If the circuit is protected by the course excess current protection, then the current capacity of the cable will be ....

(a) 12 A (b) 13 A

(c) 16 A (d) 20 A

(I.T.I. August 2015, Wireman)

60. Types of cleats available are ....

(a) 4 (b) 3

(c) 2 (d) 1

(I.T.I. August 2015, Wireman)

61. Which wiring is suitable for temporary installations?

(a) P.V.C. conduit wiring

(b) Concealed wiring

(c) Cleat wiring

(d) Matal conduit wiring

(I.T.I. August 2019, Electrician)

62. The load on each sub circuit shall be restricted to ....

(a) 800 watts (b) 1000 watts

(c) 1200 watts (d) 1500 watts

(I.T.I. August 2015, Wireman)

63. The cross sectional area of copper conductor for power wiring shall be not less than ....

(a) 0.5 mm^2 (b) 1.15 mm^2

(c) 1.5 mm^2 (d) 2.5 mm^2

(I.T.I. August 2015, Wireman)

64. The minimum size of aluminium conductor used in a power circuit is ....

(a) 1 sq mm (b) 1.5 sq mm

(c) 2.5 sq mm (d) 4 sq mm

(I.T.I. August 2015, Wireman)

65. The clearance between the conductors of opposite polarity or phase and between the conductors shall be at least ....

(a) 5.0 cm (b) 3.5 cm

(c) 2.5 cm (d) 2.0 cm

(I.T.I. August 2015, Wireman)

66. If the working voltage of an installation is 220 volts, then insulation tester rated voltage is ....

(a) 2000 V (b) 1500 V

(c) 1000 V (d) 500 V

(I.T.I. August 2015, Wireman)

67. As per NEC, the phase apparatus (output) A.C. system in Aplha-numeric natation is represented as ....

(a) L_1, L_2, L_3, N (b) R, Y, B, N

(c) A, B, C, N (d) U, V, W, N

(I.T.I. August 2015, Wireman)

68. As per IE rules, the voltage at the point of commencement of supply in the case of high or extra high voltage, shall not vary more than ....

(a) 12.5 percentage (b) 10.0 percentage

(c) 5.0 percentage (d) 2.5 percentage

(I.T.I. August 2015, Wireman)

69. The non metallic conduit pipe used in wiring, should have a minimum size in diameter is ....

(a) 15 mm (b) 18 mm
(c) 20 mm (d) 25 mm

(I.T.I. August 2015, Electrician)

70. The size of P.V.C. conduit pipe in diameter to be preferred for domestic wiring is ....

(a) 21 mm (b) 19 mm
(c) 18 mm (d) 17 mm

(I.T.I. August 2015, Electrician)

71. The threads on the conduit pipe in all cases shall be between ....

(a) 10 mm to 27 mm long
(b) 12 mm to 28 mm long
(c) 13 mm to 29 mm long
(d) 14 mm to 30 mm long

(I.T.I. February 2016, Wireman)

72. The conduit elbows are used at ....

(a) tapping the connection
(b) sharp ends
(c) straight runs
(d) away from walls

(I.T.I. February 2016, Electrician)

73. The minimum depth of junction boxes used in roof slabs shall be ....

(a) 75 mm (b) 70 mm
(c) 65 mm (d) 60 mm

(I.T.I. February 2016, Electrician)

74. Minimum thickness of TW batten shall not be less than ....

(a) 8 mm (b) 10 mm
(c) 12 mm (d) 15 mm

(I.T.I. February 2016, Wireman)

75. The size of T.W. bend commercially available is ....

(a) 12 mm × 15 mm
(b) 12 mm × 14 mm
(c) 12 mm × 13 mm
(d) 12 mm × 12 mm

(I.T.I. February 2016, Wireman)

76. The size of commercially available T.W. round block single is ....

(a) 75 mm × 37 mm
(b) 75 mm × 38 mm
(c) 75 mm × 39 mm
(d) 75 mm × 40 mm

(I.T.I. February 2016, Wireman)

77. T.W. board for fixing mains are available in ....

(a) 110 mm × 80 mm
(b) 120 mm × 80 mm
(c) 130 mm × 80 mm
(d) 150 mm × 80 mm

(I.T.I. February 2016, Electrician, Wireman)

78. Which of the following is not a disadvantage of the loop in system of wiring?

(a) More material used
(b) No junction boxes required
(c) More expensive
(d) High copper loss

(I.T.I. February 2017, Electrician)

79. Two pin sockets should not be used in domestic wiring unless the appliance to be corrected as ....

(a) Double earthed
(b) Double insulated
(c) Controlled by ELCB
(d) Controlled by MCB

(I.T.I. February 2018, Electrician)

80. The wiring system generally employed in public building (offices):

(a) Cleat wiring
(b) Cassing-Capping wiring
(c) C.T.S. wiring
(d) Conduit wiring

(I.T.I. August 2016, Electrician)

81. What type of wiring will you suggest in a cold storage?

(a) Batten wiring PVC/CTS
(b) Conduit wiring
(c) PVC/lead covered wiring
(d) Cleat wiring

(I.T.I. August 2016, Wireman)

82. Before working on any electric conductor, it should be discharged and also to prevent the same from being charged by mistake, the I.E.R. is .....

(a) I.E.R. 1956 Rule No. 38
(b) I.E.R. 1956 Rule No. 36
(c) I.E.R. 1956 Rule No. 35
(d) I.E.R. 1956 Rule No. 34

(I.T.I. August 2016, Wireman)

83. A factory requires power points for its loads, 2 nos. – 2.5 kW, 2 nos. – 1.5 kW, 10 nos. – 0.9 kW. The optimal no. of power sub-circuit will be –

(a) 5 (b) 6
(c) 7 (d) 8

(I.T.I. February 2017, Wireman)

84. The insulation resistance of an installation is more than.....

(a) 1 mega ohm (b) 1 ohm
(c) 2 kilo ohms (d) 5 mega ohms

(I.T.I. August 2016, Wireman)

85. The maximum load can be taken from a 5 ampere fuse distribution board is ....

(a) 1000 watts (b) 800 watts
(c) 3000 watts (d) 5000 watts

(I.T.I. August 2015, Wireman)

86. Each appliance in a sub circuit is operated by a separate fuse in .....

(a) Looping system
(b) Tree system
(c) Ring system
(d) Distribution system

(I.T.I. February 2017, Wireman)

87. Type of wiring system in which wires sheathed in tough rubber, clipped on wooden battens with brass clips is .....

(a) Cleat wiring (b) C.T.S.
(c) Casing cappling (d) All of these

(I.T.I. February 2017, Wireman)

88. The wiring of electrical motors in the industries is carried out according to requirement of I.E. Rule No. .....

(a) 40, 41 (b) 50, 51
(c) 55, 57 (d) 71, 72

(I.T.I. February 2017, Wireman)

89. According to I.E. rule, maximum allowable power and number of power points in a power sub circuit are ....

(a) 3000 W, 2 (b) 2000 W, 3
(c) 2000 W, 1 (d) 3000 W, 3

(I.T.I. August 2016, Electrician, Electroplator, Wireman)

90. In order to design a domestic connection with four ceiling fans, eight lamps and six power points, the minimum number of light fan sub circuit and power sub circuit are ....

(a) 4, 2 (b) 2, 3
(c) 2, 6 (d) 4, 3

(I.T.I. February 2017, Wireman)

91. What is the maximum load that can be connected in a circuit connecting lighting load only ....

(a) 500 W (b) 750 W
(c) 800 W (d) 1000 W

(I.T.I. February 2017, Electrician)

92. What is maximum number of lighting points that can be connected in a circuit?

(a) 5 (b) 10
(c) 8 (d) 12

(I.T.I. February 2017, Electrician)

93. Following is/are true about neutral wiring:

(a) Neutral grounding eliminates that need to earth point in three/five point sockets
(b) An open neutral path indicates no current flow through the respective current/sub circuit
(c) An open neutral path indicates balanced current through the three phase load
(d) All of these

(I.T.I. February 2017, Wireman)

94. The operation of a Megger is based on :

(a) moving-iron meter
(b) electrodynamic meter
(c) electrostatic meter
(d) moving coil meter

95. In open circuit, megger reading is ....

(a) Zero (b) kilo ohms
(c) infinity (d) ohm

(I.T.I. August 2016, Wireman)

96. For measuring a very high resistance we should use :

(a) Kelvin's double bridge
(b) Double Wheatstone bridge
(c) Wheatstone bridge
(d) Megger

(I.T.I. February 2019, Electrician)

97. Which instrument is used to measure insulation resistance in wiring installation?

(a) Megger
(b) Multimeter
(c) Shunt type Ohm meter
(d) Series type Ohm meter

(I.T.I. August 2019 Electrician)

98. What is the test to be carried out by using megger?

(a) Polarity test
(b) Insulation resistance test
(c) Earth electrode resistance test
(d) Earth conductor continuity test

(I.T.I. August 2019 Electrician)

99. ...... type of wiring uses insulated cable supported in porcelain cleats.

(a) Metal (b) P.V.C. conduit
(c) Cleat (d) C.T.S. / R.T.S.

(I.T.I. August 2019 Electrician)

100. Cleat wiring should be always above .... from the floor level.

(a) 5 metres (b) 8 metres
(c) 10 metres (d) 3 metres

(I.T.I. August 2019 Electrician)

101. Link clips are packed in card board boxes at the rate of ..... link clips per box.

(a) 200 (b) 300
(c) 100 (d) 50

(I.T.I. August 2019 Electrician)

102. How the conduit pipe size is specified?

(a) Length of pipe
(b) Thickness of pipe
(c) Inner diameter of pipe
(d) Outer diameter of pipe

(I.T.I. August 2019 Electrician)

103. What are the types of conduits available?

(a) Flexible conduit
(b) Heavy gauge steel screwed conduit
(c) P.V.C. conduit
(d) All of these

(I.T.I. August 2019 Electrician)

104. What is the maximum P.V.C. conduit size to make safe cold bending?

(a) 12 mm (b) 19 mm
(c) 50 mm (d) 25 mm

(I.T.I. August 2019 Electrician)

105. Ferrules are used for .....

(a) Over load
(b) For circuit break
(c) Rectification
(d) Numbering of complex wiring

(I.T.I. August 2019 Electrician)

106. The domestic installation is connected in ......

(a) Series (b) Series parallel
(c) Parallel series (d) Parallel

(I.T.I. August 2019 Electrician)

107. .... are used to take diversion from the mainline either to the switch points or light

(a) Points (b) Couplers
(c) Bends (d) Tees

(I.T.I. August 2019 Electrician)

108. Where the phase conductor is looped in looping system of wiring?

(a) Switch box
(b) Junction box
(c) Distribution box
(d) Socket connection

(I.T.I. August 2019 Electrician)

109. What is the reason of lamp glowing dim and motor running slow in a domestic wiring circuit?

(a) High value series resistance fault
(b) Short circuit between conductors
(c) Open circuit in the neutral line
(d) Open circuit in the earth conductor

(I.T.I. August 2019 Electrician)

110. What is the application of the wiring circuit?

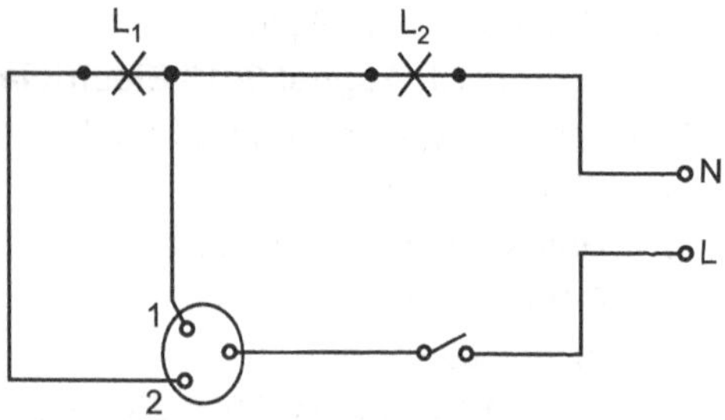

(a) Two lamps dim operation only
(b) Two lamps controlled by one switch
(c) Two lamps controlled by two switches
(d) One lamp bright and two lamps dim operation

(I.T.I. August 2019 Electrician)

111. What is the reason for home theatre wiring not to run along with power wiring?
(a) Avoid leakage current in home theatre wiring
(b) Control temperature in home theatre wiring
(c) Avoid electrical interference in audio, video system
(d) Reduce the power consumption in power supplies

(I.T.I. August 2019 Electrician)

KEY TO MULTIPLE CHOICE QUESTIONS

Electrical Accessories

1. (c)	**2.** (c)	**3.** (b)	**4.** (b)	**5.** (b)	**6.** (a)	**7.** (c)	**8.** (c)	**9.** (b)	**10.** (b)
11. (c)	**12.** (a)	**13.** (a)	**14.** (a)	**15.** (b)	**16.** (d)	**17.** (b)	**18.** (a)	**19.** (d)	**20.** (a)
21. (b)	**22.** (a)	**23.** (a)	**24.** (b)	**25.** (b)	**26.** (d)	**27.** (c)	**28.** (b)	**29.** (c)	**30.** (a)
31. (c)	**32.** (d)	**33.** (c)	**34.** (b)	**35.** (b)	**36.** (d)	**37.** (d)	**38.** (a)	**39.** (a)	**40.** (d)
41. (c)	**42.** (b)	**43.** (a)	**44.** (b)	**45.** (a)	**46.** (a)	**47.** (a)	**48.** (c)	**49.** (b)	**50.** (d)
51. (b)	**52.** (b)	**53.** (c)	**54.** (d)	**55.** (d)	**56.** (d)				

Domestic and Industrial Wiring

1. (d)	**2.** (a)	**3.** (b)	**4.** (b)	**5.** (d)	**6.** (d)	**7.** (a)	**8.** (a)	**9.** (c)	**10.** (b)
11. (b)	**12.** (c)	**13.** (b)	**14.** (d)	**15.** (a)	**16.** (b)	**17.** (c)	**18.** (c)	**19.** (b)	**20.** (d)
21. (d)	**22.** (a)	**23.** (d)	**24.** (c)	**25.** (c)	**26.** (d)	**27.** (b)	**28.** (c)	**29.** (d)	**30.** (b)
31. (b)	**32.** (c)	**33.** (c)	**34.** (a)	**35.** (c)	**36.** (c)	**37.** (c)	**38.** (c)	**39.** (c)	**40.** (d)
41. (c)	**42.** (c)	**43.** (d)	**44.** (b)	**45.** (c)	**46.** (c)	**47.** (a)	**48.** (d)	**49.** (d)	**50.** (a)
51. (a)	**52.** (a)	**53.** (b)	**54.** (d)	**55.** (a)	**56.** (b)	**57.** (b)	**58.** (d)	**59.** (b)	**60.** (b)
61. (c)	**62.** (a)	**63.** (c)	**64.** (c)	**65.** (c)	**66.** (d)	**67.** (d)	**68.** (a)	**69.** (c)	**70.** (b)
71. (a)	**72.** (b)	**73.** (c)	**74.** (b)	**75.** (d)	**76.** (d)	**77.** (d)	**78.** (b)	**79.** (b)	**80.** (d)
81. (c)	**82.** (b)	**83.** (d)	**84.** (a)	**85.** (b)	**86.** (c)	**87.** (b)	**88.** (b)	**89.** (a)	**90.** (b)
91. (c)	**92.** (b)	**93.** (b)	**94.** (d)	**95.** (c)	**96.** (d)	**97.** (a)	**98.** (b)	**99.** (c)	**100.** (a)
101. (c)	**102.** (d)	**103.** (d)	**104.** (d)	**105.** (d)	**106.** (d)	**107.** (a)	**108.** (d)	**109.** (a)	**110.** (d)
111. (c)									

Module 9

Earthing and Wiring Installation

Multiple Choice Questions

Earthing

1. The size of the earth or ground wire is based on the:
   (a) maximum fault current carrying through the ground wire
   (b) rated current carrying capacity of the service line
   (c) depends on the soil resistance
   (d) both (a) and (c)
   (I.T.I. February 2017, Electrician, Electroplator, Wireman)
2. Earth wire or ground wire is made of:
   (a) copper (b) aluminium
   (c) iron (d) galvanized steel
   (I.T.I. February 2019, Electrician, Electroplator, Wireman)
3. Average resistance of human body is:
   (a) 500 ohms (b) 1000 ohms
   (c) 1500 ohms (d) 2000 ohms
   (I.T.I. February 2019, Electrician, Electroplator, Wireman)
4. Generally earthing is provided for:
   (a) only for the safety of the equipment
   (b) only for the safety of the operating personnel
   (c) both (a) and (b)
   (d) none of the above
   (I.T.I. February 2019, Electrician, Electroplator, Wireman)
5. Ground resistance should be designed such that:
   (a) grounding resistance should be as low as possible
   (b) grounding resistance should be as high as possible
   (c) grounding resistance should be always zero
   (d) none of the above
   (I.T.I. February 2019, Electrician, Electroplator, Wireman)
6. When earth fault occurs:
   (a) voltage potential at the earth mat increases due to earthing
   (b) voltage potential at the earth mat decreases due to earthing
   (c) voltage potential at the earth mat remains zero irrespective of fault
   (d) none of the above
7. Moisture content in the soil .... the earth soil resistance:
   (a) increase
   (b) decrease
   (c) does not affect
   (d) none of the above
   (I.T.I. February 2019, Electrician, Electroplator, Wireman)

8. According to I.E. rules, the supplier shall provide and maintain on the consumer's premises, a suitable ......

(a) earth terminal (b) phase terminal
(c) neutral terminal (d) live terminal

(I.T.I. August 2014, Wireman)

9. Why does the earth conductor provide a path to ground?

(a) For over current
(b) For high voltage
(c) For leakage current
(d) For low voltage

10. Which colour wire is recommended for earth connection as per ISI code?

(a) Red (b) Green
(c) Black (d) Blue

(I.T.I. August 2014, Wireman)
(I.T.I. August 2015, Electrician, Wireman)

11. The resistance of the earth is:

(a) 1 MΩ (b) 20 MΩ
(c) infinite (d) about zero

12. In an earth-point, charcoal and salt are used:

(a) To protect the plate against rust
(b) To absorb the moisture of the surrounding soil
(c) To increase the moisture of the surrounding soil
(d) To hold the plate firmly in the soil.

13. Which one of the following is the minimum size of a copper earth continuity conductor for a domestic installation having 5 amperes capacity?

(a) 1.5 mm^2 (b) 2.5 mm^2
(c) 5 mm^2 (d) 25 mm^2

14. As a rule, the minimum distance between two electrodes shall not be less than ......

(a) One time the length of electrode
(b) Two times the length of electrode
(c) Three times the length of electrode
(d) Four times the length of electrode

15. Earth tester is used for measuring earth electrode resistance in which generated DC is converted into AC by current reverser. The reason for using AC supply for earth resistance measurement is ......

(a) AC resistance is more than DC resistance
(b) To reduce error due to electrolytic emf
(c) AC wave form is pulsating in nature
(d) AC contains sine wave form

(I.T.I. February 2015, Electrician)

16. Which method can be suggested to bring down the earth electrode resistance to an acceptable value?

(a) By reducing the size of the earth electrodes
(b) By increasing the size of the earth electrodes
(c) By pouring water in the earth pit at repeated intervals
(d) By connecting a number of earth electrodes in series

(I.T.I. February 2015, Electrician)

17. The minimum dimension of copper earth plate as per ISI code is .....

(a) 60 cm × 30 cm × 6.38 m
(b) 60 cm × 60 cm × 6.38 mm
(c) 60 cm × 30 cm × 3.18 mm
(d) 30 cm × 30 cm × 3.18 mm

(I.T.I. February 2014, Wireman)

18. As per ISI, for pipe earthing, the minimum internal diameter of galvanised iron or steel pipe is ......

(a) 12.5 mm (b) 18 mm
(c) 20 mm (d) 38 mm

(I.T.I. February 2014, Wireman)

19. Rod electrodes of copper used for earthing should be atleast 12.5 mm in diameter and those of steel or galvanised iron shall be at least in diameter ......

(a) 18 mm (b) 16 mm
(c) 12.5 mm (d) 10 mm

20. For plate earthing the thickness of the G.I. or steel plate should not be less than ......

(a) 3.30 mm (b) 4.30 mm
(c) 5.30 mm (d) 6.30 mm

(I.T.I. August 2014, Wireman)

21. At equal intervals of time the earth resistance to be measured periodically. Which instrument is used for this?

(a) Megger
(b) Ohm meter
(c) Wheat stone bridge
(d) Earth tester

(I.T.I. February 2015, Wireman)

22. How many times the faulty current will be more than the fuse rating current?

(a) At least 2.0 times
(b) At least 3.0 times
(c) At least 3.5 times
(d) At least 4.0 times

(I.T.I. February 2015, Wireman)

23. For pipe earthing the minimum length of pipe electrode should not be less than.....

(a) 2.0 m (b) 2.5 m
(c) 3.0 m (d) 3.5 m

(I.T.I. August 2015, Wireman)

24. The potential of earth is ....

(a) low (b) zero
(c) medium (d) high

(I.T.I. February 2016 Wireman)

25. The potential of neutral is ....

(a) low (b) high
(c) medium (d) zoro

(I.T.I. February 2016 Wireman)

26. The earth tester works on the principles of ....

(a) potential dividing method
(b) fall of potential method
(c) fall of resistance method
(d) current dividing method

(I.T.I. August 2015, Wireman)

27. For finding the earth resistance of an earth electrode 20 V test voltage is used. Four tests are conducted under different distance / position of spikes. Current measured in each test is 3.8A, 1.9A, 2.6A and 3.1A.

Which one of the calculated value of resistance to be considered?

(a) Lowest resistance value
(b) Highest resistance value
(c) Average of the four values
(d) Average of highest and lowest values

(I.T.I. August 2015, Wireman)

28. You have taken permit to work on a over head line and the overhead line is switched 'off'. Before undertaking the repair work on the OH line, you have to ....

(a) equip yourself with insulated gloves
(b) earth the line conductors
(c) measure the line to ground insulation resistance
(d) earth metal O.H. line poles

(I.T.I. August 2015, Wireman)

29. Following factors influence measurement of earth resistance:

(a) Temperature of soil
(b) Moisture content of soil
(c) Depth at which the electrode is embedded
(d) All of these

(I.T.I. February 2017, Wireman)

30. In a four point earth tester (P_1, C_1, P_2, C_2) the points shorted and connected to fixed earth point are ....

(a) P_1 C_1 (b) $P_2 - C_2$
(c) $P_1 - C_2$ (d) $C_1 - P_2$

(I.T.I. February 2017, Wireman)

31. The earth wire should not be thinner than a ....

(a) 20 SWG wire (b) 16 SWG wire
(c) 10 SWG wire (d) 8 SWG wire

(I.T.I. February 2018, Electrician)

32. PMEN System stands for ______________.

(a) Permanent Magnified Earth Neutral System
(b) Protective Multiple Earthed Neutral System
(c) Protective Mutual Earthed Neutral System
(d) Permanent Mutual Earthed Neutral System

(I.T.I. February 2019, Electrician)

33. What should be the value of earthing resistance for large power stations?

(a) 1 Ω (b) 0.5 Ω

(c) 2 Ω (d) 5 Ω

(I.T.I. August 2019 Electrician)

34. Which method of earth resistance measurement is illustrated?

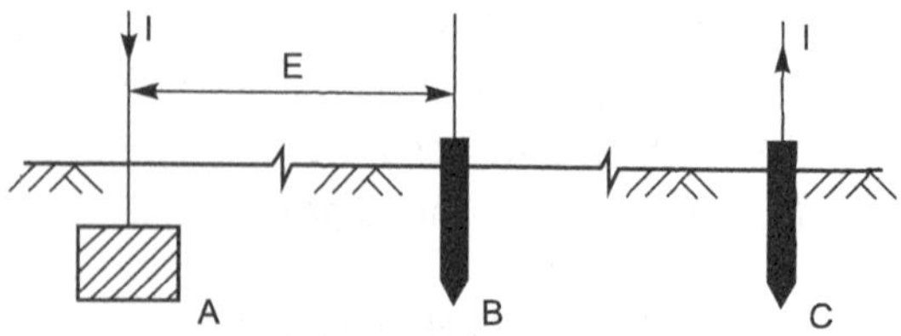

(a) Fall of potential

(b) Current dividing

(c) Fall of current

(d) Potential dividing

(I.T.I. August 2019 Electrician)

35. For which wiring installation the system earthing is to be done?

(a) Commercial wiring

(b) Godown wiring

(c) Domestic wiring

(d) Substations

(I.T.I. August 2019 Electrician)

36. How to control harmonic distortions in neutral connections as per the I.E. rule?

(a) Increasing conductor size

(b) Providing by plate earthing

(c) Earthing through impedance

(d) Providing parallel earthing

(I.T.I. August 2019 Electrician)

Wiring Installation

Choose the Correct Answer :

1. The risk of dangerous electrical shock in a building can be adequately met by earthing:

(a) All switch boxes and fuse panels

(b) All motor housing

(c) All non current carrying metal parts of the entire system

(d) All of the above

2. The conductor for final sub-circuit should have a nominal cross-sectional area not less than .... mm^2 if copper cable is used and .... mm^2 if aluminium cable is used:

(a) 1.5; 2.0 (b) 1.0; 2.0

(c) 1.0; 1.5 (d) 1.5; 1.0

3. Which of the following wiring systems is commonly used for light/fan load in domestic and commercial buildings:

(a) casing capping wiring

(b) TRS wiring

(c) Lead sheated wiring

(d) PVC wiring

4. Current rating of cables is determined based upon:

(a) Full load current

(b) Starting current

(c) 3/4th full load current

(d) Half full load current

5. The system of wiring adopted for industrial power wiring is:

(a) concealed metal conduit

(b) concealed PVC conduit

(c) surface metal conduit

(d) surface PVC conduit

6. The starter used for three phase squirrel cage induction motors of sizes ranging between 0.75kW and 11.0 kW:

(a) D.O.L. starter

(b) star delta starter

(c) 3-point starter

(d) Auto transformer starter

7. The starter used for 3-phase squirrel cage induction motors of sizes above 11 kW is:

(a) D.O.L. starter

(b) Star delta starter
(c) Auto transformer starter
(d) 3-point starter

8. The total connected load of a house wiring is 2400 watts at 240 volts supply. What is the full load current at unity power factor (u.p.f.)?

(a) 24 amperes (b) 15 amperes
(c) 10 amperes (d) 8 amperes

(I.T.I. August 2015, Electrician)

KEY TO MULTIPLE CHOICE QUESTIONS

Earthing

1. (d)	**2.** (d)	**3.** (b)	**4.** (c)	**5.** (a)	**6.** (a)	**7.** (b)	**8.** (a)	**9.** (c)	**10.** (b)
11. (d)	**12.** (c)	**13.** (a)	**14.** (b)	**15.** (b)	**16.** (c)	**17.** (c)	**18.** (d)	**19.** (b)	**20.** (d)
21. (d)	**22.** (c)	**23.** (b)	**24.** (b)	**25.** (a)	**26.** (b)	**27.** (b)	**28.** (b)	**29.** (d)	**30.** (a)
31. (d)	**32.** (b)	**33.** (b)	**34.** (a)	**35.** (d)	**36.** (c)				

Wiring Installation

1. (d)	**2.** (c)	**3.** (d)	**4.** (a)	**5.** (c)	**6.** (b)	**7.** (c)	**8.** (c)

Module 10

Illumination, Source of Light and Decoration Lighting

Multiple Choice Questions

Illumination

1. A 250 V lamp has a total flux of 3000 lumens and takes a current of 0.8 ampere from 250 V mains. Its luminous efficiency is
   (a) 12 lumens per watt
   (b) 9.6 lumens per watt
   (c) 15 lumens per watt
   (d) 240 lumens per watt

2. A lamp has a total luminous flux of 1570 lumens. Its mean spherical candle power is
   (a) 1570 (b) 500
   (c) 250 (d) 125

3. The illumination at a surface due to a source of light placed at a distance "*d*" from the surface varies as
   (a) $\frac{1}{d^2}$ (b) $\frac{1}{d}$
   (c) d (d) d^2

4. The illumination at various points on a horizontal surface illuminated by the same source varies as
   (a) $\cos\theta$ (b) $\cos^2\theta$
   (c) $\cos^3\theta$ (d) $\frac{1}{\cos q}$

5. The luminous-flux emitted in unit solid angle by a light source having a luminous-intensity of one candela is called :
   (a) illumination (b) luminance
   (c) one lumen (d) one lux.

6. The illumination required for a workplace where fine work is performed will be in the order of ....
   (a) 1000 lumens/ metre
   (b) 1000 lumens/sq. metre
   (c) 1500 lumens/sq. metre
   (d) 2000 lumens/sq. metre

   (I.T.I August 2014, Electrician)

7. The illumination required for a workplace where rough or unskilled work is performed will be in the order of ....
   (a) 100 lumens/sq. metre
   (b) 150 lumens/sq. metre
   (c) 200 lumens/sq. metre
   (d) 250 lumens/sq. metre

   (I.T.I August 2014, Wireman; I.T.I February 2015, Electrician)

8. Unit of illuminance is .....
   (a) lux (b) lux per second
   (c) lumen (d) lumer per watt

   (I.T.I. February 2017, Wireman)

9. candela is the unit of ....
   (a) Luminous flux
   (b) Luminmous intensity
   (c) Wave length
   (d) None of these

   (I.T.I. February 2019, Electrician)

10. Unit of luminous intensity is ....
(a) lumen second
(b) lumen
(c) candela
(d) candela per square metre
(I.T.I. February 2017, Wireman)

11. The unit of luminuous flux is .....
(a) steradian (b) candela
(c) lumen (d) lux
(I.T.I. August 2019, Electrician)

12. Lumen per watt is the unit of ....
(a) Light flux
(b) Luminous intensity
(c) Brightness
(d) Luminous efficiency
(I.T.I. February 2019, Electrician)

13. The unit of solid angle is ....
(a) solid angle (b) radian
(c) steradian (d) candela
(I.T.I. February 2018, Electrician)

14. Illumination is measured using the meter .....
(a) Stroboscope (b) Photo meter
(c) Ammeter (d) Energy meter
(I.T.I. August 2016, Wireman)

15. Practical ligthing schemes based on lumen method take into account :
(a) Utilization factor
(b) Maintenance factor
(c) Space/height ratio
(d) All of these
(I.T.I. February 2017, Electrician)

16. Radiant efficiency of the lumninous source depends on –
(a) Shape of the source
(b) Temperature of the source
(c) Wavelength of light rays
(d) All of these
(I.T.I. February 2019, Electrician)

17. The wave length of colour with maximum luminous intensity within visible spectrum is .....
(a) Orange (b) Green
(c) Greenish yellow (d) Red
(I.T.I. February 2017, Wireman)

Sources of Light

1. The lumen output of a 230 volts, 100 watts gas-filled lamp is approximately :
(a) 700 (b) 1000
(c) 1200 (d) 1500.

2. H.P.M.V. lamps are available in a rating of :
(a) 40, 60 watts (b) 100, 200 watts
(c) 80, 125 watts (d) 150, 300 watts.

3. When a fluorescent lamp shows a dense blackening at each end, it could possibly mean that the :
(a) gas filled in the lamp was wrong
(b) lamp was started infrequently
(c) starter is working improperly
(d) lamp is new.

4. How many lamps of 3 volts rating are required to be connected in series to a supply of 200 V for satisfactory operation ?
(a) 67 (b) 85
(c) 135 (d) 33

5. In a H.P.M.V. lamp, the mercury vapour discharges at :
(a) a higher pressure
(b) a medium pressure
(c) a low pressure
(d) none of the above

6. What is the purpose of the choke after the fluorescent tube is just lighted to its full brightness ?
(a) For supplying a higher oporating current
(b) Smoothing the sinusoidal operating current

(c) Limiting the current flowing through the fluorescent tube filaments

(d) Heating the filaments continuously during operation

(I.T.I. August, 2014, Electrician)

7. As shown in the figure given below one part of starter is labelled as 'A'. Which among the following is this part.

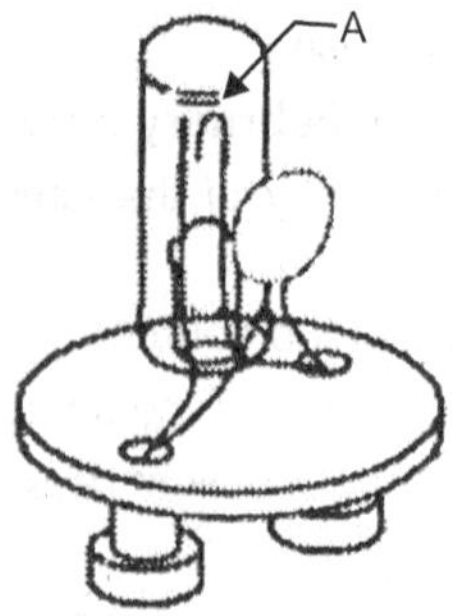

(a) Base with pins

(b) Radio frequency suppressor

(c) Bimetallic switching contact

(d) sealed glass tube fitted with gas (Neon)

8. The starter of a fluorescent tube lamp has been removed from its holder after the tube is lighted what will happen to the function of the tube lamp?

(a) Tube lamp will begin to flicker

(b) Tube lamp will go out immediately

(c) Tube will operate normally

(d) Tube will begin to hum very much

(I.T.I. February 2018, Electrician)

9. The function of choke in tube light is to induce..........

(a) low voltage and limit current

(b) low voltage and increase current

(c) high voltage and limit current

(d) high voltage and increase current

(I.T.I. February 2014, Wireman)

10. As per IE regulations when calculating the incandescent lamp power for load purpose, every lamp holder is taken as ....

(a) 40 watts (b) 60 watts

(c) 75 watts (d) 100 watts

(I.T.I. July 2014, Electrician)

11. When two 250 V, 100 W bulbs are connected is series to 250 V supply, they will give ....

(a) Bright light

(b) dim light

(c) one dim and the other bright light

(d) no light

12. Which one of the lamps mentioned below has more time delay when compared to other lamps ?

(a) Tube lamp

(b) Mercury vapour lamp

(c) Sodium vapour lamp

(d) Neon lamp

(I.T.I. February 2015, Electrician)

13. Name the part indicated as 'X' in sodium vapour lamp in the figure shown below ?

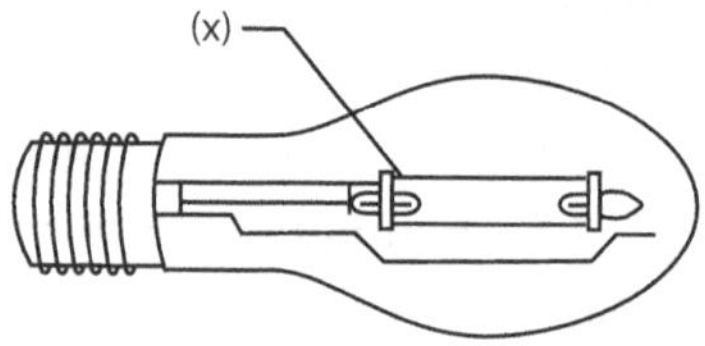

(a) Base (b) Sodium vapour

(c) Arc tube (d) Electrode

(I.T.I. August 2014, Electrician)

14. What is the colour of the light emitted by sodium vapour lamp ?

(a) bluish light (b) yellowish light

(c) reddish light (d) greenish light

(I.T.I. August 2014, Electrician)

15. The ignition voltage of sodium lamp veries from ....

(a) 400 to 600 V (b) 400 to 550 V

(c) 400 to 500 V (d) 400 to 450 V

(I.T.I. February 2015, Wireman)

16. Name the part indicated as 'X' of gas discharge lamp shown in the figure ?

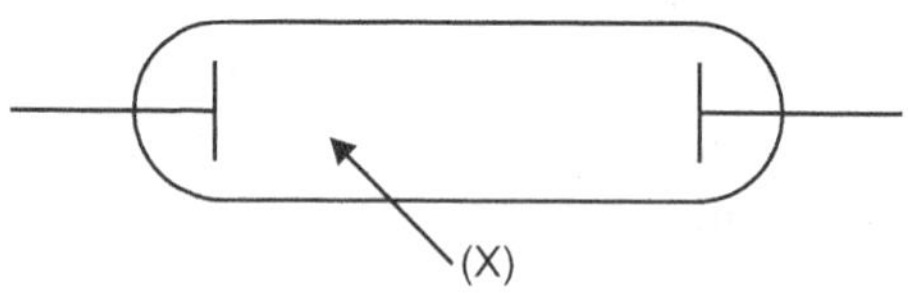

(a) glass shell

(b) lead in wire
(c) electrodes
(d) gas and metal vapour

(I.T.I. February 2015, Electrician)

17. The life of MA (mercury argon) type lamps is ....

(a) 9000 hrs (b) 8500 hrs
(c) 8000 hrs (d) 7500 hrs

(I.T.I. February 2015, Wireman)

18. The average life of sodium vapour lamp is about ....

(a) 1000 hours (b) 2000 hours
(c) 3000 hours (d) 6000 hours

(I.T.I. February 2016, Electrician)

19. Light used in the front yard of the house should not attract insects in view of safety. Which type of lamp you will avoid?

(a) Mercury vapour lamp
(b) Sodium vapour lamp
(c) Halogen lamp
(d) GLS lamp

(I.T.I. February 2018, Electrician)

20. Name the part Indicated as 'X' of incandescent lamp shown in the figure?

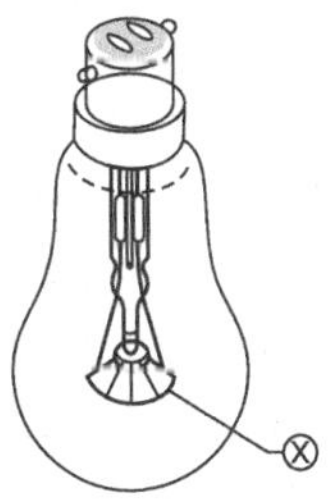

(a) Glass steam
(b) Phosphor bronze
(c) Lead in wire
(d) Tungsten filament

(I.T.I. August 2015, Electrician)

21. Carbon filament lamp operating temperature is ....

(a) 1800°C (b) 2000°C
(c) 2100°C (d) 2200°C

(I.T.I. August 2015, Wireman)

22. The purpose of the capacitor shown in the fluorescent tube lamp circuit is to ....

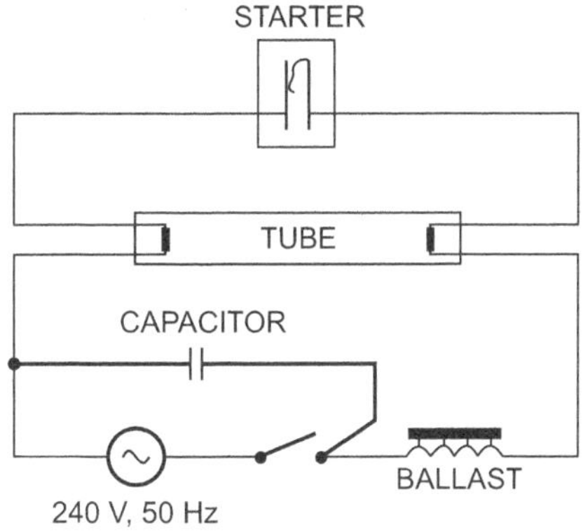

(a) compensate the reactive power of the circuit
(b) provide the voltage for the firing process
(c) limit the operating current flowing through the tube
(d) smooth the AC voltage of the power supply

(I.T.I. August 2015, Electrician)

23. Why fluorescent lamp is most commonly used?

(a) Better illumination
(b) High intensity of light
(c) High efficiency & shadowless light
(d) Instant starting, no delay as 11 PML

(I.T.I. August 2016, Wireman)

24. The flicker effect of flourescent lamp is more pronounced at .....

(a) Lower voltages
(b) Higher voltages
(c) Lower frequencies
(d) Higher frequencies

(I.T.I. August 2016, Electrician)

25. The frequency of flickers in a fluorescent lamp at 220 V, 50 Hz supply will be ......

(a) 25 per second (b) 50 per second
(c) 100 per second (d) 220 per second

(I.T.I. February 2018, Electrician)

26. The purpose of providing a choke in a tube light is ....

(a) To eliminate corona effects
(b) To avoid radio interference
(c) To improve power factor
(d) To limit current to appropriate value

(I.T.I. August 2016, Electrician)

27. The purpose of providing choke in the tube light is to .....
(a) Eliminate corona effect
(b) Avoid radio interference
(c) Limit the current to the appropriate level
(d) Improve the power factor

(I.T.I. February 2017, Electrician)

28. After starting, the voltage across the tube light is about ......
(a) 230 V (b) 110 V
(c) 200 V (d) 220 V

(I.T.I. August 2016, Wireman)

29. Luminous efficiency of a fluorescent tube is –
(a) 5-10 lumens / watt
(b) 15-20 lumens / watt
(c) 30-40 lumens / watt
(d) 60- 65 lumnens / watt

(I.T.I. February 2019, Electrician)

30. The main applications of arc lamp is .....
(a) Cinema projects
(b) Search light
(c) Heating & Illumination
(d) (a) and (b) of the given

(I.T.I. August 2016, Wireman)

31. Carbon arc lamps are commonly used in ......
(a) Domestic lighting
(b) Street lighing
(c) Cinema projectors
(d) Photography

(I.T.I. August 2019, Electrician)

32. In the given lamps, which one is not a discharge lamp?
(a) Sodium vapour lamp
(b) Low pressure mercury vapour lamp
(c) High pressure mercury vapour lamp
(d) Incandescent metal filament lamp

(I.T.I. August 2016, Wireman)

33. The best material for making filament of incandecent lamp is .....
(a) Carbon (b) Tantlum
(c) Tungsten (d) Molybdenium

(I.T.I. August 2016, Wireman)

34. Melting temperature of tungsten is ......
(a) 2000°K (b) 2500°K
(c) 2655° K (d) 3655°K

(I.T.I. February 2019, Electrician)

35. The power factor of an incandescent lamp is ....
(a) unity
(b) around 0.5 lagging
(c) around 0.5 leading
(d) zero

(I.T.I. August 2016, Electrician)

36. Which gas is sometimes used in filament lamps?
(a) Argon (b) Krypton
(c) Nitrogen (d) Carbon dioxide

(I.T.I. February 2019, Electrician)

37. What is the temperature at which a gas filled tungsten filament lamp operates?
(a) 1035°C (b) 1200°C
(c) 1500°C (d) 2300°C

(I.T.I. February 2018, Electrician)

38. The output of a tungsten filament lamp depends on –
(a) Size of lamp
(b) Size of shell
(c) Temperature of filament
(d) All of these

(I.T.I. February 2019, Electrician)

39. Very high efficiency, long life, very good lumen maintenance lamp used for outdoor yards and street lights is ......
(a) High pressure sodium lamp
(b) Low pressure sodium lamp
(c) Mercury vapour lamp
(d) Carbon arc lamp

(I.T.I. August 2016, Wireman)

40. A mercury vapour lamp gives :
(a) Pink light (b) Yellow light
(c) Greenish blue (d) White light
(I.T.I. February 2017, Electrician)

41. LED lamp has a power factor .....
(a) unity (b) leading
(c) lagging (d) None of these
(I.T.I. August 2017, Electrician)

42. For the same wattage, which lamp is cheapest?
(a) Sodium vapour lamp
(b) Mercury vapour lamp
(c) Fluorescent tube
(d) GLS lamp
(I.T.I. February 2019, Electrician)

43. Red coloured Neon sign lamps for adversiting and decoration purpose is obtained by filling .....
(a) Neon with mixture of mercury
(b) Helium
(c) Argon
(d) None of these
(I.T.I. August 2016, Wireman)

44. Which of the following lamps give nearly mono chromatic light?
(a) Sodium vapour
(b) Tube light
(c) GLS lamp
(d) Mercury vapour lamp
(I.T.I. February 2017, Electrician)

45. How the rate of evaporation in vacuum bulb is reduced?
(a) Reducing filament resistance
(b) Filling inert gas
(c) Producing arc in bulb
(d) Increasing filament resistance
(I.T.I. August 2019 Electrician)

46. Which device provides ignition voltage and acts as choke in a H.P.S.V. lamp?
(a) Leak transformer
(b) Sodium vapour
(c) Arc tube
(d) High pressure aluminium oxide
(I.T.I. August 2019 Electrician)

47. What is the name of lamp?

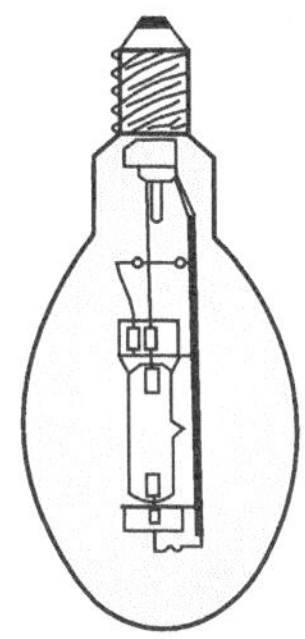

(a) M.A.T. type M.V. lamp
(b) H.P. metal halide lamp
(c) M.B. type H.P.M.V. lamp
(d) M.A. type H.P.M.V. lamp
(I.T.I. August 2019 Electrician)

48. Why the gas discharge lamps must not be operated with broken condition?
(a) Possible to get injuries
(b) To avoid electric hazard
(c) Possible to get electric shock
(d) Lead to exposure of ultraviolet rays
(I.T.I. August 2019 Electrician)

49. Which is the cold cathode lamp?
(a) Neon sign lamp
(b) Fluorescent lamp
(c) Halogen lamp
(d) Mercury vapour lamp
(I.T.I. August 2019 Electrician)

50. What is the main advantage of coiled coil lamp?
(a) High melting point
(b) Higher light output
(c) Low operating voltage
(d) Low power consumption
(I.T.I. August 2019 Electrician)

Decoration Lighting

1. In a serial set of bulb, the flasher is used. What is the function of the flasher?
   (a) To break and make the circuit but it gives no light
   (b) To break and make the circuit and it gives light
   (c) To break the circuit only but gives no light
   (d) To make the circuit only and also gives light
   (I.T.I. August 2014, Electrician)
2. To operate 6 volt bulb in a 240 volt supply, number of bulbs required is ....
   (a) 30 (b) 32
   (c) 40 (d) 42
   (I.T.I. February 2015, Electrician)
3. In a decorative series lamp connection of 6 V lamp is connected to a 240 V, A.C. supply. The number of lamps required is ....
   (a) 40 (b) 35
   (c) 30 (d) 25
   (I.T.I. February 2015, Wireman)
4. To keep the continuity of serial lamp set circuit even across the fused bulb, it is customary to connect a device across each bulb as precaution.
   What is the name of such device?
   (a) NTC thermistor
   (b) PTC thermistor
   (c) Voltage dependent resistor
   (d) Flasher
   (I.T.I. February 2018, Electrician)
5. The purpose of the drum switch used for decorative lamps circuit is to ....
   (a) give more brightness of light
   (b) use sequential switching 'ON' the lights
   (c) display the decorative lights constantly
   (d) keep the lights in their correct position
   (I.T.I. August 2015, Electrician)
6. Which type of reflector is used for yard lighting purpose?
   (a) Dispersive type (b) Mirror type
   (c) Parabolic type (d) Soft light type
   (I.T.I. February 2016, Electrician)
7. What is the name of light?

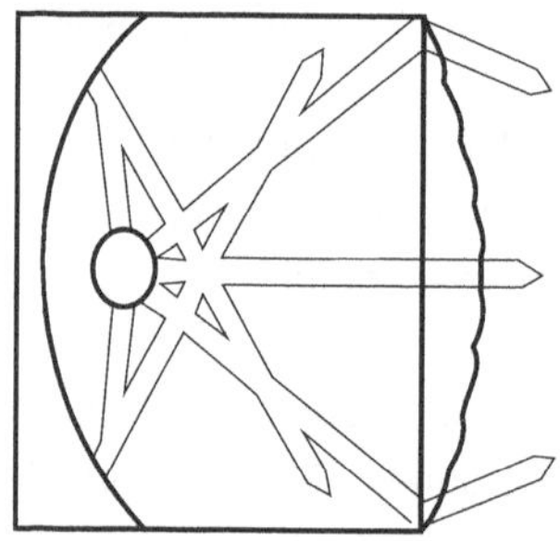

   (a) Flash light (b) Bulk light
   (c) Flood light (d) Spot light
   (I.T.I. August 2019 Electrician)
8. What is the name of the reflector?

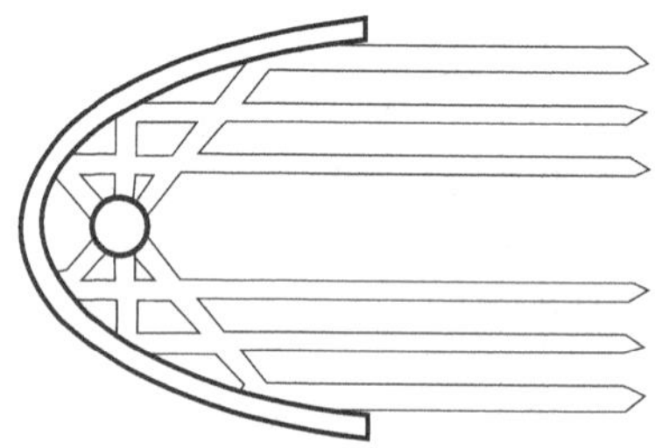

   (a) Mirror type (b) Parabolic type
   (c) Soft light type (d) Dispersive type
   (I.T.I. August 2019 Electrician)
9. Which type of lighting system is used for flood and industrial lighting?
   (a) Semi direct lighting
   (b) Indirect lighting
   (c) Direct lighting
   (d) Semi indirect lighting
   (I.T.I. August 2019 Electrician)

KEY TO MULTIPLE CHOICE QUESTIONS

Illumination

1. (c) **2.** (d) **3.** (a) **4.** (c) **5.** (c) **6.** (c) **7.** (b) **8.** (a) **9.** (b) **10.** (c)
11. (c) **12.** (d) **13.** (c) **14.** (b) **15.** (d) **16.** (b) **17.** (c)

Sources of Light

1. (c) **2.** (c) **3.** (c) **4.** (a) **5.** (a) **6.** (c) **7.** (c) **8.** (c) **9.** (c) **10.** (b)
11. (b) **12.** (c) **13.** (d) **14.** (b) **15.** (a) **16.** (d) **17.** (a) **18.** (d) **19.** (a) **20.** (d)
21. (b) **22.** (a) **23.** (c) **24.** (c) **25.** (c) **26.** (d) **27.** (c) **28.** (b) **29.** (d) **30.** (d)
31. (c) **32.** (d) **33.** (c) **34.** (d) **35.** (a) **36.** (a) **37.** (d) **38.** (c) **39.** (c) **40.** (c)
41. (c) **42.** (d) **43.** (a) **44.** (a) **45.** (b) **46.** (a) **47.** (d) **48.** (d) **49.** (a) **50.** (b)

Decoration Lighting

1. (a) **2.** (c) **3.** (a) **4.** (a) **5.** (b) **6.** (b) **7.** (d) **8.** (b) **9.** (c)

Module 11

Measuring Instruments

Multiple Choice Questions

1. The force acting in a M.C. instrument is equal to :
   (a) *B.l.v.* (b) *B.l.v. sin q*
   (c) *B.l.i.* (d) *B.l.i. sin q.*
2. The force which produces movement of the pointer of an instrument is called :
   (a) deflecting force
   (b) controlling force
   (c) damping force
   (d) distracting force.

   (I.T.I. August 2014, Wireman)
3. What is the purpose of providing a mirror behind the pointer in a measuring instrument ?
   (a) The scale is illuminated through the mirror
   (b) To observe that the pointer is bend or not
   (c) Being semi-transparent, a mirror helps in the study of inner construction of the instrument
   (d) To make the reading free from an error by removing parallax between the pointer and its image.
4. Which one of the following statements is true about the instrument in the given figure?
   (a) It is a high frequency measuring instrument
   (b) It is designed to work in horizontal position only
   (c) It is designed to work in vertical position only
   (d) Its current coil can be loaded upto 5 A.

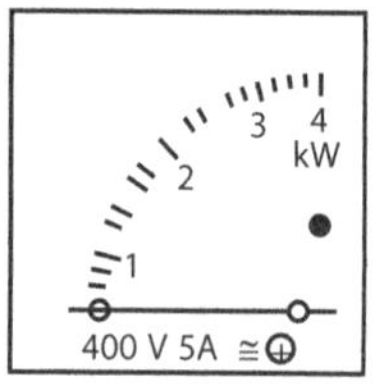

5. Which one of the following statements is true about the internal resistance of an ammeter and a voltmeter?
   (a) The internal resistance of an ammeter should be very low and that of a voltmeter should be very high
   (b) The internal resistance of an ammeter should be very high and that of a voltmeter should be very low
   (c) The internal resistance of an ammeter and a voltmeter should be very low
   (d) The internal resistance of an ammeter and a voltmeter should be very high.
6. The reliability of measuring instrument means :
   (a) the life of the instrument
   (b) the extent to which its characteristics remain linear

(c) the degree to which repeatability continues to remain within specific limits

(d) all of these.

7. The sensitivity of a measuring instrument means :

(a) the maximum value of a physical quantity that can be measured

(b) the highest reading of the scale

(c) the minimum reading of the scale

(d) the smallest amount of a physical quantity that can be detected by the instrument.

8. In order to increase the range of a voltmeter:

(a) a low resistance is connected in series

(b) a low resistance is connected in parallel

(c) a high resistance is connected in series

(d) a high resistance is connected in parallel.

9. To reduce the effect of a voltmeter on the circuit under test, we should :

(a) decrease the multiplier resistance

(b) use an instrument with a high internal resistance

(c) use a shunt resistor

(d) use an a.c. meter with a rectifier.

10. A M.I. type instrument has :

(a) a linear scale

(b) a non-linear scale

(c) an exponential scale

(d) a logarithmic scale.

11. If two ammeters X and Y require 40 mA and 50 mA current respectively to give a full-scale-deflection, then:

(a) X is more sensitive

(b) Y is more sensitive

(c) both are sensitive

(d) sensitivity cannot be judged with the given information.

12. The value of a shunt resistance in an ammeter is usually:

(a) equal to the meter resistance

(b) greater than the meter resistance

(c) lessor than the meter resistance

(d) of any value.

13. Which torque is not essential for working of an indicating instrument?

(a) Break up torque

(b) Deflecting torque

(c) Controlling torque

(d) None of these

(I.T.I. February 2019, Wireman)

14. A multimeter consists of :

(a) voltmeter and ohmmeter

(b) voltmeter and ammeter

(c) voltmeter, ammeter and ohmmeter

(d) none of these.

15. A multimeter cannot measure :

(a) current

(b) potential difference

(c) capacitance

(c) resistance.

16. Tong tester is used to measure the current in a.c. circuits without interrupting the circuit. For measurement, the split core is clamped on to the insulated current carrying conductor.

Looping of condutor passing through the split core is done sometimes.

What will be the reading for a selected range if the conductor is looped (double pass)?

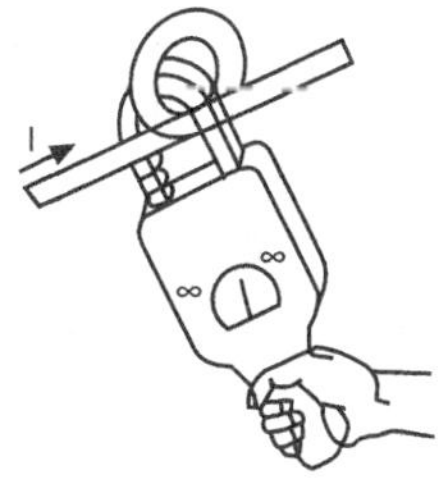

(a) Double the current which passes through the conductor

(b) 4 times (*i.e.*, 2^2) the current which passes through the conductor

(c) Half the current which passes through the conductor

(d) Same as the conductor current.

(I.T.I. February 2015, Electrician)

17. Power drawn by a welding shop which has 3-phase, 415 V, 50 Hz supply has to be measured.

The following loads are connected :

(*a*) 2 welding sets each of 5 kVA across L_1, L_2 and L_3.

(*b*) 1 set of 5 kVA across L_2 and L_3.

What is the type of wattmeter you would choose for this measurement ?

(a) 3 element, 3 wire type, 3-phae wattmeter

(b) One single phase wattmeter connected between any two phases and multiply reading by 3

(c) 3 element, 4 wire type 3-phase wattmeter

(d) 2 element type 3-phase wattmeter.

(I.T.I. February 2018, Electrician)

18. One screw is provided in M.C. instrument which is attached to the pointer. What for it is provided ?

(a) To adjust the sensitivity of the meter

(b) To correct the zero position of the pointer

(c) To tight the moving part of the meter

(d) To change the measuring range of the meter.

19. A voltmeter with a sensitivity of 1 kilo ohm per volt will have a resistance of .... ohms at 50 V range.

(a) 1000 (b) 5000

(c) 50,000 (d) 50

20. Which measuring instrument is an indicating instrument?

(a) Read type frequency meter

(b) Energy meter

(c) Ammeter

(d) Temperature recording instrument

(I.T.I. February 2014, Wireman)

21. Which type of control force is an indicating instrument, which can be kept in any position?

(a) Gravity control

(b) Spring control

(c) Fluid friction control

(d) Hydraulic friction control

(I.T.I. February 2014, Wireman)

22. The meter installed at your house to measure electrical energy is an example of ....

(a) indicating type instrument

(b) recording type instrument

(c) integrating type instrument

(d) indicating and recording type instrument

(I.T.I. February 2014, Wireman)

23. Phosphor bronze alloy is used for springs in indicating instruments because the alloy has ....

(a) high temperature co-efficient of expansion

(b) low specific resistance

(c) low resistance to fatigue

(d) good magnetic property

(I.T.I. February 2015, Electrician)

24. When an instrument is using gravity control, it can read correctly ....

(a) if it is used in horizontal position only

(b) if it is used in vertical position only

(c) if it is used in an inclined position only

(d) if it is used in an inclined and horizontal position

25. As a general rule before connecting the meters to the circuit, we should see whether pointer is at zero or not. This rule is not applicable in case of ....

(a) multimeter in voltage range

(b) wattmeter

(c) milli ammeter

(d) megger

(I.T.I. August 2014, Electrician)

26. The use of .... instrument is merely confined within laboratories standardising instrument.

(a) Absolute (b) Indicating

(c) Recording (d) Integrating

(I.T.I. February 2017, Electrician)

27. Ammeters used for current measurements in circuits with short overload time, differs from other application. Ammeters used in short overload time circuit have....

(a) non-linear scale (b) extended scale
(c) linear scale (d) fine scale

(I.T.I. February 2018, Electrician)

28. Which part of the PMMC instrument produces eddy current damping ?

(a) Moving coil
(b) Aluminium former
(c) Permanent magnet
(d) Soft iron cylindrical core.

29. Instruments which must not be used in the circuit, when the power is ON are ....

(a) wattmeter and power factor meter
(b) frequency meter and tong tester
(c) phase sequence meter and energy meter
(d) ohm meter and megger

(I.T.I. February 2018, Electrician)

30. You are testing an energy meter of 240 volts, 50Hz 5 amperes, 750 revolutions per kWh for its performance with some load. When the disc rotates 100 revolutions in 10 minutes, what is the energy supplied?

(a) 13.33 kWh (b) 1.33 kWh
(c) 0.133 kWh (d) 0.0133 kWh

(I.T.I. August 2014, Electrician)

31. You have to connect an ammeter to monitor the current drawn by a 3-phase, 5 H.P, 415 V, 50 Hz squirrel cage induction motor, connected through a DOL starter. The range of ammeter, suitable for connecting in motor circuit is ....

(a) 0 to 30 ampers (extended scale)
(b) 0 to 15 amperes (linear scale)
(c) 0 to 10 amperes (non. linear scale)
(d) 0 to 5 amperes (extended scale)

(I.T.I. January 2007, Electrician)

32. In a dynamometer type wattmeter, the deflecting torque is proportional to ....

(a) the square of the current in fixed coil
(b) the current in the moving coil
(c) the current in the fixed coil and in the moving coil
(d) the current in the fixed coil

33. What precautions you have to take when disconnecting voltmeter from potential transformers?

(a) Potential transformers secondary terminals which are open should be insulated
(b) Ground the secondary terminals on pen circuit
(c) The terminals of the potential transformer secondary have to be short circuited.
(d) After disconnecting, connect a low-resistance shunt between the terminal of the secondary

(I.T.I. July 2009, Electrician)

34. What you have to do when disconnecting the ammeter from the terminals of the CT?

(a) Earthing the secondary of CT.
(b) Short-circuiting the terminals of the CT before disconnecting the ammeter.
(c) Switching a high resistance shunt between terminals of CT after disconnecting the ammeter.
(d) Disconnecting primary from supply.

(I.T.I. July 2011, Electrician)

35. With the help of multimeter, we can measure ....

(a) voltage, current and resistance
(b) power, energy and power factor
(c) insulation resistance and A.C. curent only.
(d) ampere hour capacity and D.C. current only.,

(I.T.I. February 2014, Wireman)

36. For measuring the current, voltage (or) resistance by multimeter, which control switch to be set?

(a) Range switch
(b) Selector switch

(c) Power ON - OFF switch
(d) Function switch

(I.T.I. February 2016, Electrician)

37. How many segments are in both LCD and LED, Digital Multi Meter (DMM) display?

(a) 9 (b) 8
(c) 7 (d) 6

(I.T.I. February 2016, Electrician)

38. Two types of multimeter are in use. Analog type and Digital type.

Choose the correct statements applicable for both the types. Both the meters ....

(a) have moving elements, hence must be used in horizontal position
(b) can be used for current and voltage measurement without battery
(c) needs manual zero setting for resistance measurement
(d) needs battery for resistance measurements

(I.T.I. February 2018, Electrician)

39. Which one is a disadvantage of a PMMC instrument?

(a) Consumes less power
(b) Has efficient damping
(c) Can be used in DC only
(d) Has high torque/weight ratio

(I.T.I. February 2016, Wireman)

40. The examples for integrating instruments is ....

(a) voltmeter
(b) ammeter
(c) ampere hour meter
(d) wattmeter

(I.T.I. August 2015, Wireman)

41. Which type of instrument belongs to the integrating instrument type?

(a) Energy meter
(b) Voltmeter
(c) Ammeter
(d) Electro cardio graph (ECG)

(I.T.I. February 2016, Electrician)

42. The error of false adjustment of the instruments is ....

(a) influence errors (b) switching errors
(c) device errors (d) human errors

(I.T.I. February 2016, Wireman)

43. The error caused by the effect of vibrations is ....

(a) device errors (b) switching errors
(c) influence errors (d) human errors

(I.T.I. February 2016, Wireman)

44. The measuring instruments indicating error may be ....

(a) ± 1.5% (b) ± 2.0%
(c) ± 2.2% (d) ± 2.3%

(I.T.I. February 2016, Wireman)

45. Fine scale instruments are used for precision with an accuracy from ....

(a) 0.1 to 0.15 (b) 0.1 to 0.2
(c) 0.1 to 0.3 (d) 0.1 to 0.4

(I.T.I. February 2016, Wireman)

46. Controlling force always acts ....

(a) opposite to the deflection force
(b) same direction of the deflection force
(c) not to allow to move
(d) to keep the pointer in the same place

(I.T.I. August 2015, Wireman)

47. The material used for spring control in instrument is ....

(a) Manganin (b) Eureka
(c) Phosphor bronze (d) Silver

(I.T.I. August 2015, Electrician)

48. The material used for shunt resistor for the extension range of ammeter is ....

(a) Brass (b) Manganin
(c) Bronze (d) Copper

(I.T.I. August 2015, Electrician)

49. The unit of sensitivity of the instrument is ....

(a) ohms/volt (b) volt/ohm
(c) volt-amp (d) ampere/sec

(I.T.I. August 2015, Electrician)

50. If the pointer moves quickly to its final deflected position without any sort of oscillation, that damping is called as ....

(a) Under damping
(b) Over damping
(c) Critical damping
(d) Air friction damping

(I.T.I. February 2016, Electrician)

51. The deflecting force of repulsion type M.I. instruments is directly proportional to ....

(a) the current
(b) the voltage
(c) the square of the current
(d) the magnetic field

(I.T.I. August 2015, Electrician)

52. When 1 milli ampere is flowing through the meter coil is having 1000 ohms, and is causing F.S.D., the voltage developed across the coil resistance will be ....

(a) 1000 volt (b) 100 volt
(c) 10 volt (d) 1 volt

(I.T.I. February 2016, Electrician)

53. Which one is application of current transformer (CT) ?

(a) Auto transformer
(b) Tong Tester
(c) Earth leakage circuit breaker (ELCB)
(d) Variac

(I.T.I. August 2015, Electrician)

54. The working principle of tong tester is ....

(a) Self induction
(b) Mutual induction
(c) Fleming's right hand rule
(d) Same principle of D.C. motor

(I.T.I. August 2015, Electrician)

55. You have to select a PF meter suitable for connecting in a circuit for PF measurements. On what basis you will select its range ?

(a) Suitability for the type of load (inductive/capacitive/Resistive)
(b) PF range of meter (lagging/leading/unity)
(c) Voltage and current range of meter suitable to load
(d) Current and PF range of meter matching to load

(I.T.I. February 2018, Electrician)

56. What happens if the link in an energy meter is removed and supply is switched 'ON'?

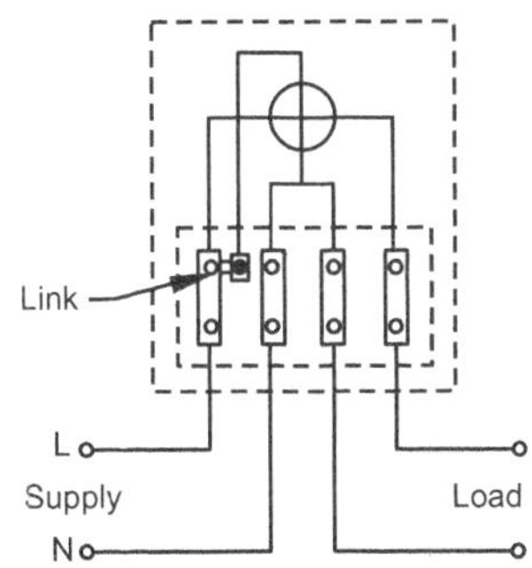

(a) Supply to the load is available. Meter will damage due to current flow in cc
(b) Supply will be available to the load. The meter run on only cc.
(c) No supply will be available to load. Meter stops
(d) Supply at load is available. Meter will not run

(I.T.I. August 2015, Electrician)

57. In energy meter the rotary disc is made of ....

(a) Iron (b) Copper
(c) Aluminium (d) Tungsten

(I.T.I. February 2019, Wireman)

58. In an energy meter breaking torque is proportional to......

(a) Speed of rotating disc
(b) Strength of the permanent magnet
(c) Position of the permanent magnet
(d) None of these

(I.T.I. August 2016, Wireman)

59. Induction type energy meter is used for measuring .....

(a) D.C. Energy (b) D.C. Power
(c) A.C. Power (d) A.C. Energy

(I.T.I. August 2016, Wireman)

60. Induction type single phase energy meters measure electric energy in ......

(a) kW (b) Wh

(c) VAR (d) kWh

(I.T.I. February 2019, Electrician)

61. In a single phase energy meter, the four terminals starting from left are designated as 1, 2, 3 and 4. The current coil is connected internally between ......

(a) 1–2 (b) 1–4

(c) 2–3 (d) 2–4

(I.T.I. February 2017, Wireman)

62. For a 3-phase unbalanced load, the energy can be measured by :

(a) 2 phase 3 wire energy meter

(b) Three single phase energy meters

(c) 3 phase 4 wire energy meter

(d) None of these

(I.T.I. August 2016, Wireman)

63. A three phase energy meter has ....

(a) 3 pressure coils and 3 current coils connected between separate phases and neutral

(b) 2 pressure coils connected between separated phases and neutral

(c) 2 pressure coils connected in series between two phases

(d) 2 current coils connected in series between separate phases and neutral

(I.T.I. February 2017, Wireman)

64. One unit of electrical energy consumed is equal to ......

(a) 36 joules (b) 3500 joules

(c) 3.6 kilo joules (d) 3.6 mega joules

(I.T.I. February 2017, Wireman)

Hint:

watt × second = watt second = joule

Actually unit of electrical energy = 1 kilo watt hour

∴ 1 kilo watt hour

$= 1000 \times \text{watt} \times 3600 \text{ seconds}$

$= 36 \times 10^5 \text{ watt seconds}$

$= 3.6 \times 10 \times 10^5 \text{ watt seconds}$

$= 3.6 \times 10^6 \text{ watt seconds}$

$= 3.6 \times 10^6 \text{ joules}$

= 3.6 mega joules

65. Accuracy of a measuring instrument indicates the .....

(a) Ratio of output value to the input value

(b) Closeness of the output reading to the true value

(c) Change in output with each change in input

(d) Degree of freedom from random errors

(I.T.I. February 2019, Electrician)

66. An instrument with high precision implies ......

(a) Does not imply anything about measurement accuracy

(b) Low accuracy

(c) High accuracy

(d) None of these

(I.T.I. February 2019, Electrician)

67. Continuity of electrical circuit is checked by .....

(a) Volt meter

(b) Ammeter

(c) Ohm meter

(d) Frequency meter

(I.T.I. August 2016, Wireman)

68. Deflection in a M.C. galvanometer is due to the torque based on the principle of ....

(a) Right hand thumb rule

(b) Fleming's right hand rule

(c) Fleming's left hand rule

(d) All of these

(I.T.I. February 2017, Wireman)

69. The controlling torque .....

(a) Brings the pointer back to zero position

(b) Deflects the needle

(c) Both (a) and (b)

(d) None of these

(I.T.I. February 2019, Wireman)

70. The function of shunt in ammeter is to ......

(a) Increase the resistance of ammeter

(b) Increase the sensitivity of the ammeter

(c) Bypass the current

(d) None of these

(I.T.I. February 2019, Elactrician)

71. Which of the following instrument indicates the instantaneous value of electrical quantity being measured at the time it is being used?

(a) Absolute (b) Indicating

(c) Recording (d) Integrating

(I.T.I. February 2017, Electrician)

72. The moving system in the indicating instruments is subjected to ....

(a) Deflecting torque

(b) Controlling torque

(c) Damping torque

(d) All of these

(I.T.I. February 2019, Electricain)

73. Which among the following is not a type of digital voltmeter?

(a) Ramp type

(b) Integrating

(c) Potentiometric type

(d) None of these

(I.T.I. February 2019, Electrician)

74. An ammeter is connected in __________ in the circuit.

(a) Series

(b) Parallel

(c) Either series or parallel

(d) None of these

(I.T.I. August 2016, Electrician)

75. A moving coil Ammter can be used for measuring.........

(a) A.C. current

(b) D.C. current

(c) Both A.C. and D.C

(d) None of these

(I.T.I. August 2016, Wireman)

76. A moving iron instrument can be used for –

(a) D.C. only

(b) A.C. only

(c) Both D.C. and A.C.

(d) None of these

(I.T.I. August 2016, Electrician)

77. In permanent magnet moving coil instrument, damping torque is provided by–

(a) Air friction

(b) Eddy current

(c) Fluid friction

(d) Either (a), (b), or (c)

(I.T.I. August 2016, Wireman)

78. Which instrument uses be vibrating reads?

(a) Power factormeter

(b) Frequency meter

(c) Tachometer

(d) Synchronoscope

(I.T.I. August 2016, Wireman)

79. The following .... type is not a wattmeter.

(a) Dynamometer

(b) Induction

(c) Electro mechanical

(d) Electrostatic

(I.T.I. February 2017, Electrician)

80. In wattmeter, current coil is ....

(a) Connected in parallel

(b) Connected in series

(c) If connected in parallel coil gets burnt

(d) Both (b) and (c)

(I.T.I. February 2019, Wireman)

81. Why damping force is required in a moving coil instrument?

(a) Makes the needle movement factor

(b) Helps the deflecting force to act fast

(c) Brings the needle to its zero position

(d) Arrests the needle without oscillations

(I.T.I. August 2019 Electrician)

82. Why the scale of moving iron instrument is havingun-uniform scale?

(a) Deflection of force is directly proportional to the square of the current

(b) Deflection force is inversely proportional to the current

(c) Deflection force is directly proportional to the current

(d) Deflection force is inversely proportional to the square of the current.

(I.T.I. August 2019 Electrician)

83. Which parameter is the cause for loading effect on measuring instruments?

(a) Low accuracy
(b) High sensitivity
(c) Low sensitivity
(d) Low influence error

(I.T.I. August 2019 Electrician)

84. Which meter is used to measure revolution per minute (R.P.M.) of a motor?

(a) Tachometer
(b) Energy meter
(c) ampere hour meter
(d) Centre zero ammeter

(I.T.I. August 2019 Electrician)

85. Which source of measuring error is caused by the effect of magnetic fields?

(a) Influence error (b) Human error
(c) Device error (d) Switching error

(I.T.I. August 2019 Electrician)

86. Which instrument is an example of an absolute instrument?

(a) A.C. voltmeter
(b) Energy meter
(c) D.C. voltmeter
(d) Tangent galvanometer

(I.T.I. August 2019 Electrician)

87. What is the purpose of phase sequence meter?

(a) To control the speed of 3 phase motor
(b) To ensure the correct phase sequence of 3-phase system
(c) To indicate the incorrect phase sequence of 3-phase system
(d) To protect motor against short circuit fault.

(I.T.I. August 2019 Electrician, Electroplator, Wireman, Lift & Escalator Mechanic)

88. How the creeping error is controlled in energy meter?

(a) By reducing rated voltage
(b) By increasing the inductive load
(c) By adjusting the brake magnet position
(d) By drilling two holes diametrically opposite on disc

(I.T.I. August 2019 Electrician)

89. Name the type of instrument.

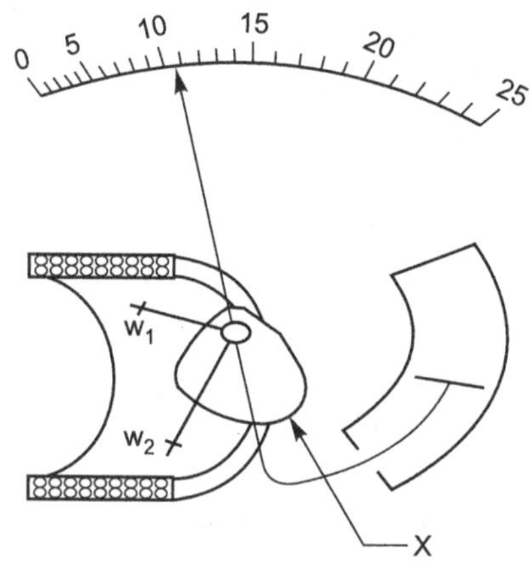

(a) Attraction type moving iron
(b) Repulsion type moving iron
(c) Permanent magnet moving coil
(d) Dynamo meter type moving coil

(I.T.I. August 2019 Electrician)

90. What is the name of the instrument?

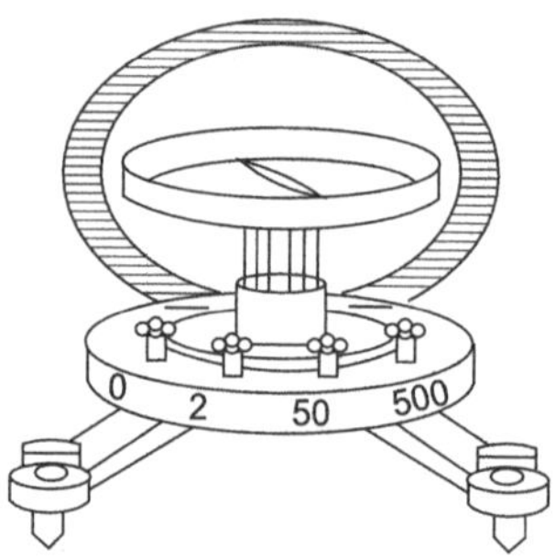

(a) Absolute instrument
(b) Indicating instrument
(c) Recording instrument
(d) Integrating instrument

(I.T.I. August 2019 Electrician)

91. What is the name of meter?

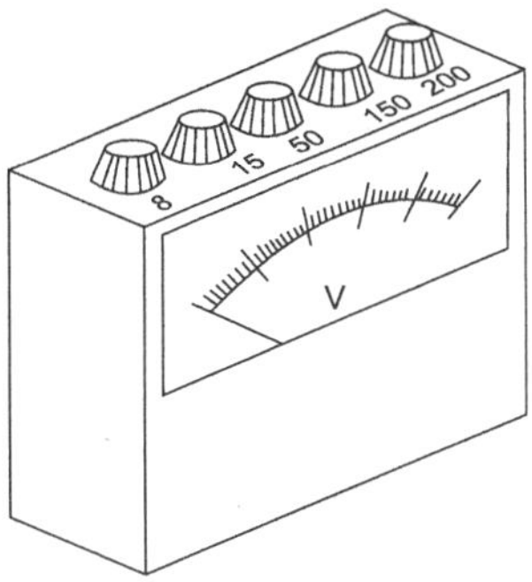

(a) Multi range DC ammeter
(b) Multi range DC voltmeter
(c) Single range AC ammeter
(d) Single range AC voltmeter

(I.T.I. August 2019 Electrician)

KEY TO MULTIPLE CHOICE QUESTIONS

1. (c)	**2.** (a)	**3.** (d)	**4.** (d)	**5.** (a)	**6.** (c)	**7.** (d)	**8.** (c)	**9.** (b)	**10.** (b)
11. (a)	**12.** (c)	**13.** (a)	**14.** (c)	**15.** (c)	**16.** (a)	**17.** (d)	**18.** (b)	**19.** (c)	**20.** (c)
21. (b)	**22.** (c)	**23.** (b)	**24.** (b)	**25.** (d)	**26.** (a)	**27.** (b)	**28.** (d)	**29.** (d)	**30.** (c)
31. (a)	**32.** (c)	**33.** (a)	**34.** (b)	**35.** (a)	**36.** (d)	**37.** (c)	**38.** (d)	**39.** (c)	**40.** (c)
41. (a)	**42.** (c)	**43.** (c)	**44.** (a)	**45.** (c)	**46.** (a)	**47.** (c)	**48.** (b)	**49.** (a)	**50.** (c)
51. (c)	**52.** (d)	**53.** (b)	**54.** (b)	**55.** (c)	**56.** (d)	**57.** (c)	**58.** (a)	**59.** (d)	**60.** (d)
61. (b)	**62.** (c)	**63.** (c)	**64.** (d)	**65.** (b)	**66.** (a)	**67.** (c)	**68.** (c)	**69.** (a)	**70.** (c)
71. (b)	**72.** (d)	**73.** (d)	**74.** (a)	**75.** (b)	**76.** (c)	**77.** (b)	**78.** (b)	**79.** (c)	**80.** (b)
81. (d)	**82.** (a)	**83.** (c)	**84.** (a)	**85.** (a)	**86.** (d)	**87.** (b)	**88.** (d)	**89.** (a)	**90.** (a)
91. (b)									

Module 12

Domestic Appliances

MULTIPLE CHOICE QUESTIONS

1. The operating temperature range of an electrical stove with a pure element is :
   (a) 300°C to 400°C
   (b) 500°C to 600°C
   (c) 750°C to 900°C
   (d) 1100°C to 1300°C
2. A 1000 W, 230 V kettle is connected to a 15 A plug-socket using a power cord rated to carry 15 A. What should be the fuse rating of the appliance ?
   (a) 4.0 A (b) 5.0 A
   (c) 15.0 A (d) 16.0 A
3. The heat proof insulating material used in heaters etc. is :
   (a) mica (b) asbestos
   (c) glass wool (d) porcelain.
4. In an automatic electric iron, a thermostat is used to:
   (a) reduce the current for controlling the temperature
   (b) change the direction of current flow for controlling the temperature
   (c) regulate the heat and thus the temperature by switching the element to ON-OFF
   (d) switch ON-OFF the indicator lamp.
5. In an electric iron, the function of pressure plate is :
   (a) to cover the heating element
   (b) to exert pressure on the garments while pressing them
   (c) to obtain a close fitting of the heating element with the sole plate
   (d) to act as a base of the equipment.
6. What is the function of a fusible safety plug in a geyser ?
   (a) It protects the geyser against blasting due to excessive steam pressure
   (b) It protects the heating element against burning.
   (c) It releases the extra water out of the geyser
   (d) It protects the thermostat of the geyser.
7. What should be the resistance of a new thermostat?
   (a) Zero or few ohms
   (b) In kilo ohms
   (c) Infinity
   (d) In mega ohms.
8. Operating temperature range of heating elements in hot plate is ....
   (a) 550°C to 700°C
   (b) 700°C to 850°C
   (c) 750°C to 900°C
   (d) 800°C to 850°C

   (I.T.I. February 2015, Electrician)
9. The insulating material used in connecting leads of the electric heater is ....
   (a) Bakelite material
   (b) Porcelain material

(c) Rubber material
(d) Asbestos material

(I.T.I. August 2014, Electrician)

10. Calculate the current and the resistance of the heating element for an Electric heater of 250 V, 1500 W, working at 500°C.

(a) 4 amps and 62.5 ohms
(b) 5 amps and 55.5 ohms
(c) 6 amps and 41.66 ohms
(d) 7.5 amps and 31.25 ohms

(I.T.I. August 2014, Electrician)

11. Which among the following is the temperature regulating component of an automatic electric iron ?

(a) Heating element (b) Sole-plate
(c) Thermostat (d) Pressure plate

12. What is the material used in between the outer casing and the inner tank of a geyser?

(a) Air (b) Mica
(c) Glass wool (d) Asbestos

(I.T.I. February 2014, Electrician)

13. Glass wool is provided in between the outer casing and inner container in storage water heater is to ....

(a) reduce the water leakage
(b) reduce the heat loss
(c) increase the life of water heater
(d) limit the load current

(I.T.I. August 2014, Wireman)

14. Electric irons are available in standard power rating ....

(a) 1000 Watts (b) 750 Watts
(c) 250 Watts (d) 100 Watts

(I.T.I. February 2014, Electrician)

15. The function of the thermostat in an Automatic electric Iron is to ....

(a) reduce the current to control temperature
(b) limit the current flow through the element
(c) regulate the heat by switching ON and OFF
(d) control the ON and OFF switch function of indicating lamp

(I.T.I. February 2015, Wireman)

16. Which type washing machine contains two tubs ?

(a) Ordinary without timer type
(b) Ordinary with timer type
(c) Semi-automatic type
(d) Automatic type

(I.T.I. February 2015, Electrician)

17. The capacitor used in fan motor is connected in ....

(a) parallel with winding
(b) parallel with starting running winding
(c) series with starting winding
(d) series with starting and running winding

18. Pumps can be classified mainly into .... categories.

(a) 2 (b) 3
(c) 4 (d) 5

(I.T.I. February 2015, Electrician)

19. Reciprocating pumps is also known as :

(a) jet pump
(b) piston pump
(c) centrifugal pump
(d) submersible pump

20. A submersible pump can lift water from any depth as the:

(a) length of the unit is more
(b) efficiency is more
(c) suction pipe is not there
(d) diameter of the unit is smaller.

21. A pump motor has an incorrect direction of rotation because of wrong connections. How do you notice this fault?

(a) The pump supplies too much water
(b) The pump does not supply water
(c) The pump motor runs too hot
(d) The fuse in the motor circuit blows out

(I.T.I. July 2014, Electrician)

22. Which one is NOT a disadvantage of the submersible motor?

(a) Initial cost high
(b) Diameter is small
(c) Requires skill for erection
(d) In case of defect entire unit is to be lifted

(I.T.I. February 2016, Wireman)

23. Heating elements wires are made of ....
(a) flat nichrome (b) tungsten
(c) eureka (d) manganin
(I.T.I. August 2015, Wireman)

24. In the geyser, the fuse is blowing repeatedly, the cause will be ....
(a) leakage around thermostat
(b) thermostat setting is too low
(c) grounded heating element
(d) thermostat setting is too high
(I.T.I. August 2015, Electrician)

25. The efficiency of heater is ....
(a) $\frac{\text{Heat utilised}}{\text{Heat generated}}$
(b) $\frac{\text{Heat generated}}{\text{Heat utilised}}$
(c) Heat generated × Heat utilised
(d) Heat generated and Heat utilised
(I.T.I. August 2015, Wireman)

26. Number of canopy fitted to a ceiling fan is ....
(a) 1 (b) 2
(c) 3 (d) 4
(I.T.I. February 2016, Wireman)

27. According to the IS recommendations, the clearance between the bottom part of the ceiling fan and the floor should not be less than ....
(a) 1.3 m (b) 2.0 m
(c) 2.25 m (d) 2.40 m
(I.T.I. February 2018, Electrician)

28. The type of motor used in food mixer is ....
(a) shaded pole motor
(b) universal motor
(c) capacitor start motor
(d) capacitor run motor
(I.T.I. February 2016, Wireman)

29. The element of electric heater is made of ....
(a) Copper (b) Aluminium
(c) Nichrome (d) Carbon
(I.T.I. February 2019, Wireman)

30. In automobiles the sound is produced by horn due to
(a) Magnetostriction
(b) Vibrating diaphragm
(c) Moving coil
(d) Oscillating coil
(I.T.I. February 2018, Electrician)

31. Inside the geyser there is a ....
(a) Filament (b) Immersion rod
(c) Both (a) & (b) (d) None of these
(I.T.I. August 2016, Electrician)

32. Following domestic appliances use thermostats for tripping purpose :
(a) Electric geyser
(b) Electric press
(c) Microwave heater
(d) All of these
(I.T.I. February 2017, Wireman)

33. In the given appliances, which one does not have a power point:
(a) Fridge (b) Electric grinder
(c) Air conditioner (d) Pedestal fan
(I.T.I. August 2016, Wireman)

34. If the ceiling fan, when switched on runs at slow speed in the reverse direction. It can be concluded that :
(a) Winding has burnt out
(b) Bearings are worn out
(c) Capacitor is ineffective
(d) None of these
(I.T.I. August 2016, Electrician)

35. For ceiling fans, generally the single phase motor used is –
(a) Spilt phase type
(b) Capacitor start type
(c) Capacitor start and run type
(d) Permanent capacitor type
(I.T.I. February 2019, Electrician)

36. A ceiling fan of 1400 mm sweep (56 inches, will have a motor of rating :
(a) 10 to 15 watts (b) 50 to 70 watts
(c) 120 to 180 watts (d) 250 to 500 watts
(I.T.I. February 2018, Electrician)

37. What is the symbol indicates?

(a) Ceiling fan (b) Table fan
(c) Bracket fan (d) Exhaust fan

(I.T.I. August 2019 Electrician)

38. What is the name of the circuit?

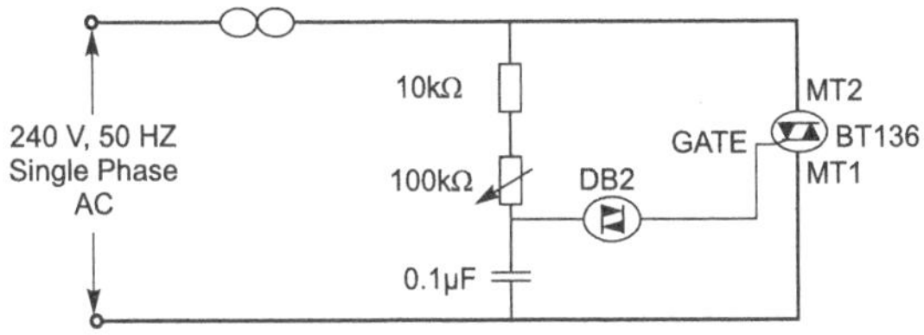

(a) Electronic triggering circuit of S.C.R.
(b) Electronic voltage multiplier
(c) Electronic voltage stabilizer
(d) Electronic fan regulator

(I.T.I. August 2019 Electrician)

39. Which type of motor is used in the wet grinder?

(a) Universal motor
(b) Capacitor start capacitor run motor
(c) Repulsion motor
(d) Capacitor start induction run motor

(I.T.I. August 2019 Electrician)

40. What is the name of the part marked 'X' of the wet grinder?

(a) Motor and pulley
(b) Male grinding stone
(c) Vertical stand holder
(d) Female grinding stone

(I.T.I. August 2019 Electrician)

41. Which is the cause for the fault that food mixer runs at slow speed.

(a) Dry bearing
(b) Partially grounded winding
(c) Loose mounting screws
(d) Rotor rubbing against stator

(I.T.I. August 2019 Electrician)

42. Which speed control method is used in food mixture motors?

(a) Voltage control method
(b) Field diverter control method
(c) Armature diverter method
(d) Series field tappping method

(I.T.I. August 2019 Electrician)

43. What falult occur due to excessive heat produced in automatic electric iron?

(a) Sticks to clothes
(b) Iron give shock
(c) Tears the clothes
(d) Blisters on sole plant

(I.T.I. August 2019 Electrician)

44. What is the purpose of pressure plate in non-automatic electric iron?

(a) Transfer the heat from the heating element
(b) Serve as a shield and protection from head
(c) Keep the heating element firmly against sole plate
(d) Give support as a resting point to its back and handle

(I.T.I. August 2019 Electrician)

45. What is the function of neutral path in A.C. supply system for appliances?

(a) Maintains load current constant
(b) Provides voltage level constant
(c) Reduces voltage drop in wiring
(d) Provides current return path

(I.T.I. August 2019 Electrician)

46. Which factor depends on the amount of generated heat in heating elements?

(a) Voltage
(b) Current
(c) Resistivity
(d) Square of the current

(I.T.I. August 2019 Electrician)

47. Which is the possible cause if "oven will not turn off" in cooking range?

(a) Loose connection
(b) Incorrect setting of timer
(c) Open circuit in element
(d) Incorrect connection

(I.T.I. August 2019 Electrician)

48. What is the function of the thermo cut out in microwave oven?

(a) Provides a current path to the stirrer motor
(b) Completes the circuit to the coil of the cook relay
(c) Prevents damage to magnetron at overheat condition
(d) It provides the holding circuit energises the coil

(I.T.I. August 2019 Electrician)

49. What is the function of magnetron tube in a microwave oven?

(a) Oscillates and produce cooking frequency
(b) Changes the polarity in every half cycle
(c) Amplifies the micro wave signal
(d) Converts microwave energy to electrical energy

(I.T.I. August 2019 Electrician)

50. Name the part marked 'X' of the geyser.

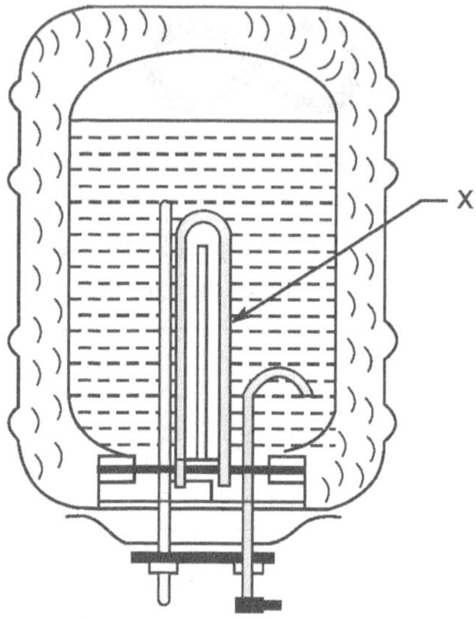

(a) Thermostat
(b) Heating element
(c) Drain pipe cover
(d) Pressure release valve

(I.T.I. August 2019 Electrician)

KEY TO MULTIPLE CHOICE QUESTIONS

1. (c)	**2.** (b)	**3.** (d)	**4.** (c)	**5.** (c)	**6.** (a)	**7.** (b)	**8.** (a)	**9.** (b)	**10.** (c)
11. (c)	**12.** (c)	**13.** (b)	**14.** (a)	**15.** (c)	**16.** (c)	**17.** (c)	**18.** (a)	**19.** (b)	**20.** (c)
21. (b)	**22.** (b)	**23.** (a)	**24.** (c)	**25.** (a)	**26.** (b)	**27.** (d)	**28.** (b)	**29.** (c)	**30.** (b)
31. (b)	**32.** (d)	**33.** (d)	**34.** (c)	**35.** (d)	**36.** (c)	**37.** (a)	**38.** (d)	**39.** (d)	**40.** (b)
41. (b)	**42.** (d)	**43.** (d)	**44.** (c)	**45.** (d)	**46.** (d)	**47.** (b)	**48.** (c)	**49.** (a)	**50.** (b)

Module 13

TRANSFORMER

MULTIPLE CHOICE QUESTIONS

1. On removing the cores of a transformer, the production of magnetic flux :
   (a) is reduced
   (b) is increased
   (c) remains the same
   (d) none of these.

2. In a transformer, the linkage of magnetic flux takes place from :
   (a) primary to secondary
   (b) secondary to primary
   (c) primary to secondary and from secondary to primary
   (d) none of these.

3. Since, the cores of a transformer are made of silicon-steel, therefore :
   (a) copper-loss remains low
   (b) eddy-current-loss remains low
   (c) hysteresis-loss remains low
   (d) both copper and eddy-current losses remain low.

4. The capacity of a star-connected transformer as compared to a delta-connected transformer will be:
   (a) 3 times (b) 1/3 times
   (c) 1/2 times (d) equal.

5. In the breather of a transformer, silica-jelly is used to stop the ______ to reach upto the insulating oil :
   (a) rain water (b) dry air
   (c) moisture (d) steam.

6. The efficiency of a transformer is normally in the range of:
   (a) 50% to 70% (b) 60% to 75%
   (c) 80% to 90% (d) 90% to 98%.

7. What type of core is used in a transformer designed to work at high frequencies (MHz range) ?
   (a) Aluminium core
   (b) Iron core
   (c) Air core
   (d) Open type iron core.

8. If the secondary turns of a transformer are doubled and at the same time the primary voltage is reduced to half then the secondary voltage will :
   (a) be halved
   (b) be doubled
   (c) not change
   (d) be four times high.

9. The no-load current of a transformer in terms of full-load current is usually :
   (a) 1% to 2% (b) 2% to 5%
   (c) 5% to 12% (d) 12% to 20%.

10. The two transformers operated in parallel will share the load depending upon their :
   (a) rating in kVA
   (b) efficiency
   (c) leakage reactance
   (d) per unit impedance.

11. What is the formula used to find out the full load current in three phase transformer ?

(a) $\frac{\sqrt{3}.kVA \times 1000}{V}$ (b) 3.V.I cos θ

(c) $\frac{kVA \times 1000}{\sqrt{3} \times V}$ (d) $\frac{kVA}{\sqrt{3}} \times \cos q$

(I.T.I. Mock Test 2013, Electrician)

12. In a transformer, the number of turns per volt is 6. The primary and secondary voltages are 240 V and 32 V respectively. Find the number of primary and secondary winding turns?

(a) 880 turns; 200 turns
(b) 1200 turns; 192 turns
(c) 1440 turns ; 192 turns
(d) 1550 turns; 200 turns

13. Results of a transformer with a rating of 15 kVA, 1100V/220V, 50Hz are obtained, while testing.

Short circuit test power input = 510 watts

Open circuit test power input = 252 watts

What is the efficiency of the transformer at full load, 0.8 p.f.?

(a) 92% (b) 93%
(c) 96% (d) 94%

14. What is the maximum load upto which an open delta transformer can safely maintain supply to load?

(a) 33% (b) 50%
(c) 58% (d) 66%

(I.T.I. July 2011, Electrician)

15. No load loss of a transformer has two components. They are eddy current and hysteresis loss. In which part of a transformer hysteresis loss will occur?

(a) Core (b) Breather
(c) Transformer oil (d) Conservator

16. Which material is used for the construction of the power transformer core?

(a) Mild steel
(b) High carbon steel
(c) Silicon steel (d) Stainless steel.

(I.T.I. July 2014, Electrician)

17. What is the formula to determine the transformation ratio 'K'?

(a) $K = \frac{N_2}{N_1} = \frac{E_2}{E_1} = \frac{I_2}{I_1}$

(b) $K = \frac{N_1}{E_1} = \frac{E_2}{E_2} = \frac{I_2}{I_1}$

(c) $K = \frac{N_2}{I_1} = \frac{E_2}{N_1} = \frac{I_2}{E_1}$

(d) $K = \frac{N_2}{N_1} = \frac{E_2}{E_1} = \frac{I_1}{I_2}$

(I.T.I. August 2014, Electrician; I.T.I February 2015, Wireman)

18. The percentage of voltage regulation (V_R) of a transformer is determined by using the formula ....

(a) $V_R = \frac{\text{No load voltage} - \text{fullload voltage}}{\text{no load voltage}} \times 100$

(b) $V_R = \frac{\text{Full load voltage} - \text{No load voltage}}{\text{Full load voltage}} \times 100$

(c) $V_R = \frac{\text{Full load voltage} - \text{No load voltage}}{\text{Full load voltage}} \times 100$

(d) $V_R = \frac{\text{No load voltage} - \text{Full load voltage}}{\text{Full load voltage}} \times 100$

(I.T.I. February 2015, Electrician)

19. Which one of the following can be increased by using a step-up transformer?

(a) Power (b) Voltage
(c) Current (d) Frequency

20. In 1 kVA transformer the number of turns per volt is 8. What is the number of turns in high voltage side, If the voltage is 110 V?

(a) 1920 (b) 880
(c) 192 (d) 125

(I.T.I. February 2015, Electrician)

21. What is the full load current of 1 kVA, 240V/120V single phase transformer at 0.8 P.F.?

(a) 8.33 amps (b) 6.66 amps
(c) 4.16 amps (d) 3.33 amps

(I.T.I. August 2014, Electrician, I.T.I February 2015, Wireman)

22. As compared to an amplifier, a transformer cannot ....
(a) decrease the output power
(b) increase the output power
(c) increase the output current
(d) increase the output voltage

23. Small type transformer core for shell type generally use ....
(a) E & I type of core.
(b) F & F type of core.
(c) E & E type of core.
(d) T & U type of core.

(I.T.I. February 2015, Electrician)

24. What principle makes the instrument transformer working?
(a) Self induction
(b) Lenz's law
(c) Mutual induction
(d) Fleming's right hand rule

25. Current transformers are classified into five types. Which type of current transformer (C.T.) principle is used in clipon or clamp on Ammeter?
(a) Oil immersed C.T.
(b) Dry type C.T.
(c) Ring type C.T.
(d) Wound type C.T.

26. What is the formula to be used to measure the actual power consumed, when the wattmeter is used along with C.T. and P.T.?
(a) Wattmeter reading × C.T. ratio
(b) Wattmeter reading × P.T. ratio
(c) Wattmeter reading × C.T. ratio × P.T. ratio
(d) $\dfrac{\text{Wattmeter reading}}{\text{C.T. ratio}} \times \text{P.T. ratio}$

27. What is the standard rating of secondary of potential transformers?
(a) 110 V (b) 220 V
(c) 330 V (d) 440 V

28. You find a C.T. of 10 : 1 ratio and P.T. of 100 : 1 ratio are available in the stores. Ammeter of 0 to 5 amperes and voltmeter of 0 to 110 V are also available. The maximum high value of voltage and current that could be measured by using the above is ....
(a) 1100 V, 50 A (b) 11 kV, 50 A
(c) 11 kV, 500 A (d) 110 kV, 500 A

(I.T.I. February 2015, Electrician)

29. The standard rated value of secondary current of current transformer as per IS 2705 (part I) - 1981 shall be either ....
(a) 5 ampere or 10 ampere
(b) 2.5 ampere or 5 ampere
(c) 1.5 ampere or 6 ampere
(d) 1 ampere or 5 ampere

(I.T.I. February 2018, Electrician)

30. The standard values of rated secondary current of the current transformer is ....
(a) 2 Amp (b) 3 Amp
(c) 5 Amp (d) 10 Amp

(I.T.I. February 2016, Electrician)

31. The rated secondary voltage of a single phase potential transformer is ....
(a) 240 V (b) 230 V
(c) 115 V (d) 110 V

(I.T.I. February 2018, Electrician)

32. In which material, core of the transformer is constructed ?
(a) Silicon iron steel (b) Mild steel iron
(c) Carbon steel iron (d) Pig iron steel

(I.T.I. February 2016, Electrician)

33. The input side of the single phase transformer is called as ....
(a) secondary side (b) primary side
(c) high voltage side (d) low voltage side

(I.T.I. February 2016, Electrician)

34. If the two winding transformers, the load (out put) side is known as ....
(a) primary side
(b) secondary side

(c) low voltage side
(d) high voltage side

(I.T.I. August 2015, Electrician)

35. Calculate the current in low voltage side of the 1 KVA, 240 V/ 120 V single phase transformer.
(a) 4.16 amperes (b) 6.72 amperes
(c) 8.33 amperes (d) 9.12 amperes

(I.T.I. August 2015, Electrician)

36. A 100 KVA, 2400V/240V, 50Hz transformer has 300 turns on the secondary winding. Calculate the number of turns in primary winding?
(a) 1500 turns (b) 1800 turns
(c) 2400 turns (d) 3000 turns

(I.T.I. February 2016, Electrician)

37. What will be the colour of the silicon gel after it absorbs moisture from air?
(a) Blue (b) Brown
(c) Violet (d) Pink

(I.T.I. February 2016, Electrician)

38. Which one method of cooling is generally adopted for low capacity distribution transformer upto 100 kVA ?
(a) Oil blast cooling method
(b) Air blast cooling method
(c) Forced circulation of cooling method
(d) Natural air cooling method

(I.T.I. August 2015, Electrician)

39. Buchholz's relay is connected in high capacity power transformers.
The purpose of the Buchholz's relay is to ....
(a) arrest the flow of moisture into the tank
(b) disconnect the transformer, when dead short circuit fault occur
(c) flow insulating oil from the conservator tank
(d) control the level of oil tank

(I.T.I. February 2018, Electrician)

40. What will happen to 415 V/240 V, 1 kVA transformer when it is connected to a D.C. supply ?
(a) The output will be 240 V
(b) The output will be 415 V
(c) The transformer may burn
(d) Remains as it is

(I.T.I. August 2015, Electrician)

41. Which method of group of transformer for 3 phase transformer's connection employed at the sub-station end of the transmission line?
(a) Primaries in Δ, Secondaries in Δ
(b) Primaries in Y, Secondaries in Y
(c) Primaries in Y, Secondaries in Δ
(d) Primaries in Δ, Secondaries in Y

(I.T.I. August 2015, Electrician)

42. The rating of a transformer is measured in ....
(a) kVAR (b) kW
(c) kVA (d) kWH

(I.T.I. February 2019, Electrician)

43. Transformers are rated in terms of kVA instead of kW, because ......
(a) Load factor is often not known
(b) kVA is fixed whereas kW depends on the load power factor
(c) The total transformer loss depends upon the volt ampere
(d) It has become customary

(I.T.I. February 2017, Electricain)

44. A transformer is used to –
(a) change A.C. to D.C.
(b) change D.C. to A.C.
(c) Step up or down D.C. voltage
(d) Step up or down A.C. voltages

(I.T.I. August 2016, Electrician)

45. Total core loss is also called as .....
(a) Magnetic loss
(b) Eddy current loss
(c) Hysteresis loss
(d) Copper loss

(I.T.I. February 2019, Electrician)

46. Open circuit test on transformer gives information of .....
(a) core loss at rated voltage and frequency

(b) core loss resistance and magnetizing reactance
(c) turns ratio of the transformer
(d) All of these

(I.T.I. February 2017, Wireman)

47. The iron losses in a transformer can be determined by .....
(a) Open circuit test
(b) Short circuit test
(c) Earth test
(d) Oil test

(I.T.I. August 2016, Wireman)

48. Hysteresis loss of a transformer occurs due to –
(a) Core loss
(b) Eddy current loss
(c) Alternating flux in the core
(d) Copper loss

(I.T.I. February 2017, Wireman)

49. When load on the transformer is decreased ---- losses decrease.
(a) Eddy current (b) Hysteresis
(c) Copper (d) Friction

(I.T.I. February 2017, Electrician)

50. The copper losses in transformer can be reduced ....
(a) By changing the core materials
(b) By laminating the core
(c) By reducing the resistance of the windings
(d) All of these

(I.T.I. August 2016, Electrician)

51. Eddy current loss of a transformer is proporational to ....
(a) $f \cdot B_m^2$ (b) $f \cdot B_m$
(c) $f^2 B_m$ (d) $f^2 \cdot B_m^2$

(I.T.I. February 2017, Wireman)

52. Eddy current loss within a transformer can be reduced by –
(a) Laminating the core
(b) Running transformer above 90% of full kVA
(c) Using it as an autotransformer
(d) Increasing the number of turns

(I.T.I. February 2017, Wireman)

53. The CRGO silion steel laminations are used for core of a transformer, CRGO stands for .....
(a) Cold Rolled Grain Obtained
(b) Cold Rolled Green Obtained
(c) Cold Rolled Grain Oriented
(d) Clear Red Green Oriented

(I.T.I. February 2017, Wireman)

54. Use of superior quantity of laminated CRGO silicon iron sheet reduces ......
(a) Hysteresis loss
(b) Mechanical, friction and windage losses
(c) Copper losses due to load currents
(d) Other thermal and mechanical losses

(I.T.I. August 2016, Wireman)

55. The efficiency of a transformer will be maximum when :
(a) Copper losses = iron losses
(b) Hyteresis losses = eddy current losses
(c) Eddy current losses = copper losses
(d) Copper losses = hysteresis losses

(I.T.I. February 2019, Electrician)

56. The effective value of induced voltage in transformer is represented by :
(a) $4.44 f \phi_m N$ (b) $4.44 f . N^2 \phi_m$
(c) $4.44 f^2 . N\phi_m$ (d) $4.44 f . N . \phi_m$

(I.T.I. February 2017, Electrician)

57. What will happen to 415 V/ 240, I kVA transformer when it is connected to a D.C. supply?
(a) The output will be 240 volts
(b) The transformer may burn
(c) The output will be 415 volts
(d) Remain as it is

(I.T.I. February 2018, Electrician)

58. Power tansformer secondary side neutral is earthed due to :
(a) For testing the phase at any point
(b) For the safety of the machine
(c) For balancing the load on transformer
(d) All of these

(I.T.I. August 2016, Wireman)

59. Current Transformer and Potential transformer are used to increase the range of :

(a) AC Ammeter and AC voltmeter respectively

(b) DC Ammeter and AC voltmeter respectively

(c) DC Ammeter and DC voltmeter respectively

(d) AC Ammeter and DC voltmeter respectively

(I.T.I. February 2017, Electrician)

60. A 100 : 5 current transformer is used to measure a maximum of 1000 amperes at 500 volts. The secondary reading under rated condition is .....

(a) 50 A, 10 kV (b) 500 A, 25 V

(c) 100 A, 5 V (d) 10 A, 100 V

(I.T.I. February 2017, Wireman)

61. In which of the transformer is the secondary nearly short circuited under normal operating conditions?

(a) CT (b) PT

(c) Distribution Transformer

(d) Power Transformer

(I.T.I. August 2016, Electrician)

62. Transformer oil is used for ....

(a) Cooling, insulation

(b) Cooling, breathing

(c) Heating, cooling

(d) All of these

(I.T.I. February 2017, Wireman)

63. The method used for cooling power transformer is/are .....

(a) Forced air cooling

(b) Forced oil cooling

(c) Natural air cooling

(d) All of these

(I.T.I. August 2016, Wireman)

64. In an Auto Transformer, the Primary and Secondary are ...... Coupled.

(a) Only Magnetically

(b) Only Electrically

(c) Magnetically as well as Electrical

(d) None of these

(I.T.I. February 2018, Electrician)

65. Applications of Auto Transformer is/are ....

(a) As starter of induction motor/ synchronous motor

(b) Continuously variable supply in electrical/testing laboratories

(c) As regulating booster transformers

(d) All of these

(I.T.I. August 2016, Wireman)

66. In a distribution transformer, voltage ratio can be adjusted to suit the loading conditions as ......

(a) On load tap changing

(b) Off load tap changing

(c) Controlled by an AVR

(d) All of these

(I.T.I. August 2016, Wireman)

67. Breathers ......

(a) are used as dehydrating agent

(b) contain silica gel

(c) contain 50% oil in cylindrical vessel

(d) both (a) and (b)

(I.T.I. February 2017, Wireman)

68. A sort of a drum mounted on the top of the transformer with a level indicator is .....

(a) Conservator (b) Breather

(c) Explosion vent (d) Buchholz Relay

(I.T.I. August 2016, Wireman)

69. The following is/are true about a conservation and breather:

(a) When transformer losses are there, the conservator breaths out

(b) When conservator oil cools, it breaths in

(c) Silica gel of breather absorbs moisture during breathing to turn blue

(d) All of these

(I.T.I. February 2017, Wireman)

70. Buchholz's relay senses –

(a) One fault (b) Two faults

(c) Three faults (d) Four faults

Hint: The two faults are:

(i) Minor fault (Heat is Produced)

(ii) Severe fault (Large volume of gas is evolved)

(I.T.I. February 2017, Wireman)

71. Scott connections are used for ?
(a) 1ϕ to 3ϕ transformation
(b) 3ϕ to 2ϕ transformation
(c) 3ϕ to 1ϕ transformation
(d) 2ϕ to 3ϕ transformation
(I.T.I. February 2017, Electrician)

72. In which transformer, tertiary winding is used?
(a) Star-Delta (b) Star-Star
(c) Delta-Delta (d) Delta-Star
(I.T.I. February 2017, Electrician)

73. One or more of the following conditions are essential for parallel operation of transformers–
(a) Polarity of the transformers should be same
(b) Equal voltage ratio should be provided
(c) Percentage impedance should be same
(d) All of these
(I.T.I. February 2017, Wireman)

74. Which cooling method is used in pole mounting distribution transformer?
(a) Air natural
(b) Oil natural air blast
(c) Oil forced air forced
(d) Oil natural air natural
(I.T.I. August 2019 Electrician)

75. How does the moisture is controlled in breather filled on power transformers?
(a) Using silica gel
(b) Using sodium chloride
(c) Using transformer oil
(d) Using Ammonium jelly
(I.T.I. August 2019 Electrician)

76. Which is determined by the crackle test of transformer oil?
(a) Acidity
(b) Moisture
(c) Viscosity
(d) Dielectric strength
(I.T.I. August 2019 Electrician)

77. Which type of transformer oil test is illustrated?

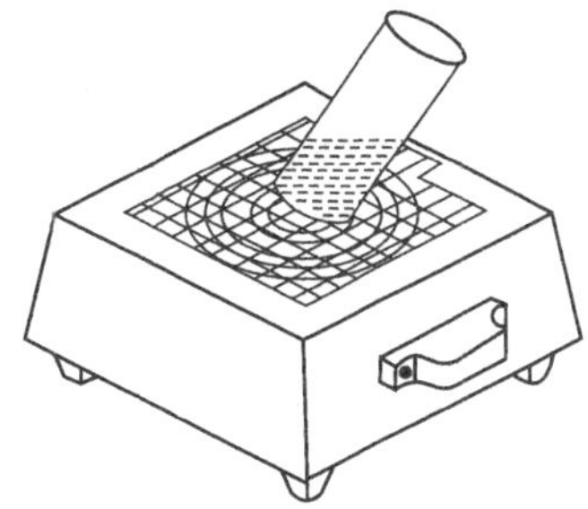

(a) Field test
(b) Acidity test
(c) Crackle test
(d) Dielectric strength test
(I.T.I. August 2019 Electrician)

78. Which loss is determined by conducting short circuit test in transformer?
(a) Stray loss
(b) Eddy current loss
(c) Hysteresis loss
(d) Copper loss
(I.T.I. August 2019 Electrician)

79. Which construction technique is used to reduce copper loss in larger transformers?
(a) Use of laminated core
(b) By reducing core thickness
(c) Use of stepped core arrangement
(d) By using grain oriented core
(I.T.I. August 2019 Electrician)

80. What is the purpose of tap changing in power transformers?
(a) Maintain primary voltage constant
(b) Change voltage ratio in distribution
(c) Maintain secondary voltage constant
(d) Load the transformer for maximum efficiency
(I.T.I. August 2019 Electrician)

81. What is the composition of steel and silicon in transformer core?
(a) Steel 97% and silicon 3%
(b) Steel 95% and silicon 5%
(c) Steel 93% and silicon 7%
(d) Steel 90% and silicon 10%
(I.T.I. August 2019 Electrician)

82. Which winding of a tranformer has more number of turns?

(a) Low voltage winding
(b) High voltage winding
(c) Primary winding
(d) Secondary winding

(I.T.I. August 2019 Electrician)

83. Which transformer is used to measure high voltage installations?

(a) Pulse tranformers
(b) Potential transformers
(c) Ignition transformers
(d) Constant voltage transformers

(I.T.I. August 2019 Electrician)

84. What is the name of transformer?

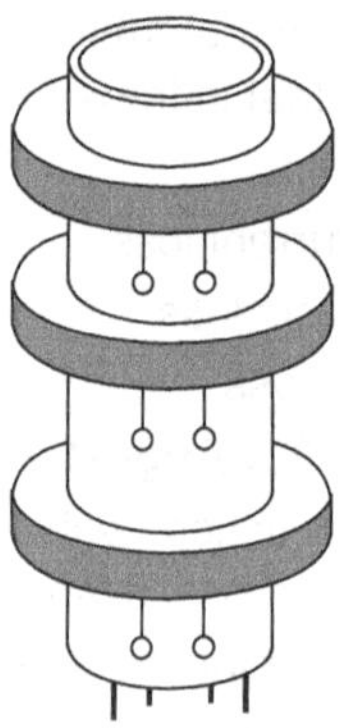

(a) Iron core transformer
(b) Air core transformer
(c) Ring core transformer
(d) Ferrite core transformer

(I.T.I. August 2019 Electrician)

85. What is the disadvantage of auto transformer?

(a) More losses
(b) Heavier in weight
(c) Poor voltage regulation
(d) Cannot isolate secondary winding

(I.T.I. August 2019 Electrician)

86. 3 phase transformers as compared to single phase transformers are ......

(a) More economical
(b) Not so easy in construction
(c) Easy to construct
(d) Easy to handle

(I.T.I. August 2019 Electrician)

KEY TO MULTIPLE CHOICE QUESTIONS

1. (a)	**2.** (c)	**3.** (c)	**4.** (b)	**5.** (c)	**6.** (d)	**7.** (c)	**8.** (c)	**9.** (b)	**10.** (d)
11. (c)	**12.** (c)	**13.** (d)	**14.** (c)	**15.** (a)	**16.** (c)	**17.** (d)	**18.** (d)	**19.** (b)	**20.** (b)
21. (a)	**22.** (b)	**23.** (a)	**24.** (c)	**25.** (c)	**26.** (c)	**27.** (a)	**28.** (b)	**29.** (d)	**30.** (c)
31. (d)	**32.** (a)	**33.** (b)	**34.** (b)	**35.** (c)	**36.** (d)	**37.** (d)	**38.** (d)	**39.** (b)	**40.** (c)
41. (c)	**42.** (c)	**43.** (b)	**44.** (d)	**45.** (a)	**46.** (d)	**47.** (a)	**48.** (c)	**49.** (c)	**50.** (c)
51. (d)	**52.** (a)	**53.** (c)	**54.** (a)	**55.** (a)	**56.** (d)	**57.** (b)	**58.** (b)	**59.** (a)	**60.** (a)
61. (a)	**62.** (a)	**63.** (b)	**64.** (c)	**65.** (d)	**66.** (d)	**67.** (d)	**68.** (a)	**69.** (d)	**70.** (b)
71. (b)	**72.** (d)	**73.** (d)	**74.** (d)	**75.** (a)	**76.** (b)	**77.** (c)	**78.** (d)	**79.** (c)	**80.** (b)
81. (a)	**82.** (b)	**83.** (b)	**84.** (b)	**85.** (d)	**86.** (a)				

1st Year

Workshop Calculation and Science

1. Units, Fractions, Square Roots

Units

1. Convert 120°C into degree Fahrenheit.

(a) 250°F (b) 248°F

(c) 240°F (d) 238°F

2. Convert 98.6°F into degrees centigrade.

(a) 22°C (b) 25°C

(c) 27°C (d) 37°C

3. 8°C is equal to ....

(a) 269 K (b) 279 K

(c) 280 K (d) 392 K

4. Normal human body temperature is 98.6° Fahrenheit. Its celsius equivalent is equal to ....

(a) 27°C (b) 37°C

(c) 40°C (d) 50°C

5. Absolute zero temperature is equal to ....

(a) −273.15°C (b) 0°C

(c) 100°C (d) 273.15°C

6. Following is unit of temperature:

(a) joule (b) pascal

(c) kelvin (d) calorie

7. 1 H.P. (metric) = ...

(a) 750 watts (b) 746 watts

(c) 735.5 watts (d) 720 watts

8. S.I. unit of force is ....

(a) kilogram (b) newton

(c) joule (d) acceleration

9. Convert 10 inches into mm:

(a) 2.54 mm (b) 25.4 mm

(c) 254 mm (d) 2540 mm

10. What are the fundamental units in all the systems?

(a) Length, mass and time

(b) Length, mass and current

(c) Mass, time and current

(d) Length, mass and temperature

Fractions

1. Find the H.C.F. of 18, 42, 24.

(a) 9 (b) 8

(c) 6 (d) 3

2. Lowest common multiple (L.C.M.) of 30, 36, 48 and 60 is ....

(a) 60 (b) 120

(c) 240 (d) 720

3. Place the missing number: $\frac{3}{4} = \frac{1}{12} \times$ ....

(a) 9 (b) 7

(c) 6 (d) 3

4. Convert $\frac{18}{16}$ into mixed fraction.

(a) $3\frac{1}{8}$ (b) $2\frac{3}{4}$

(c) $2\frac{1}{8}$ (d) $1\frac{1}{8}$

5. Add: $6\frac{1}{3}+2\frac{1}{3}+3\frac{7}{9}$ = ...

(a) $13\frac{7}{9}$ (b) $11\frac{7}{9}$

(c) $9\frac{5}{9}$ (d) $12\frac{4}{9}$

6. Simplify: $1\frac{1}{2}\times 2\frac{2}{3}\div\left(3\frac{3}{4}\times\frac{28}{105}\right)$

(a) 16 (b) 14

(c) 6 (d) 4

7. $\dfrac{3\frac{1}{3}+2\frac{1}{5}}{2\frac{1}{3}+2\frac{2}{5}}$ = ...

(a) $3\frac{2}{15}$ (b) $1\frac{1}{5}$

(c) $\frac{12}{71}$ (d) $1\frac{12}{71}$

Decimal Fractions

1. 48 seconds = ... hour

(a) 0.13 (b) 0.013

(c) 0.0013 (d) 0.00013

2. Decimal fraction of $\frac{9}{26}$ is ....

(a) 0.3614 (b) 0.3461

(c) 0.3164 (d) 0.1346

3. Simplify the following fraction:

12.836 – 6.5265 + 0.875 – 0.005

(a) 717.95 (b) 71.795

(c) 7.1795 (d) 0.71795

4. 3 + 33 + 333 + 3.33 = ...

(a) 362.3 (b) 372.33

(c) 702.33 (d) 702

5. $\frac{80.4}{20}-(-4.2)$ = ...

(a) 822 (b) 82.2

(c) 8.22 (d) 0.822

6. (0.05 × 5) – (0.005 × 5) = ...

(a) 2.25 (b) 22.5

(c) 0.225 (d) 2.025

7. Value of (0.4 × 7.5 × 0.6 × 25) is ....

(a) 45 (b) 155

(c) 115 (d) 145

Square Roots

1. $\frac{26}{\sqrt{x}}=2$ $\therefore$ x = ...

(a) 196 (b) 225

(c) 125 (d) 169

2. Find square root of 0.49.

(a) 0.5 (b) 0.6

(c) 0.7 (d) 0.8

3. Find square root of 55225.

(a) 135 (b) 235

(c) 355 (d) 435

4. Square root of 4761 is equal to ....

(a) 69 (b) 59

(c) 46 (d) 72

5. $\sqrt{0.2916}$ = ...

(a) 54 (b) 5.4

(c) 0.54 (d) 0.054

6. $\sqrt{45796}$ = ...

(a) 264 (b) 254

(c) 234 (d) 214

7. Find the value of $\sqrt{\frac{196}{225}}$

(a) $\frac{14}{15}$ (b) $\frac{15}{14}$

(c) $\frac{16}{25}$ (d) $\frac{14}{25}$

8. Simplification of $\frac{1}{\sqrt{3}-\sqrt{2}}$ is ....

(a) 31.46 (b) 3.146

(c) 0.3146 (d) 0.03146

9. Square root of 1734489 is ....
   (a) 117 (b) 1127
   (c) 1317 (d) 1147

10. Value of $\sqrt[3]{4\frac{12}{125}}$ = ?
   (a) $1\frac{2}{5}$ (b) $1\frac{3}{5}$
   (c) $1\frac{4}{5}$ (d) $2\frac{2}{5}$

Pythagoras Theorem

1. In this right angled triangle, $AC^2 = AB^2 + BC^2$. Which theorem states the above statement.

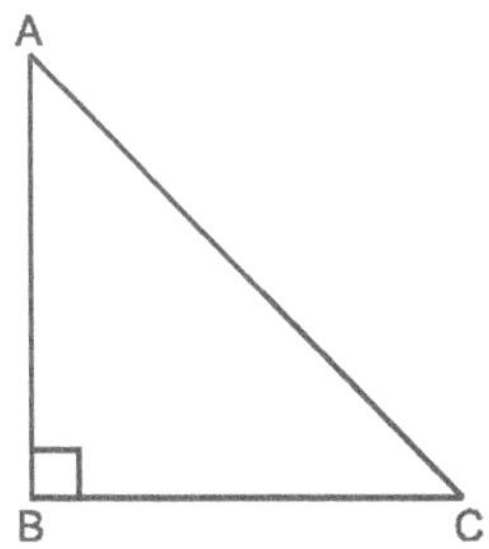

   (a) Pythagoras theorem
   (b) Lami's theorem
   (c) Bernoullis' theorem
   (d) Thevenins theorem

2. A stay wire is to be provided for a 9 metres high pole at a horizontal distance of 3 metres. Find the length of the wire required?
   (a) 12.5 metres (b) 10.42 metres
   (c) 9.486 metres (d) 7.382 metres

Solving Problems by using Calculator

1. A scientific calculator usually includes which of the following functions?
   (a) logarithmic functions, using both base 10 and base e
   (b) trigonometric functions (some including hyperbolic trigonometry)
   (c) exponential functions and roots beyond the square root
   (d) All of the above

2. Some functions may require you to press which of these buttons to access them?
   (a) Control
   (b) Function
   (c) Tab
   (d) None of the above

3. What function does the symbol ^ indicate?
   (a) Insert pi
   (b) Log base 10
   (c) Raise to the power of
   (d) Square root

4. What are the important things to keep in mind when using Trigonometric Functions?
   (a) Order of Button presses
   (b) Radians vs Dgerees
   (c) Both of the above
   (d) None of the above

5. Use of what is necessary to enter long equations requiring proper order of calculations?
   (a) Parentheses (b) Commas
   (c) Semicolons (d) Arrow

KEY TO MCQs 1. Units, Fractions, Square Roots

Units

1. (b) 2. (d) 3. (d) 4. (b) 5. (a) 6. (c) 7. (c) 8. (b) 9. (c) 10. (a)

Fractions

1. (c) 2. (d) 3. (a) 4. (d) 5. (d) 6. (d) 7. (d)

Decimal Fractions

1. (b) 2. (b) 3. (c) 4. (b) 5. (c) 6. (c) 7. (a)

Square Roots

1. (d) 2. (c) 3. (b) 4. (a) 5. (c) 6. (d) 7. (a) 8. (b) 9. (c) 10. (b)

Pythagoras Theorem

1. (a) 2. (c)

Use of Scientific Calculator

1. (d) 2. (b) 3. (c) 4. (c) 5. (a)

2. Ratio & Proportion, Percentage

Ratio

1. Find the value of "x" if x : 15 : : 3 : 2·5
   (a) 6 (b) 5
   (c) 3 (d) 2.5
2. An alloy consisting of three metals A, B and C in the proportion of 1 : 2 : 3. Weight of the alloy is 72 kg. Then weight of A, B, C are in the ratio of ....
   (a) 36 : 24 : 12 (b) 12 : 36 : 24
   (c) 24 : 12 : 36 (d) 12 : 24 : 36
3. Calculate the value of each resistance, if the total resistance of 1200 ohms of circuit is divided in the ratio of 3 : 5.
   (a) 400 ohms ; 800 ohms
   (b) 450 ohms ; 750 ohms
   (c) 500 ohms ; 700 ohms
   (d) 550 ohms ; 650 ohms
4. Three partners have capital is in the ratio 5 : 6 : 9. The share of first partner for the year's profit is ₹ 625. What is the total profit.
   (a) ₹ 5000 (b) ₹ 4000
   (c) ₹ 3000 (d) ₹ 2500
5. How many pieces of $3\frac{1}{2}$ cm long string can be cut from a length of 7 metres?
   (a) 107 (b) 160
   (c) 200 (d) 268
6. 72 metres long rope is to be cut into equal lengths. Find the number of pieces, if one piece is equal to 4 metres.
   (a) 36 (b) 24
   (c) 18 (d) 12
7. A belt passes over a 280 mm diameter pulley rotating at 300 R.P.M. How far does a point on the pulley move in 2 minutes?
   (a) 526 metres
   (b) 528 metres
   (c) 546 metres
   (d) 548 metres
8. If A : B = 5 : 7 and B : C = 6 : 11, then A : B : C is ....
   (a) 55 : 66 : 77 (b) 30 : 42 : 77
   (c) 35 : 49 : 42 (d) None of these

Proportion

1. 25 workers can complete a work in 30 hours. If the same work is carried over by 10 workers, then time taken is ....
   (a) 75 hours (b) 65 hours
   (c) 45 hours (d) 35 hours
2. 96 metres long rope is to be cut into equal lengths. Find the number of pieces, if one piece is equal to 6 metres.
   (a) 32 (b) 30
   (c) 16 (d) 10

3. The selected cable will have 1 volt drop for every 14.7 metre cable run. If voltage drop is 6.801 volts, the length of the cable is ....

(a) 100 metres (b) 102 metres
(c) 104 metres (d) 110 metres

4. A generator set is running 24 days in a months. If the consumption of diesel per month is 192 litres, calculate the number of litres per day generator is consuming.

(a) 16 (b) 12
(c) 10 (d) 8

5. On a given linear spring force of 2 newtons is required to produce displacement of 3 cm. Amount of force required to displace the same spring by 27 cm is ....

(a) 9 newtons (b) 13.5 newtons
(c) 18 newtons (d) 36 newtons

6. A construction site requires 2 Electricians to complete 5 tasks in 3 days. If the task is doubled, the No. of Electricians required to complete the task in one day is ....

(a) 3 (b) 6
(c) 12 (d) 60

Percentage

1. The cost price of 21 mangoes are equal to the selling price of 18 mangoes. Find the profit percentage.

(a) 16.66% (b) 15.66%
(c) 14.66% (d) 13.66%

2. The input of an electric motor is 600 watts and its output is 530 watts. The percentage efficiency is ....

(a) 66.33% (b) 72.33%
(c) 78.33% (d) 88.33%

3. 2% of 150 is ....

(a) 3 (b) 5
(c) 8 (d) 10

4. What percent is 200 metres is of 8 kilometres?

(a) 2.5% (b) 25%
(c) 2% (d) 4%

5. Find the total number of crank pin if 12% of it is 1080:

(a) 6000 (b) 7000
(c) 8000 (d) 9000

6. The ratio 5 : 4 expressed as percent is equal to ......

(a) 12.5% (b) 40%
(c) 80% (d) 125%

7. A T.V. is purchased at Rs. 5000 and sold at Rs. 4000. Find the loss percent.

(a) 10% (b) 20%
(c) 25% (d) 28%

8. One person gains 70 paise on Rs. 70. What is gain percent?

(a) 0.1% (b) 1%
(c) 7% (d) 10%

9. When a plot is sold for Rs. 18700, the owner loses 15%. At what price must that plot be sold in order to gain 15%?

(a) Rs. 21000 (b) Rs. 22000
(c) Rs. 25300 (d) Rs. 25800

10. How much time will it take for an amount of Rs. 450 to yield Rs. 81 as interest at 4.5% per annum of simple interest?

(a) 3.5 years (b) 4 years
(c) 4.5 years (d) 5 years

11. The compound interest on Rs. 30000 at 7% per annum is Rs. 4347. The period in years is ....

(a) 2 (b) 3
(c) 4 (d) 1

12. At a certain rate of compound interest, Rs. 15320 becomes Rs. 30239 in 6 years. What is the rate of compound interest?

(a) 12% (b) 13%
(c) 11% (d) 14%

KEY TO MCQs 2. Ratio & Proportion

Ratio

1. (a) 2. (d) 3. (b) 4. (d) 5. (c) 6. (c) 7. (b) 8. (b)

Proportion

1. (a) 2. (c) 3. (a) 4. (d) 5. (c) 6. (c)

Percentage

1. (a) 2. (d) 3. (a) 4. (a) 5. (d) 6. (d) 7. (b) 8. (b) 9. (c) 10. (b)
11. (a) 12. (a)

3. Material Science

1. Name the mechanical property of metal that can withstand a shock.
   (a) Tenacity (b) Elesticity
   (c) Brittleness (d) Toughness
2. Corrosion of metal involves ....
   (a) physical reaction
   (b) chemical reaction
   (c) Both (a) and (b)
   (d) None of these
3. Which one of the following is ferrous metal?
   (a) Aluminium (b) Brass
   (c) Lead (d) Pig iron
4. The bronze metal is made of ....
   (a) copper and tin
   (b) copper and lead
   (c) copper and aluminium
   (d) copper and zinc
5. The gun metal is made of ....
   (a) copper, tin and zinc
   (b) tin and zinc
   (c) copper and nickel
   (d) zinc and nickel
6. When heat is applied it starts melting, then the property is called ....
   (a) Ductility (b) Malleability
   (c) Fusibility (d) Brittleness
7. Which property of metal is stated under physical property?
   (a) Structure (b) Ductility
   (c) Tenacity (d) Elasticity
8. Which one of the following is a non metal?
   (a) Copper (b) Brass
   (c) Steel (d) Mica
9. Bronze contains ....
   (a) Copper 80%, Tin 20%
   (b) Copper 85%, Tin 15%
   (c) Copper 90%, Tin 10%
   (d) Copper 95%, Tin 5%
10. To make "Bronze" which metal is mixed with "Copper"?
    (a) Tin (b) Iron
    (c) Zinc (d) Manganese
11. Combination of copper and tin makes ....
    (a) Brass (b) Bronze
    (c) Cast iron (d) Wrought iron
12. .... is a ferrous metal.
    (a) Steel (b) Brass
    (c) Copper (d) Bronze
13. The ability of a material to deform without breaking is called ....
    (a) Elasticity (b) Plasticity
    (c) Ductility (d) Brittleness

14. Property of zinc is ....
(*a*) Ductile
(*b*) Malleable
(*c*) Rusting
(*d*) Melting at 400°C

15. Melting temperature of "wrought iron" is ....
(a) 1500°C (b) 150°C
(c) 555°C (d) 1555°C

KEY TO MCQs 3. Material Science

1. (d)	**2.** (b)	**3.** (d)	**4.** (a)	**5.** (a)	**6.** (c)	**7.** (a)	**8.** (d)	**9.** (c)	**10.** (a)
11. (b)	**12.** (a)	**13.** (c)	**14.** (b)	**15.** (d)					

4. Mass, Weight, Volume and Density

1. Specific gravity of water is ....
(a) 1.5 (b) 1.4
(c) 1.3 (d) 1.0

2. Specific gravity of electrolyte of a lead acid battery at fully charged condition is ....
(a) 1.28 (b) 1.50
(c) 1.40 (d) 1.35

3. Which of the following has no unit?
(a) Density (b) Specific gravity
(c) Acceleration (d) Velocity

4. The amount of matter contained in a body is known as:
(a) Weight (b) Force
(c) Mass (d) None of these

5. Density of water is maximum at ....
(a) 0°C (b) 100°C
(c) 4°C (d) 200°C

6. Density of a sustance can be given by:
(a) Its mass per unit volume
(b) Its mass per unit area
(c) Its volume per unit mass
(d) None of these

KEY TO MCQs 4. Mass, Weight, Volume and Density

1. (d)	**2.** (a)	**3.** (b)	**4.** (c)	**5.** (c)	**6.** (a)

5. Work, Power and Energy

1. Work done = force × ....
(a) distance (b) velocity
(c) speed (d) acceleration

2. 1 joule = ...
(a) 1 Nm (b) 1 N m2
(c) 1 N^2m^2 (d) 1 kg m/sec^2

3. Determine the work done in joules, required to drive a lift in raising a load of 2000 newtons at a height of 500 cm.
(a) 400 joules (b) 1000 joules
(c) 5000 joules (d) 10000 joules

4. Calculate the work done in joules for lifting 200 kg of water through a height of 6 metres.
(a) 117.72 joules (b) 1200 joules
(c) 11772 joules (d) 12000 joules

5. 20 joule of work is done by an object in 5 seconds. It's power is ....
   (a) 100 watts (b) 0.25 watt
   (c) 1.00 watt (d) 4 watts
6. 15 kN metre of work is done by a crane in 30 seconds. What is the power in kilowatts?
   (a) 2 kW (b) 30 kW
   (c) 0.5 kW (d) 7.5 kW
7. How much power is required to lift a load of 3000 kg to a height of 15 metres in 6 minutes?
   (a) 250 kg metres per second
   (b) 200 kg metres per second
   (c) 125 kg metres per second
   (d) 75 kg metres per second
8. Capacity to do work is called ....
   (a) Work (b) Power
   (c) Energy (d) Force
9. In motor, which form of input energy is converted into what form of energy?
   (a) Chemical to electrical
   (b) Mechanical to electrical
   (c) Thermal to electrical
   (d) Electrical to mechanical
10. The application of conversion of chemical energy into electrical energy is ....
    (a) generator (b) cell
    (c) soldering iron (d) alternator
11. Potential energy = ...
    (a) $\frac{1}{2}mv^2$ (b) mgh
    (c) 2 mgh (d) 2 mv^2
12. What is the potential energy in a body mass 5 kg on top of a pole 20 metres high?
    (a) 1000 joules (b) 900 joules
    (c) 980 joules (d) 800 joules
13. Calculate the potential energy in a body of mass 5 kg on top of ladder 10 metres in height.
    (a) 320.6 joules (b) 420 joules
    (c) 478.5 joules (d) 490.5 joules
14. Find the horse power (metric) of an engine to lift a weight of 2 tonnes to a height of 30 metres in 2 minutes.
    (a) 4.67 H.P (b) 5.46 H.P
    (c) 6.67 H.P (d) 7.32 H.P
15. A pump can raise 1000 litres of water per minute to a height of 50 metres. Calculate the H.P. of pump.
    (a) 9.99 H.P (b) 11.11 H.P
    (c) 13.15 H.P (d) 14.50 H.P
16. If a pump can lift 200 kg of water through a height of 6 metres in 10 seconds, then its power is ....
    (a) 1100 watts (b) 1000 watts
    (c) 1300 watts (d) 1176 watts

KEY TO MCQs 5. Work, Power and Energy

1. (a)	**2.** (a)	**3.** (d)	**4.** (c)	**5.** (d)	**6.** (c)	**7.** (c)	**8.** (c)	**9.** (d)	**10.** (b)
11. (b)	**12.** (c)	**13.** (d)	**14.** (c)	**15.** (b)	**16.** (d)				

6. Heat and Temperature

1. Which one of the following is a form of energy?
   (a) Temperature (b) Heat
   (c) Boiling (d) Fusion
2. Latent heat of ice is ....
   (a) 80 calories per gram
   (b) 40 calories per gram
   (c) 60 calories per gram
   (d) 20 calories per gram
3. How much fall in Fahrenheit temperature, if a piece of red hot M.S. rod was lowered from 900°C to 200°C by dipping in water?

(a) 1298° F (b) 1292° F
(c) 1286° F (d) 1284° F

4. Freezing point of water is ....
(a) 0°C (b) 10°C
(c) 45°C (d) 100°C

5. Boiling point of water in celsius is ....
(a) 0°C (b) 100°C
(c) 108°C (d) 273°C

6. At what temperature will centigrade and Fahranheit thermometers give the same reading?
(a) –32° (b) –40°
(c) +32° (d) +40°

7. The degree of hotness/coldness of a body is called as ....
(a) heat (b) temperature
(c) Kelvin (d) radiation

8. The quantity of heat required to raise the temperature of 1 gram of water through 1°C is called as ....
(a) British Thermal Unit
(b) specific heat
(c) calorie
(d) centigrade heat unit

9. Calorie is the quantity of heat required to raise the temperature of 1 gram of water through ....
(a) 1°C (b) 2°C
(c) 3°C (d) 3.5°C

10. How much amount of latent heat in calories is required to convert 1 gram ice into liquid?
(a) 100 calories (b) 80 calories
(c) 70 calories (d) 60 calories

11. Convert 10 Centigrade Heat Units (C.H.U.) into calories.
(a) 4536.0 calories (b) 456.6 calories
(c) 44.36 calories (d) 43.45 calories

12. Specific heat of copper is ....
(a) 2.5 (b) 2.0
(c) 1.0 (d) 0.1

13. Specific heat of water is ....
(a) 4 (b) 3
(c) 2 (d) 1

14. Heat energy is converted into mechanical energy in ....
(a) steam engine (b) electric fan
(c) induction stove (d) lamp

15. Which one of the following is the example of heat radiation made of heat transfer?
(a) Boiling of water
(b) Sun rays falls on ground
(c) Heating of iron bar
(d) Melting of lead.

16. Which instrument is used to measure the temperature above 300°C?
(a) Gas thermometer
(b) Alcohol thermometer
(c) Mercury thermometer
(d) Pyrometer

17. Radiation pyrometer is used to measure temperature upto ....
(a) 1000°C (b) 2000°C
(c) 3000°C (d) 4000°C

KEY TO MCQs 6. Heat and Temperature

1. (b)	**2.** (a)	**3.** (b)	**4.** (a)	**5.** (b)	**6.** (b)	**7.** (b)	**8.** (c)	**9.** (a)	**10.** (b)
11. (a)	**12.** (d)	**13.** (d)	**14.** (a)	**15.** (b)	**16.** (d)	**17.** (c)			

7. Basic Electricity

1. What is the value of hot resistance of the 200 watts lamp, when it is connected across 240 volts supply?

(a) 480 Ω (b) 288 Ω
(c) 240 Ω (d) 188 Ω

2. 1 kilowatt hour is equal to .... joules.

(a) 360 (b) 1800
(c) 1800×10^5 (d) 36×10^5

3. The unit of electrical energy is ....

(a) kW (b) kW hour
(c) kV (d) kVAR

4. The coil resistance of 1500 W/250 V heater is .... ohms.

(a) 46.76 (b) 41.67
(c) 38.76 (d) 39.64

5. Electric lamps of 1000 W, 60 W, 40 W and 5 W are in use in a house where mains supplied with 240 volts. Calculate the current drawn by the main in amperes.

(a) 1.971 (b) 0.9375
(c) 0.3952 (d) 0.3795

6. How can you increase the pulling strength of an electromagnet?

(a) Increase the field intensity
(b) Reduce the current in the coil
(c) Reduce the number of turns in the coil
(d) Increase the B-H curve of the material

7. The emf produced in the D.C. generator is ... induced emf.

(a) Statically (b) Dynamically
(c) Magnetically (d) Electrostatically

8. The polarity of induced e.m.f. in a generator can be reversed by reversing the direction of ....

(a) Rotation (b) Field flux
(c) Both (a) and (b) (d) None of these

9. The direction of the generated e.m.f. is determined by .....

(a) Lenz' law
(b) Faraday's law of electromagnetic induction
(c) Fleming's left hand rule
(d) Fleming's right hand rule

10. As per Ohm's law:

(a) $V = I \cdot R$ (b) $I = \frac{R}{V}$
(c) $V = \frac{I}{R}$ (d) $P = \frac{V}{I}$

11. Current is measured in ....

(a) walts (b) volts
(c) henries (d) amperes

12. Unit of e.m.f. is ....

(a) volt (b) ampere
(c) joule (d) H.P.

13. What is the effective resistance of the circuit consisting of three resistance 4Ω, 6Ω and 12Ω are connected in parallel across a supply of 44 volts?

(a) 1 ohm (b) 2 ohms
(c) 3 ohms (d) 4 ohms

14. What is the resistance between A and C?

(a) 3 ohms (b) 4 ohms
(c) 5 ohms (d) 6 ohms

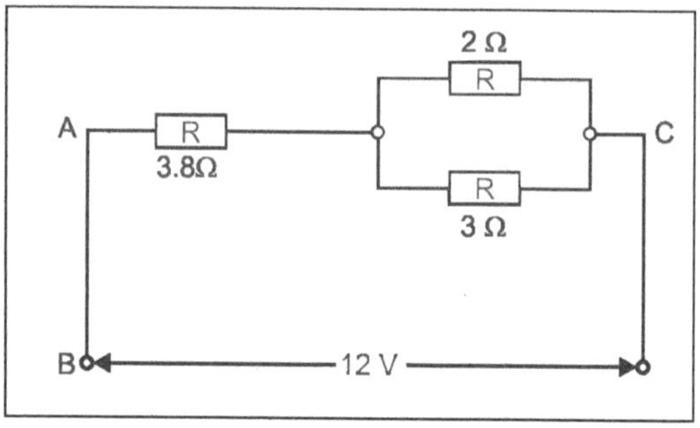

15. What is the total resistance of the circuit?

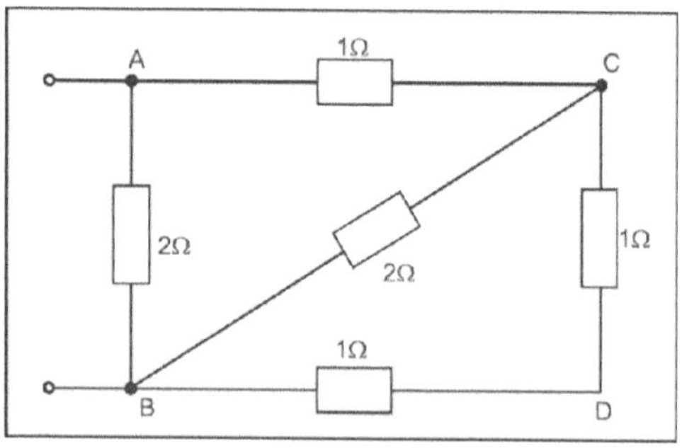

(a) 1π (b) 2π
(c) 3π (d) 4π

16. If two resistors each 20 ohms are connected in parallel, then the total value of resistance is ....
(a) 10 ohms (b) 20 ohms
(c) 40 ohms (d) 50 ohms

17. In a circuit, three resistors of 9 ohms each, are connected in parallel. What will be the equivalent resistance of the circuit?
(a) 9 ohms (b) $\frac{1}{3}$ ohm
(c) $\frac{1}{9}$ ohm (d) 3 ohms

18. If 5 resistors of 25 ohms each are connected in parallel, then the effective resistance of the circuit is ....
(a) 125 ohms (b) 75 ohms
(c) 25 ohms (d) 5 ohms

19. Ten resistors each of value 10 ohms are connected in parallel. Equivalent resistance is ....
(a) 100 Ω (b) 10 Ω
(c) 1 Ω (d) Zero

20. Three resistances 20 Ω, 30 Ω and 60 Ω are connected in parallel. Their combined resistance is given by ....
(a) 110 Ω (b) 50 Ω
(c) 20 Ω (d) 10 Ω

21. Two resistors of value 40 Ω and 60 Ω are connected in parallel to the supply of 220 volts. What is the total current in the circuit?
(a) 5.54 A (b) 9.167 A
(c) 3.667 A (d) 10.296 A

22. A certain circuit is composed of two parallel resistors.The total resistance is 1403 ohms. One of the resistors is 2 kilo ohms. The other resistor value is ....
(a) 150 kilo ohms
(b) 4.7 kilo ohms
(c) 3.5 kilo ohms
(d) 190 kilo ohms

23. A moving coil galvanometer is converted into an Ammeter by connecting a low resistance in ....
(a) Parallel (b) Series
(c) Both (a) and (b) (d) None of these

24. In series circuit, resistance ....
(a) Increases
(b) Decreases
(c) Remains constant
(d) None of these

25. One thousand resistors each of value 0.001 ohm are connected in series. Equivalent resistor is of value .....
(a) 1 Ω (b) 2 Ω
(c) 1000 Ω (d) 106 Ω

26. Three resistors of values 2 Ω, 3 Ω and 4 Ω are connected in parallel. The equivalent resistor is connected in series to a resistor of value $1\frac{1}{13}\Omega$. Final value of resistance is ....
(a) 1 Ω (b) 2 Ω
(c) 9 Ω (d) $\frac{100}{13}\Omega$

27. An example of insulator is ....
(a) Copper (b) Silver
(c) Rubber (d) Aluminium

28. Which is the conducting material?
(a) Aluminium (b) Porcelain
(c) Rubber (d) Fibre

29. Which one of the following will conduct electricity?
(a) Graphite (b) Porcelain
(c) Rubber (d) Wood

30. One of the following has free electrons in abundance ....
(a) Conductors (b) Semiconductors
(c) Insulators (d) All of these

31. .... is a semi-conductor
(a) Aluminium (b) Silicon
(c) Copper (d) None of these

KEY TO MCQs 7. Basic Electricity

1. (b)	**2.** (d)	**3.** (b)	**4.** (b)	**5.** (b)	**6.** (a)	**7.** (b)	**8.** (a)	**9.** (d)	**10.** (a)
11. (d)	**12.** (a)	**13.** (b)	**14.** (c)	**15.** (d)	**16.** (a)	**17.** (d)	**18.** (d)	**19.** (c)	**20.** (d)
21. (b)	**22.** (b)	**23.** (a)	**24.** (a)	**25.** (a)	**26.** (b)	**27.** (c)	**28.** (a)	**29.** (a)	**30.** (a)
31. (b)									

8. Mensuration

1. What is the formula to calculate area of equilateral triangle whose side is "*a*"?

(a) $\frac{\sqrt{3}}{4}a$ (b) $\frac{\sqrt{3}}{4}a^2$

(c) $\frac{4}{\sqrt{3}}a$ (d) $\frac{4}{\sqrt{3}}a^2$

2. What is the formula to calculate surface area of cube whose side is "*a*".

(a) 4a (b) $4a^2$

(c) 6a (d) $6a^2$

3. The angle which is more than 180° and less than 360° is called ....

(a) reflex angle (b) acute angle

(c) obtuse angle (d) straight angle

4. The sum of the two angles is equal to 90°, then that angles are called as ....

(a) corresponding angles

(b) complementary angles

(c) alternate interior angles

(d) right angles

5. What is the formula to find the volume of equilateral triangular prism having the side '*a*' and height "*h*" ?

(a) $\frac{3}{4}ah$ (b) $\frac{3}{4}a^2h$

(c) $\frac{\sqrt{3}}{4}a^2h$ (d) $\frac{3}{4}ah^2$

6. If two sides *a* and *b* and angle *c* is given, then the formula for area of triangle is ....

(a) $\frac{1}{2}a \cdot b \cdot \sin c$ (b) $\frac{1}{2}a \cdot \sin c$

(c) $\frac{1}{2}b \cdot \sin c$ (d) $a \cdot b \sin c$

7. Number of sides of a hexagon is ....

(a) 7 (b) 6

(c) 5 (d) 4

8. The ratio of length and breadth is 5 : 4 and the perimeter of the rectangles is 36 metres. The area is ....

(a) 79 cm^2 (b) 80 cm^2

(c) 81 cm^2 (d) 82 cm^2

9. Parallelogram base is 10 cm and the area is 120 cm^2. The height is ....

(a) 10 cm (b) 11 cm

(c) 12 cm (d) 13 cm

10. Find the height of parallelogram, whose base is 10 cm and area is 120 cm^2.

(a) 15 cm (b) 12 cm

(c) 10 cm (d) 8 cm

11. The side of rhombus is 5 cm and one of its diagonal is 8 cm. The area is ....

(a) 24 cm^2 (b) 25 cm^2

(c) 26 cm^2 (d) 27 cm^2

12. A trapezium has sum of parallel sides as 15 cm with maximum distance between the parallel sides as 4 cm. The area of the trapezium is .... square cm.

(a) 30 (b) 112.5

(c) 225 (d) 450

13. Surface area of wooden table top of diameter 1.4 meters ....
(a) 1.54 sq. metres
(b) 3.08 sq. metres
(c) 7.7 sq. metres
(d) 10.78 sq. metres

14. An equilateral triangular plot has side 20 metres. Find out its area.
(a) 143.2 square metres
(b) 153.2 square metres
(c) 163.2 square metres
(d) 173.2 square metres

15. Find the perimeter of a triangle whose sides are in the ration 3 : 4 : 5 and the longest side is 15 cm.
(a) 66 cm (b) 56 cm
(c) 36 cm (d) 46 cm

16. An equilateral triangle has each side 60 mm. Find out its area.
(a) 1550.8 mm^2 (b) 1556.8 mm^2
(c) 1558.8 mm^2 (d) 1560.8 mm^2

17. Calculate the height of right angled triangle whose area is 80 cm^2 and base is 20 cm?
(a) 8 cm (b) 10 cm
(c) 12 cm (d) 14 cm

18. In a trianglo, two equal sides are at right angle. Then other two angles are ....
(a) 45°, 60° (b) 30°, 60°
(c) 60°, 60° (d) 45°, 45°

19. The perimeter of a square is 800 cm. Find its diagonal.
(a) 254.5 cm (b) 272.6 cm
(c) 282.8 cm (d) 294.9 cm

20. Find the perimeter of a semi-circle whose radius is 14 cm $\left(\text{Take } \pi = \frac{22}{7}\right)$
(a) 22 cm (b) 44 cm
(c) 66 cm (d) 72 cm

21. The volume of a rod of 1.6 metres height, is 1.017 $metre^3$. Determine its diameter in mm ....
(a) 0.8997 (b) 8.997
(c) 89.97 (d) 899.7

22. A cylindrical tank has 22000 c.c. water. If the tank height is 70 cm, calculate the diameter of the tank.
(a) 9 cm (b) 10 cm
(c) 12 cm (d) 20 cm

23. A cylinder of radius 75 cm and height 100 cm can hold how many litres of water?
(a) 1767.38 litres (b) 1777.38 litres
(c) 1877.83 litres (d) 1978.78 litres

24. Find the volume of hollow cylinder of height 25 cm, outer diameter is 20 cm and inner diameter is 18 cm.
(a) 1945.10 cm^3
(b) 1549.15 cm^3
(c) 1491.5 cm^3
(d) 1149.15 cm^3

25. Find the volume of hollow sphere of 20 cm outer diameter and radial thickness of 2.5 cm.
(a) 2420.02 cm^3 (b) 2420.21 cm^3
(c) 2420.22 cm^3 (d) 2422.62 cm^3

26. An equilateral triangular prism having side 6 cm and length 16 cm. The volume is .... cubic cm.
(a) 249. 408 (b) 259.408
(c) 269.408 (d) 279.408

27. Equilateral triangular prism has height 40 cm, volume 4360 c.c. The side is .... cm.
(a) 21.68 (b) 17.78
(c) 16.64 (d) 15.86

28. The square prism have base of 10 cm and height 40 cm. The volume is .... cubic cm
(a) 4500 (b) 4400
(c) 4200 (d) 4000

29. The side of the cube is 4.5 metres. Its volume is .... cubic metres.
(a) 94.125 (b) 93.125
(c) 92.125 (d) 91.125

30. The volume of the largest right circular cone that can be cut out from a cube of edge 4.2 cm is .... cubic cm.
(a) 9.7 (b) 77.6
(c) 58.2 (d) 19.4

KEY TO MCQs 8. Mensuration

1. (b)	**2.** (d)	**3.** (a)	**4.** (b)	**5.** (c)	**6.** (a)	**7.** (b)	**8.** (b)	**9.** (c)	**10.** (b)
11. (a)	**12.** (a)	**13.** (a)	**14.** (d)	**15.** (c)	**16.** (c)	**17.** (a)	**18.** (d)	**19.** (c)	**20.** (d)
21. (d)	**22.** (d)	**23.** (a)	**24.** (c)	**25.** (d)	**26.** (a)	**27.** (d)	**28.** (d)	**29.** (d)	**30.** (d)

9. Levers and Simple Machines

1. Which one of the following is an example of first order lever?

(a) Lemon squeezer (b) Cutting plier
(c) Fire tongs (d) Nut craker

2. Which one of the following is an example for second order lever?

(a) lemon squeezer
(b) cutting plier
(c) crow bar
(d) fire tongs

3. Which one is the example for the combination of first and second order levers?

(a) Nut cracker
(b) Type writer
(c) Handle of tube well pump
(d) Wheel barrow

4. Which is the example of 3rd order lever. Choose the correct answer?

(a) A wheel barrow
(b) Common balance
(c) Lemon squeezer
(d) Human fore arm

5. What is third order of lever?

(a) Nut cracker
(b) Wheel Barrow
(c) Human Forearm
(d) A pair of scissors

6. A weight of 3000 kg is to be lifted by a bat of length 3 metres. Load arm is 1 metre while the effort arm is 2 metres. Find out mechanical advantage.

(a) 1 (b) 5
(c) 2 (d) 4

7. In a simple machine, effort applied is 25 kg and its mechanical advantage is 6. Find load lifted.

(a) 150 kg (b) 75 kg
(c) 50 kg (d) 175 kg

8. The ratio of load lifted to effort applied is known as

(a) efficiency
(b) velocity ratio
(c) mechanical advantage
(d) moment of couple

9. The formula for mechanical advantage in a simple machine is

(a) $\frac{\text{Effort}}{\text{Load}}$ (b) $\frac{\text{Load}}{\text{Effort}}$

(c) $\frac{\text{Distance moved by load}}{\text{Distance moved by effort}}$

(d) $\frac{\text{Distance moved by effort}}{\text{Distance moved by load}}$

10. Mechnical advatage = ...

(a) load/effort (b) load × effort
(c) output × input (d) effort/load

11. Simple machine velocity ratio is ......

(a) velocity ratio

$$= \frac{\text{Distance moved by load}}{\text{Distance moved by effort}}$$

(b) velocity ratio

$$= \frac{\text{Distance moved by effort}}{\text{Distance moved by load}}$$

(c) velocity ratio = $\frac{\text{Output}}{\text{Input}}$

(d) velocity ratio = $\frac{\text{Load}}{\text{Effort}}$

12. The ratio between mechanical advantage and velocity ratio is called ......
(a) mechanical advantage
(b) velocity ratio
(c) efficiency
(d) stress

13. Theoretically, if the mechanical advantage, velocity ratio of a simple machine are equal in magnitude, what will be its efficiency?
(a) 70% (b) 80%
(c) 90% (d) 100%

14. In a simple machine, the velocity ratio is found to be 20. An effort of 25 kg lifts a load 400 kg.
Eficiency of machine is ......
(a) 90% (b) 80%
(c) 75% (d) 70%

KEY TO MCQs 9. Levers and Simple Machines

1. (b)	**2.** (a)	**3.** (b)	**4.** (d)	**5.** (c)	**6.** (c)	**7.** (a)	**8.** (c)	**9.** (b)	**10.** (a)
11. (b)	**12.** (c)	**13.** (d)	**14.** (b)						

10. Trigonometry

1. What is the value of sin 30°?
(a) $\sqrt{3}$ (b) $\frac{1}{2}$
(c) $\frac{1}{\sqrt{3}}$ (d) $\frac{\sqrt{3}}{2}$

2. What is the ratio of cos θ?
(a) $\frac{1}{\sin\theta}$ (b) $\frac{1}{\tan\theta}$
(c) $\frac{1}{\sec\theta}$ (d) $\frac{1}{\cot\theta}$

3. The value of sin 90° × cos 90° = ...
(a) zero (b) 0.866
(c) 1 (d) 1.732

4. sin (90 + θ) = ...
(a) cos θ (b) – sin θ
(c) – cot θ (d) sec θ

5. What is the value of sin θ, if cos θ = $\frac{4}{5}$
(a) $\frac{1}{5}$ (b) $\frac{2}{5}$
(c) $\frac{3}{5}$ (d) $\frac{4}{5}$

6. What is the value of tan 30°?
(a) $\frac{2}{\sqrt{3}}$ (b) $\frac{1}{\sqrt{3}}$
(c) $\sqrt{3}$ (d) $\frac{\sqrt{3}}{2}$

7. Reverse of cos θ = ...
(a) tan θ (b) cot θ
(c) secant θ (d) cosecant θ

8. If sin θ = $\frac{1}{3}$, find the value of cosec θ.
(a) $\frac{\sqrt{8}}{3}$ (b) 3
(c) $\frac{3}{\sqrt{8}}$ (d) $\frac{1}{3}$

9. If sin θ = $\frac{3}{5}$, then find the value of cos θ.
(a) $\frac{4}{5}$ (b) $\frac{4}{3}$
(c) $\frac{5}{4}$ (d) $\frac{3}{5}$

10. $\sin 45° =$ ....

(a) 1 (b) $\frac{1}{2}$

(c) $\frac{\sqrt{3}}{2}$ (d) $\frac{1}{\sqrt{2}}$

11. $\sin^2\theta + \cos^2\theta =$ ....

(a) 1 (b) 2

(c) 3 (d) 4

12. $(\tan^2 \theta + 1) \cos^2 \theta =$ ....

(a) cosec θ (b) sec θ

(c) cot θ (d) 1

13. A ladder leans against a wall making an angle of 60° with ground. If the foot of the ladder is 5 metres away from the wall, calculate the length of the ladder.

(a) 5 metres (b) 10 metres

(c) 15 metres (d) 20 metres

14. The top of a hill makes an angle of 60° at a point 100 metres away from the foot of the hill. Calculate the height of the hill.

(a) 17.321 metres (b) 137.12 metres

(c) 153.7 metres (d) 173.21 metres

15. $\sin 90° =$ ....

(a) $\frac{1}{2}$ (b) $\frac{1}{\sqrt{2}}$

(c) $\frac{\sqrt{3}}{2}$ (d) 1

16. $\tan 45° =$ ....

(a) 1 (b) $\frac{1}{2}$

(c) 3 (d) 4

17. Find the value of $\tan\theta$, if $\cos\theta = \frac{5}{13}$

(a) $\frac{12}{5}$ (b) $\frac{12}{13}$

(c) $\frac{15}{8}$ (d) $\frac{17}{8}$

18. $\cos 2A =$ ....

(a) $1 - 2\sin^2 A$ (b) $2\sin^2 A$

(c) $2\sin^2 A - 1$ (d) $\text{cosec}^2 A$

19. $\sec^2\theta - 1 =$ ....

(a) $\tan^2\theta$ (b) $\sin^2\theta$

(c) $\cos^2\theta$ (d) $\text{cosec}^2\theta$

KEY TO MCQs 10. Trigonometry

1. (b)	**2.** (c)	**3.** (c)	**4.** (a)	**5.** (c)	**6.** (b)	**7.** (c)	**8.** (b)	**9.** (a)	**10.** (d)
11. (a)	**12.** (d)	**13.** (b)	**14.** (d)	**15.** (d)	**16.** (a)	**17.** (a)	**18.** (a)	**19.** (a)	

1st Year

Engineering Drawing

1. Introduction to Engineering Drawing and Drawing Instruments

1. Drawing sheets are designated by Indian standards according to their sizes.
   Designation of 1 m^2 surface area drawing sheet is
   (a) A3 (b) A2
   (c) A1 (d) A0
2. The trimmed size of A2 designation drawing sheet is ....
   (a) 594 × 841 (b) 420 × 594
   (c) 841 ×1189 (d) 297 × 420
3. What is the standard drawing board size of D_1 as per IS :1444 -1989?
   (a) 1500 × 1000 × 25
   (b) 1000 × 700 × 25
   (c) 500 × 350 × 15
   (d) 700 × 350 × 15
4. Set squares are used with the support of 'T' square to draw all straight lines except ....
   (a) vertical lines
   (b) inclined lines
   (c) horizontal lines
   (d) perpendicular lines
5. The HB grade pencils are ....
   (a) soft (b) hard
   (c) medium soft (d) medium hard
6. The part that doesn't belong to T-square is ________
   (a) Working edge (b) Blade
   (c) Stock (d) Ebony Wheel
7. Small bow compass can draw circles less than ______ mm radius.
   (a) 25 mm (b) 30 mm
   (c) 35 mm (d) 40 mm
8. Many factors are to be considered for the selection of particular grade of pencil. One of the factors for the selection of pencil is the -----
   (a) Size of drawing paper
   (b) Type of line work
   (c) Type of drawing
   (d) Quality of pencil
9. To draw parallel lines at angles of 105° to the horizontal how set square are to be manipulated?
   (a) 45° and 30° (b) 45° and 30°
   (c) 90° × 30° (d) 90° × 45°
10. Set squares are used with the support of 'T' square to draw all straight lines except ....
    (a) vertical lines
    (b) inclined lines
    (c) horizontal lines
    (d) Perpendicular lines
11. Several grades of pencils are available from which suitable grades are chosen for specific drafting. The pencil grade suitable for lettering and object lines are ....
    (a) 8H or 9H (b) 2H or 3H
    (c) H or 2H (d) H or HB

12. Manipulating two set squares together may give many angles to draw and measure. What is the angle 'a' formed by the 45° and 60°- 30° set-squares ....

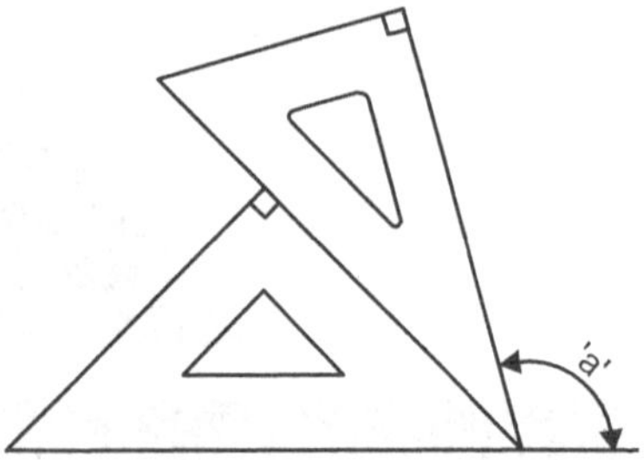

(a) 100° (b) 105°
(c) 110° (d) 120°

13. Angle is formed by the combination of two set squares Manipulating two set squares together may be forming various angles

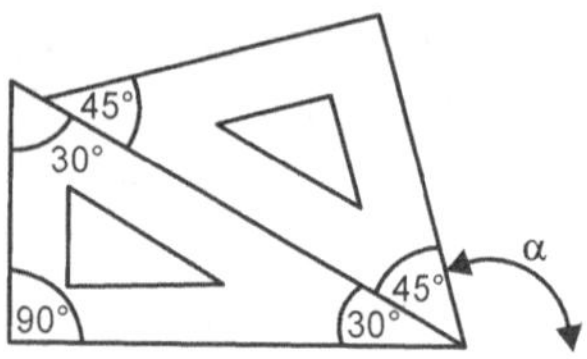

The angle 'α' formed by the combination of two set square

(a) 150° (b) 105°
(c) 90° (d) 75°

14. Identify the drawing instrument which is used to ink the small circles drawing in pencil?

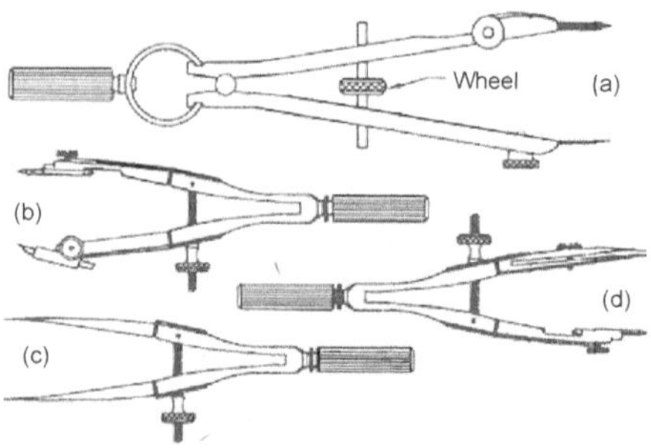

15. The areas of the two subsequent sizes of drawing sheet are in the ratio _____.

(a) 1 : 5 (b) 1 : 4
(c) 1 : 2 (d) 1 : 10

16. __________ is used to draw curves which are not circular.

(a) Compass (b) Protractor
(c) French curves (d) Pro circle

17. From the shown Instruments (a), (b), (c), (d) which instrument is used to transfer very small dimensions and also parting lines into many equal parts.

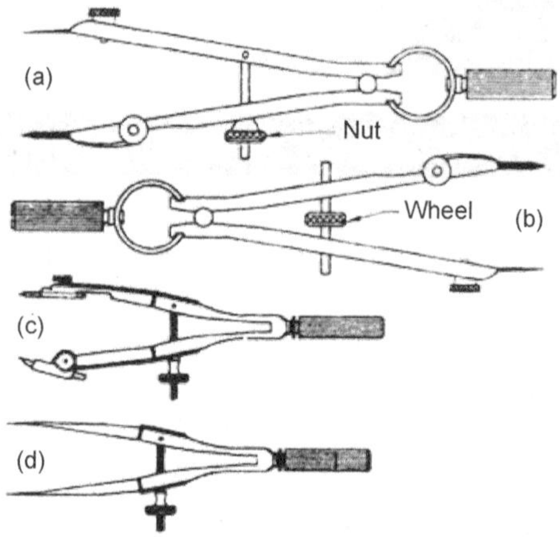

18. Which degree set square is used directly to draw a 60° line along with 'T' square ....

(a) 60° – 30° (b) 45° – 90°
(c) 30° – 45° (d) 30° – 75°

19. The head (stock) of a T-square is made out of .....

(a) Wood / celluloid
(b) metalic plate
(c) hard board
(d) ceramic material

20. Protractor is an instrument for measuring ....

(a) angle in degrees (b) angle in radians
(c) angle in minute
(d) linear measurement

21. Which type of drawing sheet has biggest area ?

(a) A0 (b) A1
(c) A2 (d) A3

22. What is the angle of inclined letter to be drawn towards right side?

(a) 75° (b) 60°
(c) 30° (d) 15°

23. Protractor is an instrument used for measuring ....
(a) angle in degrees
(b) angle in radians
(c) angle in minutes
(d) linear measurement

24. From the shown Instruments (a), (b), (c), (d) which instrument is having adjustable pencil point to draw very small circles

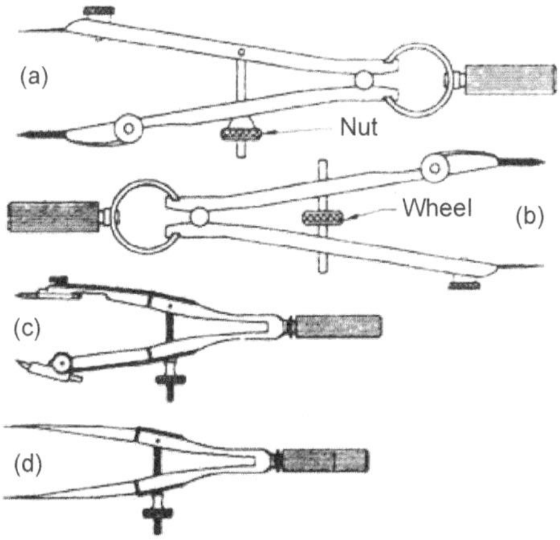

25. For drawing general lines, lettering, arrows, we use ....
(a) 9H grade (b) 7B grade
(c) H / 2H medium (d) HB grade

26. An erasing shield is used ....
(a) for shielding the drawn lines
(b) for rubbing oft (erasing) unwanted lines, curves without damaging other lines
(c) to draw lines through the shield
(d) to draw different curves

27. Mention the instrument designed for drawing multiple identical small circles:
(a) Small compass
(b) Drop spring bow pencil
(c) Bow compass
(d) Adjustable compass

28. Choose the type of lines that can be drawn only by using 'T' square is
(a) Horizontal parallel lines
(b) Vertical parallel lines
(c) Inclined parallel lines
(d) Concentric parallel lines

29. Identify the drop spring bow pencil used for drawing multiple identical small circles.

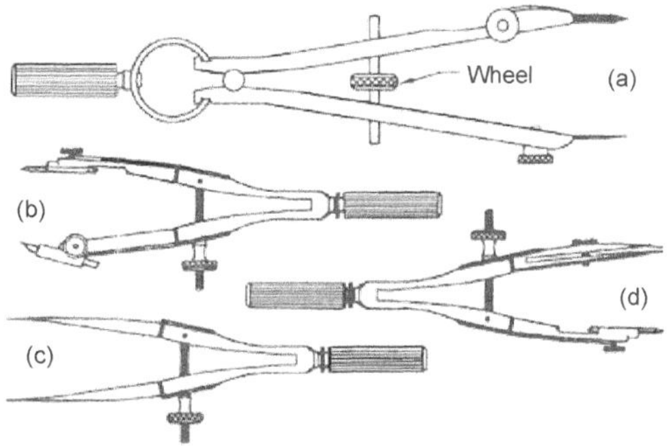

30. The angle which we can't make using both the Set-squares is ______________
(a) 15° (b) 105°
(c) 165° (d) 125°

31. Which size of trimmed drawing sheet is designed by A3?
(a) 210 mm × 297 mm
(b) 297 mm × 420 mm
(c) 420 mm × 594 mm
(d) 148 mm × 210 mm

32. How many battens will be there for a Drawing board?
(a) 1 (b) 2
(c) 3 (d) 4

KEY TO MCQs 1. Intro. to Engg. Drg. & Drg. Ins.

1. (d)	**2.** (b)	**3.** (b)	**4.** (b)	**5.** (c)	**6.** (d)	**7.** (a)	**8.** (b)	**9.** (a)	**10.** (b)
11. (d)	**12.** (b)	**13.** (b)	**14.** (d)	**15.** (c)	**16.** (c)	**17.** (d)	**18.** (a)	**19.** (a)	**20.** (a)
21. (a)	**22.** (d)	**23.** (a)	**24.** (c)	**25.** (c)	**26.** (b)	**27.** (b)	**28.** (a)	**29.** (b)	**30.** (d)
31. (b)	**32.** (b)								

2. Free Hand Drawings

1. Which quadrilateral shape is trapezium ?

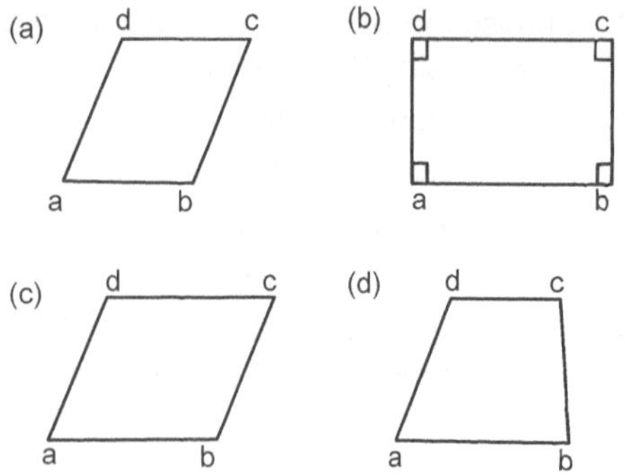

2. Degree is the unit of angle. Which one represents 1°?

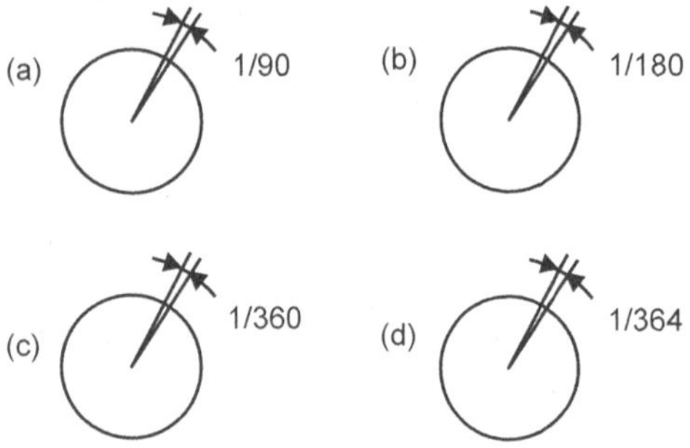

3. Which plane figure is heptagon from the following?

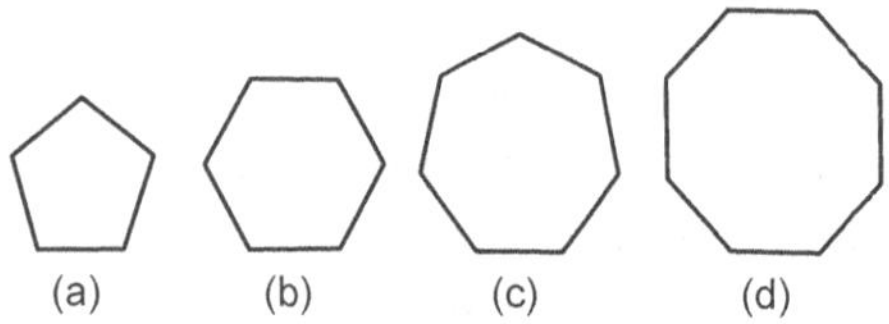

4. In the given figure which of the following construction line is drawn first?

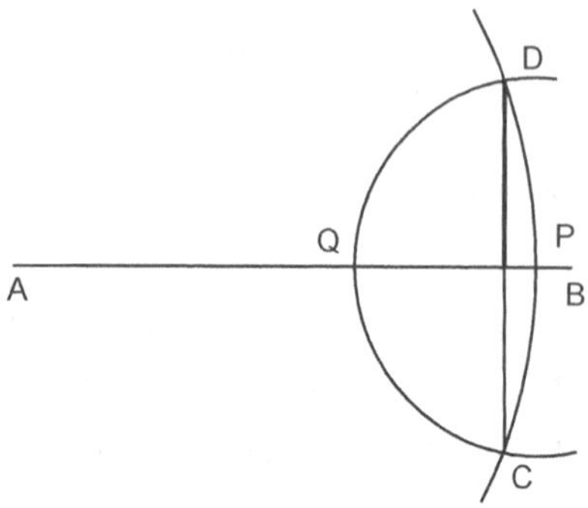

(a) Line AP (b) Arc DPC
(c) Arc DQC (d) Line DC

5. Polygons are known by the number of sides which plane figure is Nonagon? Choose the correct answer.

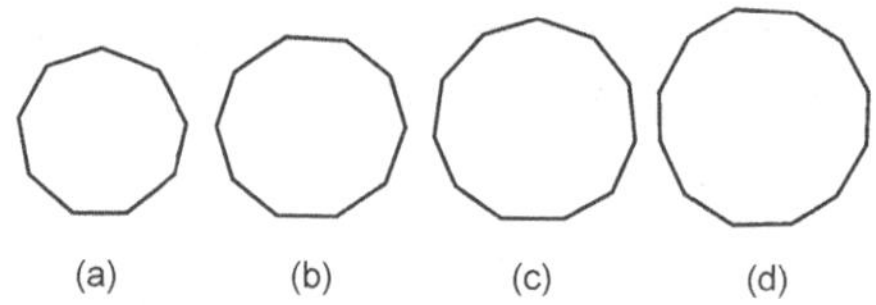

6. Which is a trapezoid?

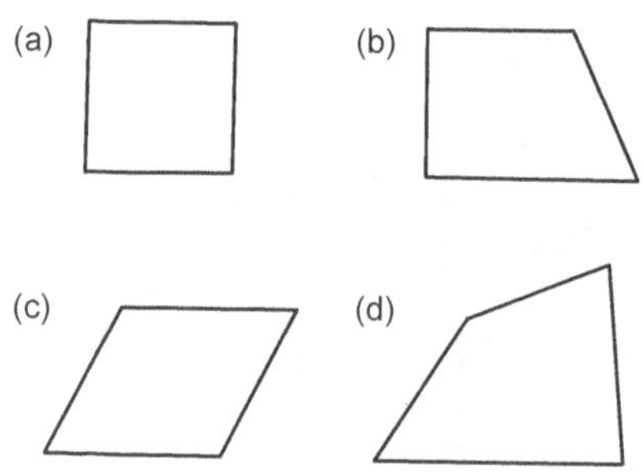

7. Quadrilaterals are known by their shapes. Which one is a rhombus ?

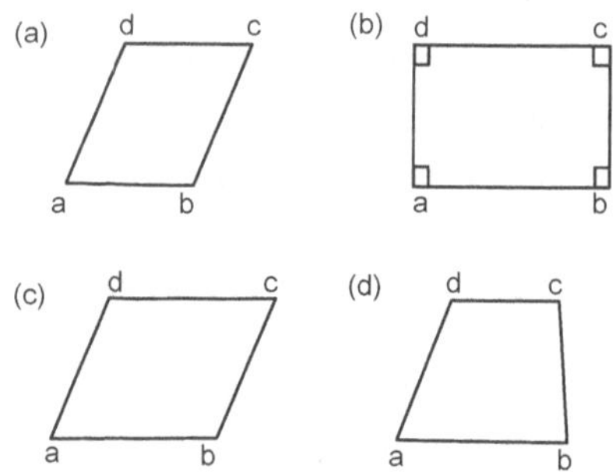

8. Identify 'octagon' from the given figure:

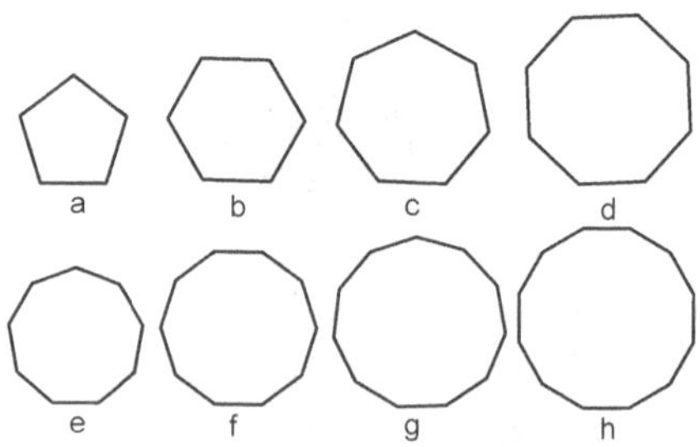

9. Development of which solid is drawn by parallel line method?

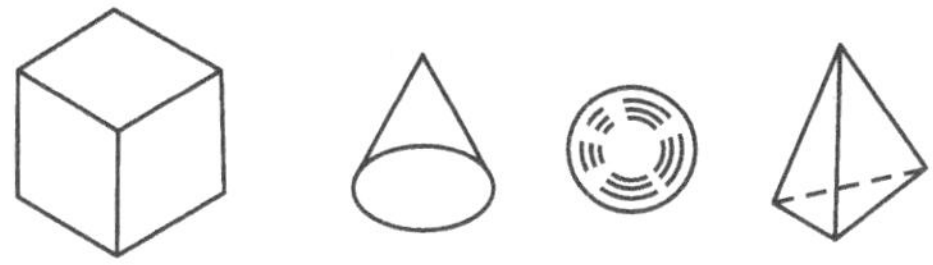

(a) (b) (c) (d)

10. Two opposite sides are equal and parallel and four angles are right angles are called as ....

(a) trapezium (b) square
(c) parallelogram (d) Rectangle

11. Name the part 'X' indicated in figure?

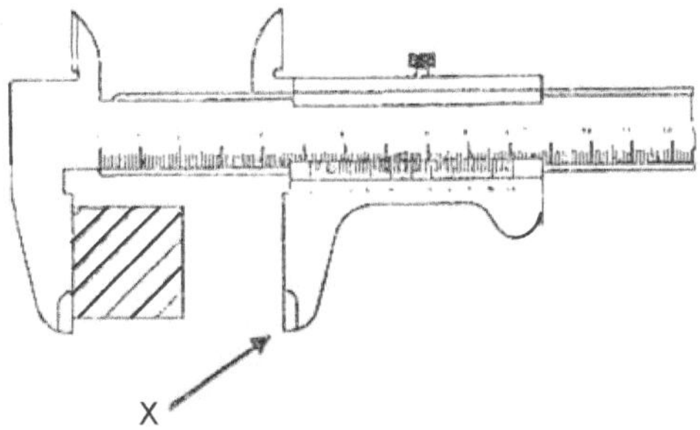

(a) Main scale divisions graduated on this part
(b) It is fixed measuring part
(c) It is moving measuring part
(d) Beam

12. Identify the name of the part which is marked as 'X'.

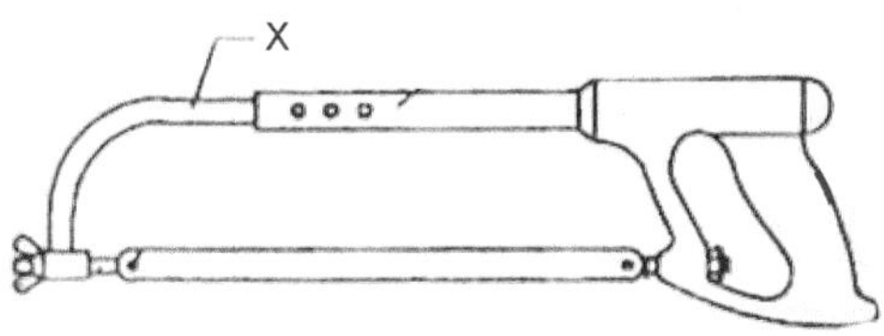

(a) Frame
(b) Blade holder
(c) Retaining pipe
(d) Frame length adjustment

13. Identify the name of the tool.

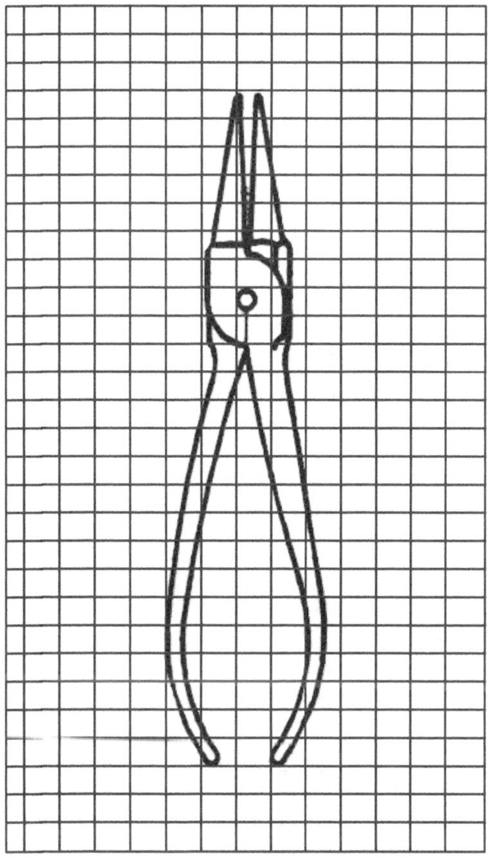

(a) Combination plier
(b) Long nose plier
(c) Criclip plier
(d) End cutting plier

KEY TO MCQs 2. Free Hand Drawings

1. (d) **2.** (c) **3.** (c) **4.** (b) **5.** (a) **6.** (d) **7.** (a) **8.** (d) **9.** (a) **10.** (d)
11. (c) **12.** (a) **13.** (c)

3. Drawing of Geometrical Figures

1. Which one of the following figures shows complementary angle?

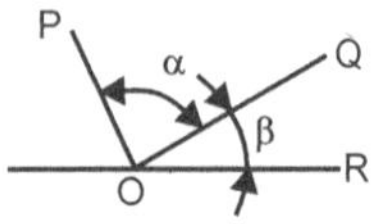

(a)

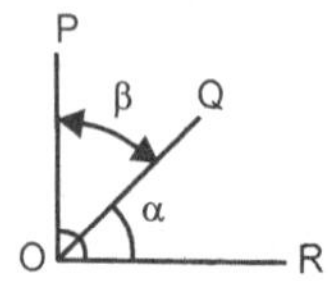

(b)

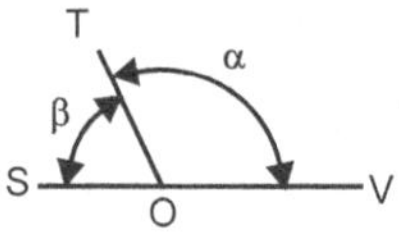

(c)

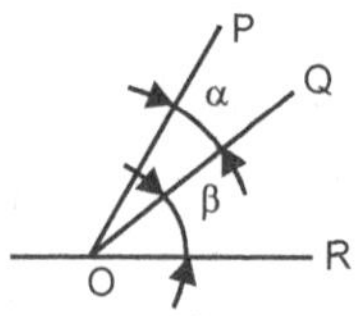

(d)

2. From the given drawing identify the pentagonal pyramid.

(a)

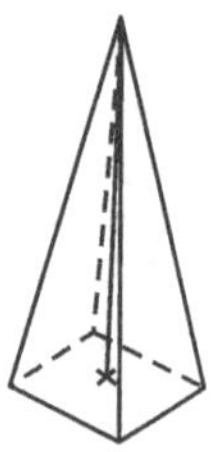
(b)

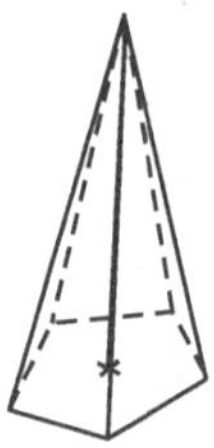
(c)

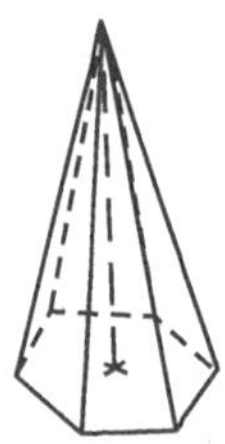
(d)

3. All object are bounded by geometrical surfaces. The surfaces may be classified in different ways. Which one solid is bounded by plane surfaces.

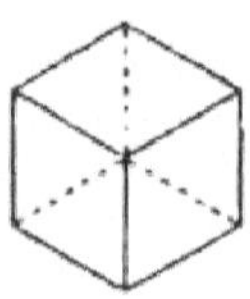
(a)

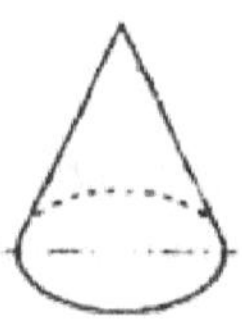
(b)

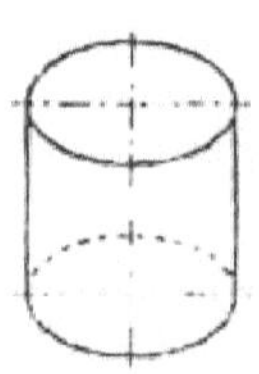
(c)

(d)

4. Given all are regular cylinders. Identify the one which is in perspective view.

(a)
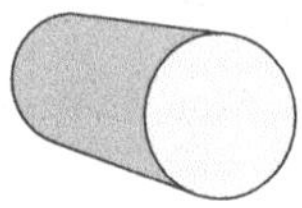

(b)
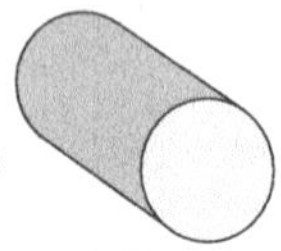

(c)
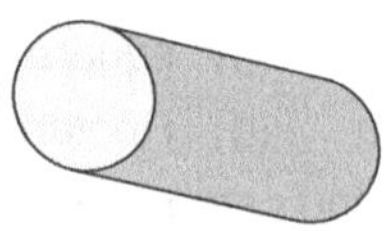

(d)
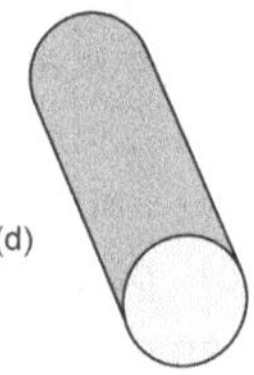

5. Identify the side view of for the below given pyramid.

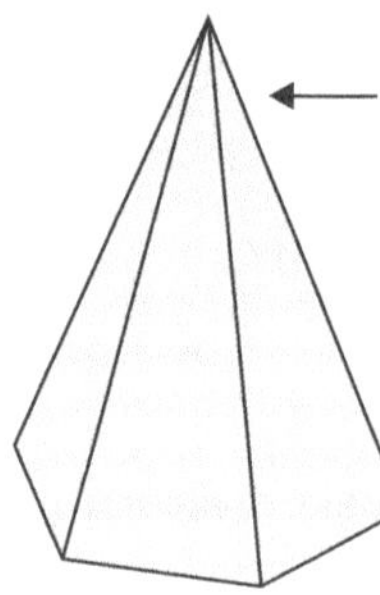

(a)
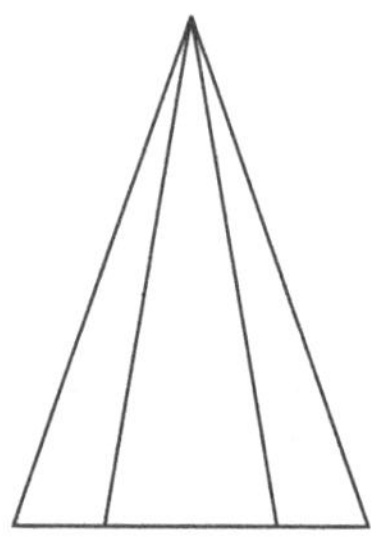

(b)
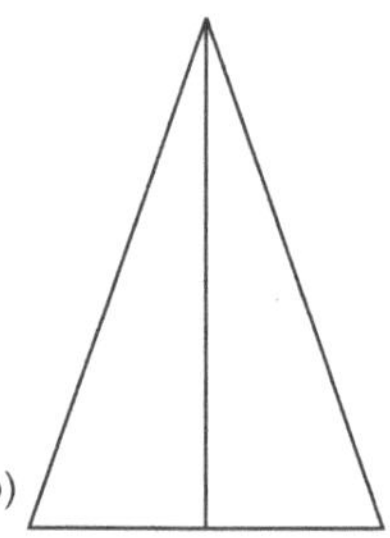

(c)
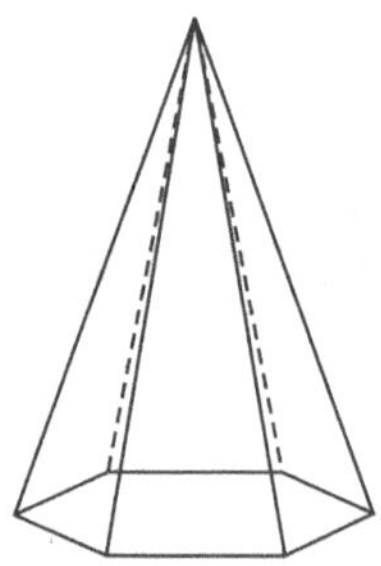

(d)
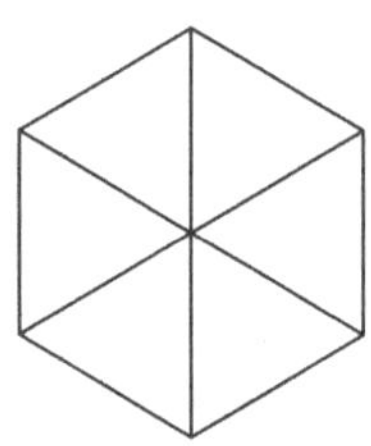

6. Angles between two lines may be any value up to 360°. Which one is an acute angle ?

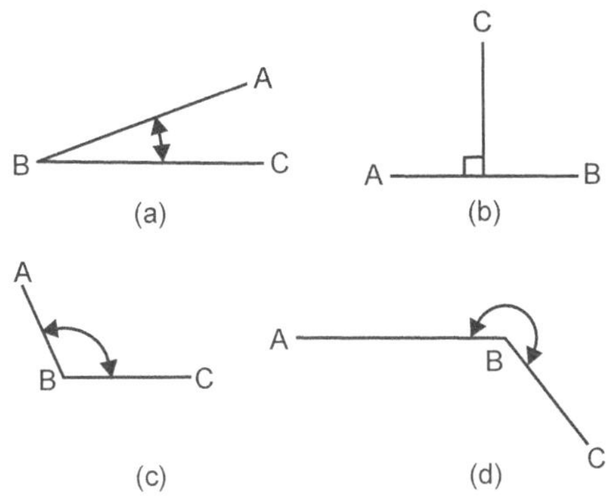

7. Which one of the solid bounded by double curved surfaces and it is formed by revolving a curve about a straight line?
   (a) Cylinder (b) Cone
   (c) Pyramid (d) Sphere

8. Identify the square pyramid?

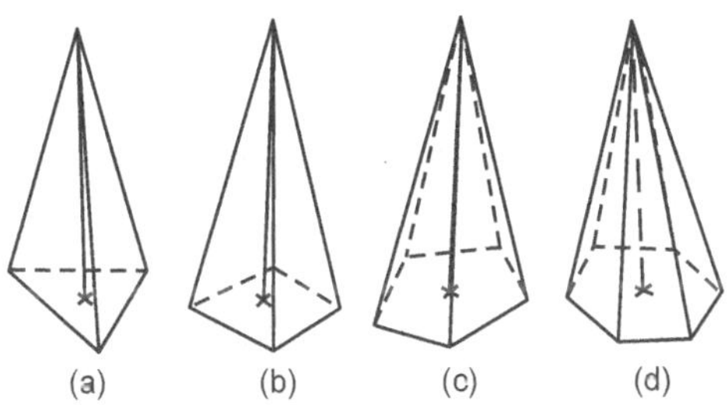

9. Which one of the solid bounded by single curved surface and it can be rolled to a plane?
   (a) Sphere (b) Cylinder
   (c) Paraboloid (d) Prism

10. There are different types of angles. An obtuse angle is one which is -----
    (a) Less than 90°
    (b) More than 90° and less than 180°
    (c) Equal to 180°
    (d) More than 180° and Less than 360°

11. Regular heptagon has .... sides
    (a) five (b) six
    (c) seven (d) eight

12. There are many styles of letter styles in Engineering drawing.
    Name the style of lettering ....

A B C D E F G H
a b c d e f g h

    (a) gothic (b) roman
    (c) italic (d) text

13. Identify the style of the letter given below?

ABCDEFGH
abcdefgh

    (a) Italic (b) Gothic
    (c) Roman (d) Text

14. All the lettering should be printed so as they are read / viewed from the ....
    (a) top of drawing
    (b) left side of drawing
    (c) bottom of drawing
    (d) right side of drawing

KEY TO MCQs 3. Drawing of Geometrical Figures

1. (b)	2. (c)	3. (a)	4. (a)	5. (b)	6. (a)	7. (d)	8. (b)	9. (b)	10. (b)
11. (c)	12. (a)	13. (a)	14. (c)						

4. Dimensioning Practice

1. Which are the two systems of placing dimensions?

i. Aligned system, ii. Break system,
iii. Unidirectional system,
iv. Directional system

(a) i, ii (b) i, iii
(c) ii, iv (d) i, iv

2. Which of the following dimension is according to the 'aligned system' of dimensioning?

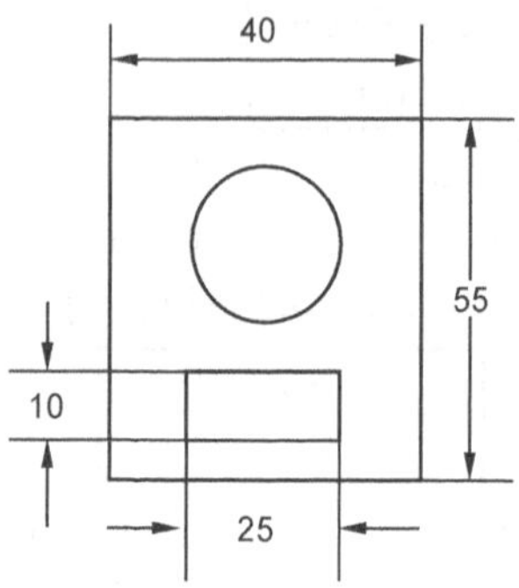

(a) 40 (b) 55
(c) 25 (d) 10

3. From the given figure, identify extension line.

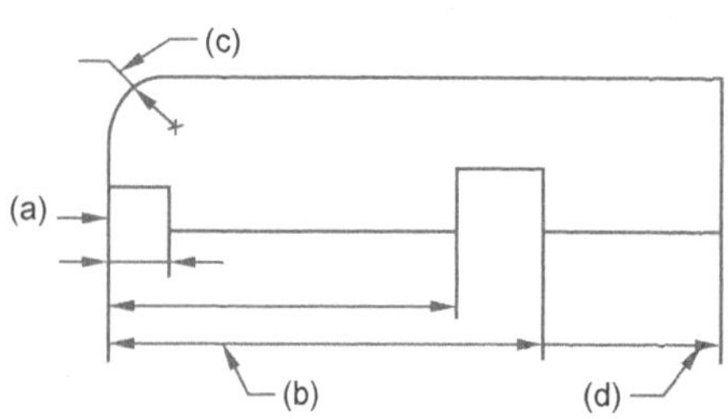

4. The wrongly represented line in the below figure is ________

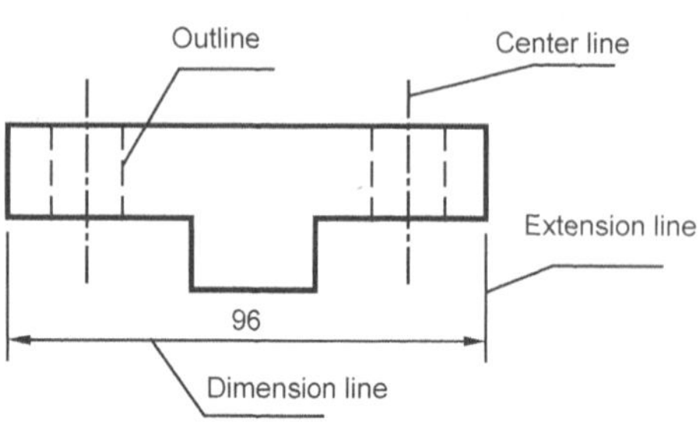

(a) dimension line
(b) extension line
(c) out line
(d) center line

5. Identify the type of dimensioning used in the example below.

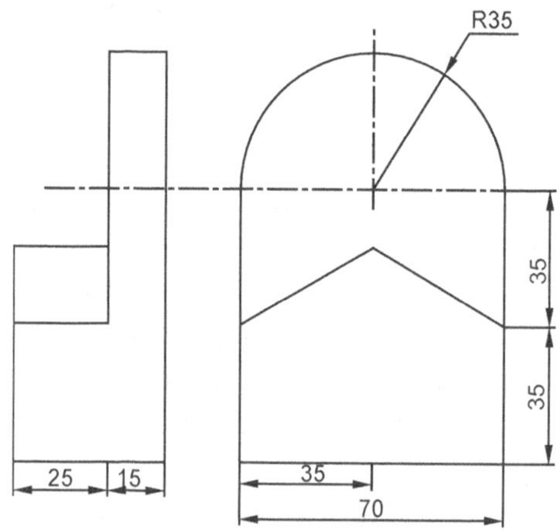

(a) Parallel and aligned dimension
(b) Chain and aligned dimension
(c) Chain and unidirectional dimension
(d) Combined and aligned dimension

6. Identify the 'a' in the given picture

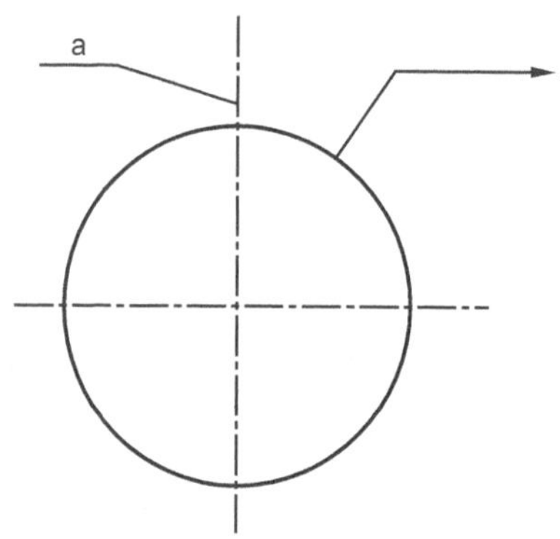

(a) dimension line
(b) hidden line
(c) centre line
(d) outline

7. From the given figure identify leader line.

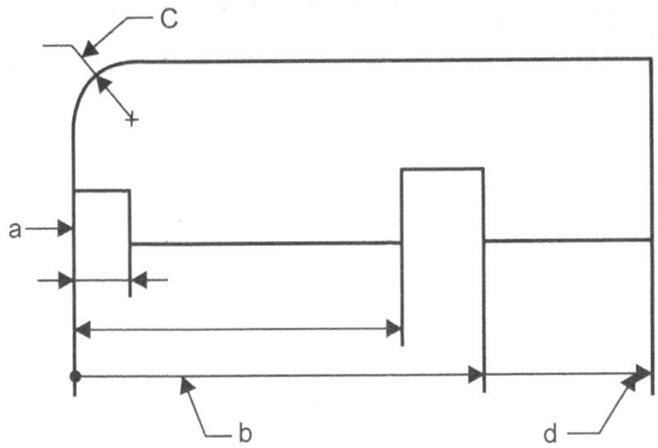

8. Name the symbol on a drawing indicated by Sϕ50

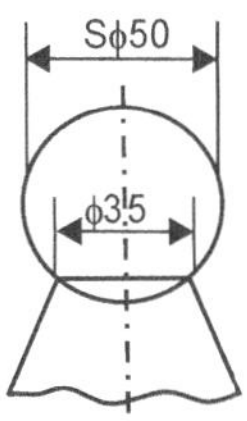

(a) Spherical diameter
(b) Spherical radius
(c) Spherical square
(d) Spherical circumference

9. Identify the type of line?

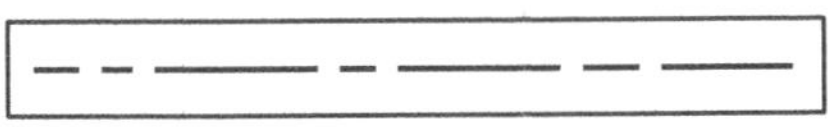

(a) Dashed thin line
(b) Chain thin line
(c) Chain thick line
(d) Chain thin double dashed

10. Identify the name of arrangement of dimensioning.

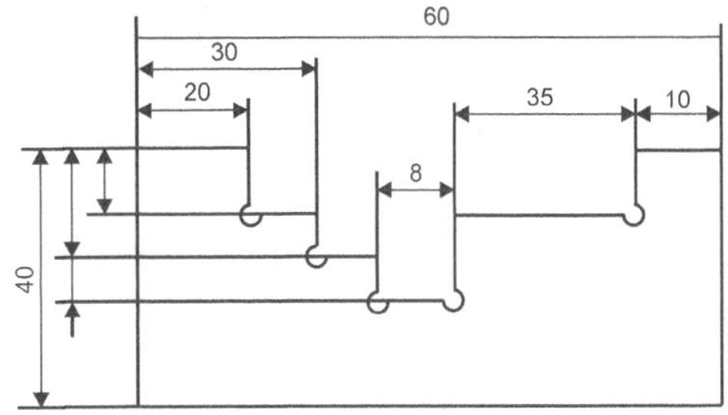

(a) Chain dimensioning
(b) Parallel dimensioning
(c) Combined dimensioning
(d) Super imposed running dimensioning

11. From the below figure, what is the name of the line X ?

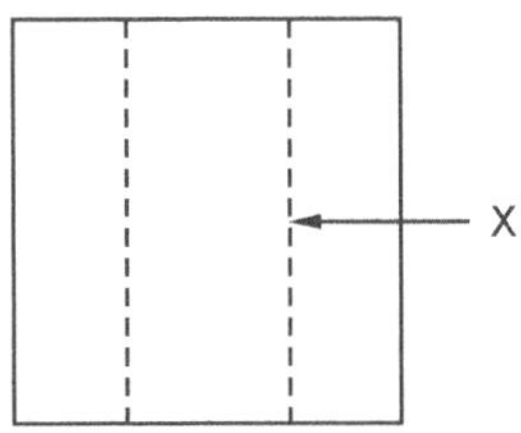

(a) Outline (b) Section line
(c) Hidden line (d) Hatching

12. Name the line marked 'A' in the given figure.

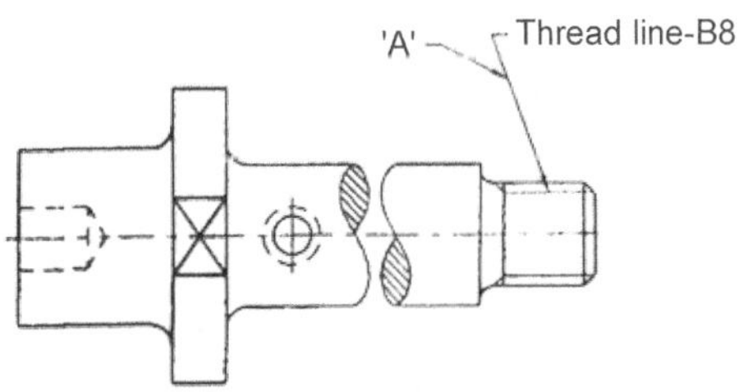

KEY TO MCQs 4. Dimensioning Practice

1. (b) **2.** (a) **3.** (a) **4.** (c) **5.** (d) **6.** (c) **7.** (c) **8.** (a) **9.** (c) **10.** (d) **11.** (c) **12.** (a)

5. Symbolic Representation, Reading of Circuit Diagrams and Layout Drawings

1. Which is the symbol for single 'V' butt weld ?

(a) (b) (c) (d)

2. Name the fitting symbol given in figure.

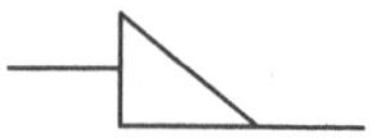

(a) Eccentric reducer
(b) Concentric reducer
(c) Elbow
(d) Pipe

3. What is the name of the BIS symbol used in electronic circuit diagram?

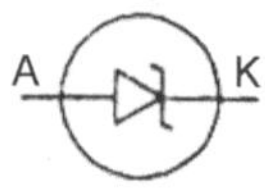

(a) Varactor diode (b) Zener diode
(c) Photo diode (d) Tunnel diode

4. Name the part shown as 'X' in the magnetic coil.

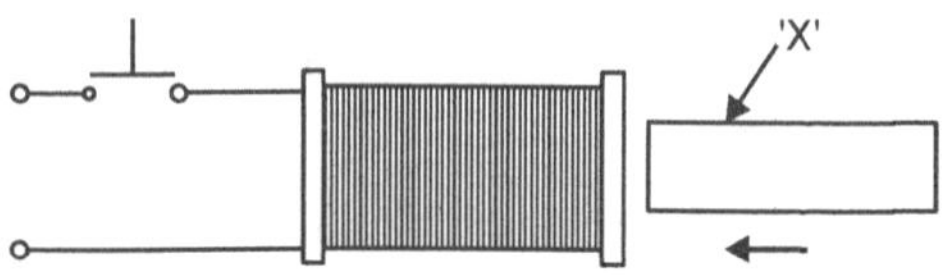

(a) core (b) coil
(c) plunger (d) armature

5. What does the symbol given in the below figure represents ?

(a) AC generator (b) DC generator
(c) AC motor (d) DC motor

6. What does the symbol marked as 'X' in the bell circuit show in figure?

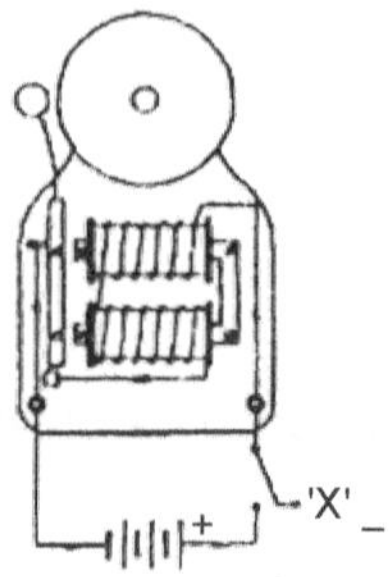

(a) Battery (b) Switch
(c) Spring (d) Screw contact

7. What is the name of the BIS symbol?

(a) DC motor (b) AC motor
(c) DC meter (d) AC meter

8. Name the B.I.S. symbol used in circuit diagram.

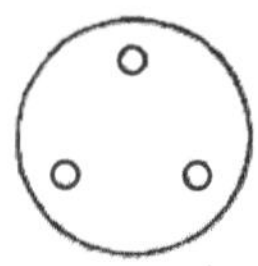

(a) Lamp
(b) Socket outlet, 6A
(c) Push button
(d) Plug and socket

9. What is the name of the BIS symbol?

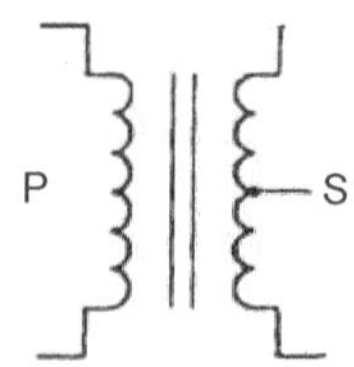

(a) Transformer tapped secondary
(b) Transformer multiple Secondar

(c) Ferrite core transformer
(d) Tuned transformer

10. Symbol shown in figure represent ....

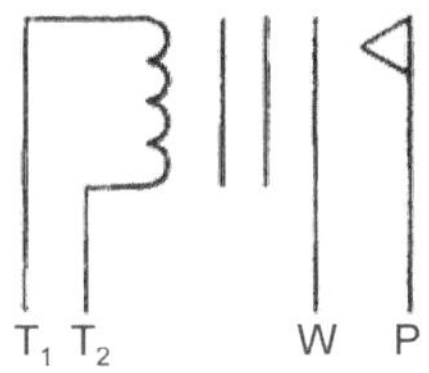

(a) Electro magnetic relay
(b) spring return switch
(c) relay contacts
(d) magnetic relay

11. Symbol shown in the figure below represents.

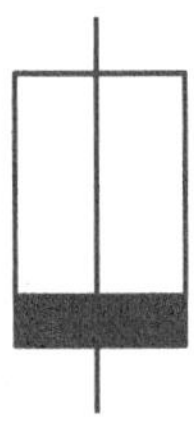

(a) fuse
(b) alternate symbol for fuse
(c) alternate symbol for fuse where supply side is indicated by a thick line
(d) alternate symbol for fuse where load side is indicated by a thick line

12. Identify the conventional representation of material?

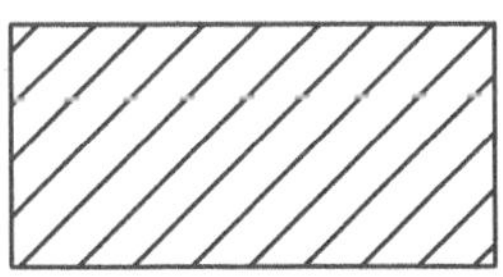

(a) Lead (b) Glass
(c) Steel (d) Wood

13. This safety sing is covered under ....

(a) Prohibitory sign (b) Mandatory sign
(c) Warning sign (d) Information sign

14. Identify the symbol used for alternating current in the circuit diagram?

(a) —— (b) ∼
(c)  (d) 

15. What is the name of BIS symbol?

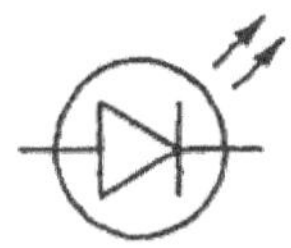

(a) LDR
(b) LED
(c) Photo diode
(d) Tunnel diode

16. What does this safety sign indicate?

(a) Sun rising
(b) Very hot do not touch
(c) Laser beam
(d) Risk of fine

17. Name the symbol used in electronic circuit diagram.

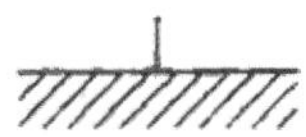

(a) Earth point
(b) Chasis earth connection
(c) Position of fault connection
(d) Frame connection

18. What is the name of the part marked 'X' in figure?

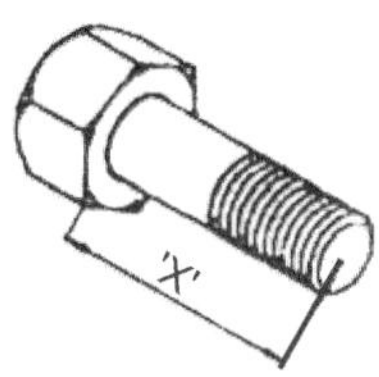

(a) Length of bolt
(b) Length of threaded portion
(c) Pitch of thread
(d) Length of unthreaded portion

19. Name the part indicated as 'X' in the DC generator as shown in figure.

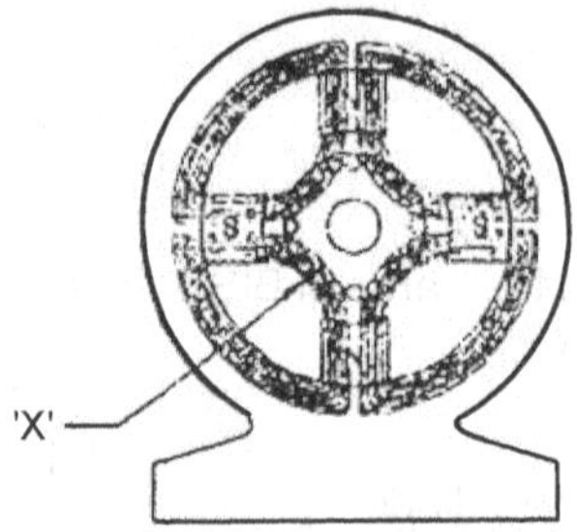

(a) armature core
(b) field coil
(c) armature conductors
(d) field pole

20. Identify two way switch symbol used in the wiring of electrical installation.

(a) 
(b) 
(c) 
(d) 

KEY TO MCQs 5. Sy,. Rep., Reading of Diagrams

1. (c)	**2.** (a)	**3.** (b)	**4.** (a)	**5.** (a)	**6.** (b)	**7.** (a)	**8.** (b)	**9.** (a)	**10.** (a)
11. (c)	**12.** (c)	**13.** (d)	**14.** (b)	**15.** (b)	**16.** (c)	**17.** (b)	**18.** (a)	**19.** (c)	**20.** (c)

ELECTRICIAN TRADE THEORY

(As Per NSQF Level-4 Compliant Competency Based Curriculum)

2nd YEAR

Module 1

D.C. Generator

Multiple Choice Questions

1. If the axis of a moving coil is kept parallel to the flux then the e.m.f. induced in the coil will be :

(a) minimum (b) maximum
(c) nil (d) one volt

2. For inducing an e.m.f. of 1 volt in 1 second time, a conductor will have to cut :

(a) 10 maxwell lines
(b) 10^8 lines
(c) 1 weber magnetic flux
(d) none of the above

3. On doubling the supply frequency, the eddy current will :

(a) become half of their former value
(b) become four times of their former value
(c) remain unaffected
(d) become eight time of their former value

4. In order to reduce hysteresis losses, the cores of electrical machines are generally made of :

(a) tungsten steel
(b) cast iron
(c) high carbon steel
(d) silicon steel

5. Whenever a conductor cuts magnetic flux, an e.m.f. is induced in the same. The statement refers to :

(a) Joule's law (b) Coulomb's law
(c) Faraday's law (d) Lenz's law

6. The given figure shows a coil with a closed iron core. What do the dotted lines represent ?

(a) magnetic potential
(b) magneto motive force
(c) magnetic field strength
(d) magnetic flux

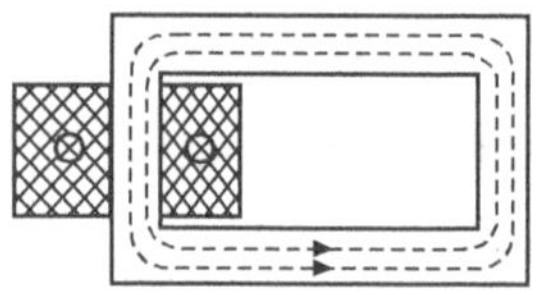

7. The magnetic flux shown in above Fig. can be increased by:

(a) increasing the magneto motive force
(b) decreasing the magneto motive force
(c) reversing the direction of current
(d) making an air-gap in the core

8. The law which states that the induced e.m.f. and current always opposes the cause producing it was presented by :

(a) Lenz (b) Faraday
(c) Maxwell (d) Ohm

9. An open-circuit curve is drawn between :

(a) E and I_F (b) N and I_L
(c) E and I_A (d) E_0 and I_F

(where E_0 = no-load voltage; I_F = field current; I_A = armature current; I_L = load-current; N = no. of armature conductors; E = load voltage.)

10. In which type of d.c. machine-poles is residual magnetism necessary ?

(a) Shunt generator (b) Motor
(c) Separately excited generator
(d) Permanent magnet type generator

11. The output voltage of a single loop generator is a :
(a) steady d.c. (b) steady a.c.
(c) pulsating a.c. (d) pulsating d.c.

12. The voltage generated by a generator is determined by the equation :
(a) $E = V - I_a \cdot R_a$ (b) $E = V + I_a \cdot R_a$
(c) $E = L\frac{di}{dt}$ (d) $E = -N\frac{df}{dt}$

13. The brushes are always placed along :
(a) MNA (magnetic neutral axis)
(b) GNA (geometric neutral axis)
(c) perpendicular to MNA
(d) perpendicular to GNA

14. In any d.c. generator, the e.m.f. generated in the armature is maximum when :
(a) rate of change of flux is minimum
(b) rate of change of flux is maximum
(c) flux linked with conductors is maximum
(d) flux linked with conductors is minimum

15. The method which can be used to improve commutation process is :
(a) to increase the resistance of brushes
(b) to reduce the no. of turns in the armature coil so as to reduce the no. of commutator segments
(c) to neutralise the reactance voltage by producing a reversing e.m.f. in the short-circuited coil under commutation
(d) all the above

16. The function of interpoles is :
(a) to neutralise the reactance voltage to help commutation process
(b) to neutralise the cross-magnetising effect of armature reaction
(c) to reduce sparking at the commutator
(d) all the above

17. Generators are run in parallel for achieving the advantage of :
(a) high reliability (b) high efficiency
(c) meeting higher load demands
(d) all the above

18. Shunt generators are used where the main requirement is a constant :
(a) voltage and current
(b) current
(c) voltage over a wide load range
(d) voltage over a narrow load range

19. Interpoles are normally connected in :
(a) parallel with the field windings
(b) series with the field windings
(c) parallel with the load
(d) series with armature winding

20. In d.c. generators, the cause of rapid brush wear may be :
(a) imperfect cantact
(b) severe sparking
(c) rough commutator surface
(d) any one of the above

21. In d.c. generators, lap winding is used for :
(a) low voltage, low current
(b) high voltage, high current
(c) low voltage, high current
(d) high voltage, low current

22. D. C. generator preferred for charging automobiles batteries is :
(a) shunt generator
(b) series generator
(c) long shunt compound generator
(d) short shunt compound generator

23. The overall efficiency of a d.c. generator lies in the range of :
(a) 60-80% (b) 75-85%
(c) 80-90% (d) 85-95%

24. The function of an equalizer ring is :
(a) to avoid unequal distribution of current at the brushes
(b) to help in reducing sparking
(c) both the above
(d) none of the above

25. A conductor is rotating within a magnetic field as shown in the given figure. At which of the positions shown does the peak voltage occur ?

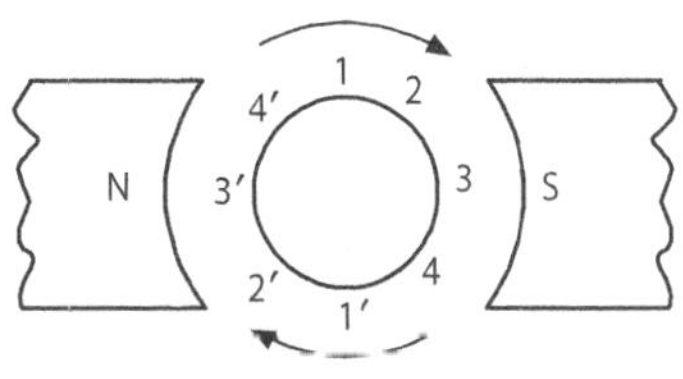

(a) At position 1 and 1′
(b) At position 2 and 2′

(c) At position 3 and 3′
(d) At position 4 and 4′

26. The essential condition for stable parallel operating of two D.C. generators having similar characteristics is that, they should have
(a) same kilowatt output ratings
(b) dropping voltage characteristics
(c) same percentage regulation
(d) same no load and full load speed

27. For parallel operation the D.C. generators normally preferred are
(a) Shunt and over compound
(b) Series
(c) Shunt and under compound
(d) Over compound

28. Equalizer bar is necessary for parallel operation of D.C.
(a) series and over compound generator
(b) under compound generator
(c) shunt and series generator
(d) over compound generator

29. A DC generator works on the principle of ....
(a) Faraday's law of electrolysis
(b) Fleming's left hand rule
(c) Lenz's law
(d) Faraday's law of electromagnetic induction

(I.T.I. February 2014, Electrician)

30. The formula to find EMF induced (E) in DC generator is ....
(a) $E = \frac{j\,ZN}{60} , \frac{P}{A}$ volts
(b) $E = \frac{j\,ZN}{60} , \frac{A}{P}$ volts
(c) $E = \frac{j\,ZP}{60}$ volts
(d) $E = \frac{j\,ZN}{120}$ volts

(I.T.I. February 2014, Electrician)

31. A pulley of DC generator is found to have completed 2000 revolutions within 1.25 minutes. The speed of the DC generator in r.p.m. will be :
(a) 1200 r.p.m (b) 1400 r.p.m
(c) 1600 r.p.m (d) 2000 r.p.m

32. The 10 pole D.C generator has 960 armature conductors and a flux per pole of 20 mwb running at 600 rpm. Calculate the e.m.f generated when the armature is connected in simplex wave winding?
(a) 960 volts (b) 720 volts
(c) 640 volts (d) 360 volts

(I.T.I. August 2014, Electrician)

33. The generators are designed to deliver full load current (I_L) and the fall of no-load voltage will be about ....
(a) 5 to 8% (b) 4 to 7 %
(c) 3 to 6% (d) 2 to 5%

(I.T.I. February 2015, Wireman)

34. Which type of D.C. generator is called as constant voltage generator ?
(a) D.C. series generator
(b) D.C. shunt generator
(c) D.C. Long shunt compound generator
(d) D.C. short shunt compound generator

(I.T.I. February 2015, Electrician)

35. Which among the following statements is **not** correct?
(a) The best method of stopping a DC shunt generator is to open the field circuit switch
(b) The existence of residual magnetism is must in a self-excited DC generator
(c) As the shunt field current increases, the induced voltage also increases initially
(d) As the speed of the DC generator increases, the induced voltage also increases up to a certain extent.

36. Which type of generator gives a constant voltage output at all loads?
(a) Level compound generator
(b) Commutative compound generator
(c) Differential compounded generator
(d) Series generator

37. The generator in which the shunt field is connected after the series field is called ....
(a) D.C series generator
(b) D.C shunt generator

(c) D.C short shunt compound generator
(d) D.C long shunt compound generator

(I.T.I. August 2015, Electrician)

38. One of the main reasons for NOT building up the voltage in D.C. generator is ....
(a) the generator runs at rated speed
(b) the field circuit resistance is more than field critical resistance
(c) the residual magnetism existing in the field cores
(d) brushes are in proper contacts

(I.T.I. August 2015, Electrician)

39. The rough commutation of the D.C. generator armature can be reduced by ....
(a) providing compensating winding in the main pole faces
(b) shifting brush position from GNA to MNA
(c) providing inter poles between the main poles
(d) using low resistance brushes

(I.T.I. August 2015, Electrician)

40. Which one of the following is the reason for loosing the residual magnetism of self excited DC generator ?
(a) The armature core is not properly laminated
(b) The generator is kept idle for a long time
(c) The armature resistance is more than 1 ohm
(d) Incorrect brush position

(I.T.I. February 2016, Electrician)

41. A typical magnetization characteristic is shown in the figure.
What 'Oa' in this curve is called ?

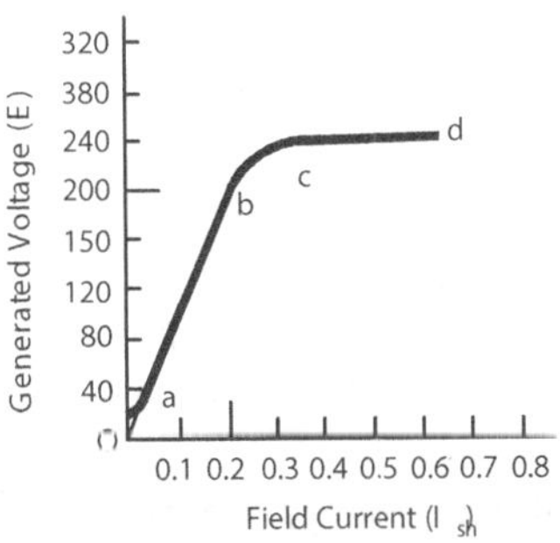

(a) Field critical resistance
(b) Knee
(c) Residual magnetism voltage
(d) Linear region

(I.T.I. February 2016, Electrician)

42. The insulating material used in between commutator segment is ....
(a) ebonite (b) mica
(c) empire (d) glass

(I.T.I. August 2015, Electrician)

43. The thickness of lamination of armature core discs is ....
(a) 0.3 mm (b) 0.4 mm
(c) 0.5 mm (d) 0.6 mm

(I.T.I. February 2016, Wireman)

44. As per BIS the shunt field winding terminals of separately excited type DC generator are marked as ....
(a) D_1 and D_2 (b) B_1 and B_2
(c) F_1 and F_2 (d) E_1 and E_2

(I.T.I. February 2016, Electrician)

45. A DC shunt generator delivers 15 amps to a load at 220 V terminal voltage. The shunt field resistance is 2200 ohms. The armature current will be ....
(a) 16.0 amps (b) 15.1 amps
(c) 1.0 amps (d) 0.1 amp

(I.T.I. August 2015, Electrician)

46. A long shunt compound generator delivers a load current of 100 A at 400 V, the shunt fiold resistance is 200 ohm. Calculate the armature current?
(a) 100 A (b) 102 A
(c) 4 A (d) 2 A

(I.T.I. February 2016, Electrician)

47. Body or yoke of a D.C. generator is made of ...
(a) Cast iron(b) Cast steel
(c) Rolled steel(d) All of these

(I.T.I. February 2017, Wireman)

48. In DC machine, yoke offers–
(a) Mechanical protection to the machine
(b) Flux path completion
(c) Produce working flux
(d) Mechanical protection to the machinc or flux path completion

(I.T.I. February 2019, Electrician)

49. The field coils of D.C. machine are made up of ....
(a) Carbon (b) Copper
(c) Steel (d) Mica
(I.T.I. February 2019, Wireman)

50. In a D.C. machine, brushes are placed ......
(a) Along geometrical neutral axis
(b) Along magnetic neutral axis
(c) Between two adjacent poles
(d) All of these
(I.T.I. February 2017, Wireman)

51. In a D.C. machine, the number of brushes in lap winding is always:
(a) Double the number of poles
(b) Same as the number of poles
(c) Half the number of poles
(d) Two
(I.T.I. August 2016, Electrician)

52. A D.C. generator converts A.C. voltages into D.C. voltage using ....
(a) Commutators (b) Carbon brushes
(c) Shunt coils (d) Series coils
(I.T.I. February 2017, Wireman)

53. The polarity of induced e.m.f. in a generator can be reversed by reversing the direction of .....
(a) Rotation (b) Field flux
(c) Both (a) and (b) (d) None of these
(I.T.I. February 2019, Wireman)

54. In Fleming's Right Hand Rule for a D.C. generator conventionally fingers have direction as ....
(a) Thumb-motion, index-field, middle-current
(b) Thumb-motion, index-current, middle-field
(c) Thumb-current, index-field, middle-motion
(d) Thumb-field, index-motion, middle-current
(I.T.I. February 2017, Wireman)

55. Which of the following generators are used in arc welding?
(a) Shunt generators
(b) Series generators
(c) Cumulative compound generators
(d) Differential compound generators
(I.T.I. August 2016, Electrician)

56. Which of the following windings are necessary in case of all types of D.C. machines?
(a) Closed type (b) Lap
(c) Wave (d) Open type
(I.T.I. February 2017, Electrician)

57. In wave winding the parallel paths 'A' is given by ....
(a) A = P (b) A = 2
(c) A = Z (d) A = 4
(I.T.I. August 2016, Electrician)

58. The number of parallel paths in wave wound armature are –
(a) Equal to the number of poles of the machines
(b) Equal to two irrespective of the number of poles
(c) Equal to number of commutator segment
(d) Equal to the number of armature conductors
(I.T.I. August 2016, Electrician)

59. Which of the following reflects the shunt field resistance of a D.C. compound generator?
(a) 1.0 Ω (b) 100 Ω
(c) 1.5 Ω (d) 1 kilo ohms
(I.T.I. February 2017, Electrician)

60. In a DC generator the initial flux is produced due to ......
(a) Saturation of the core
(b) Residual magnetism
(c) Eddy currents
(d) Hysteresis
(I.T.I. February 2017, Electrician)

61. The effect of armature reaction is ....
(a) Reduction in EMF
(b) Increase in iron losses
(c) Sparking at commutator
(d) Reduction in EMF, Increase in iron losses
(I.T.I. February 2019, Electrician)

62. The rotational losses in D.C. machines is equal to the .....
(a) Kinetic energy of armature
(b) Rate of change of kinetic energy

(c) Square of the kinetic energy of armature
(d) Half of the kinetic energy of armature

(I.T.I. February 2019, Electrician)

63. Cross magnetising component of armature reaction is...
(a) At right angles to the main field m.m.f.
(b) Changing direction of main field fluxes
(c) At right angles to the demagnetising component
(d) All of these

(I.T.I. February 2017, Wireman)

64. Demagnetizing component of armature reaction is ...
(a) In phase with opposition of main field m.m.f.
(b) In phase with main field m.m.f.
(c) Perpendicular to main field m.m.f.
(d) In phase with cross reaction component

(I.T.I. February 2017, Wireman)

65. Which energy is converted into electrical energy by generator?
(a) Heat (b) Mechanical
(c) Chemical (d) Kinetic

(I.T.I. August 2019 Electrician)

66. Which material is used to make brush in generator?
(a) Steel and graphite
(b) Aluminium and graphite
(c) Cast iron and graphite
(d) Carbon and graphite

(I.T.I. August 2019 Electrician)

67. Which metal is used to make pole core of large D.C. generator machines?
(a) Soft iron (b) Cast iron
(c) Cast steel (d) Stainless steel

(I.T.I. August 2019 Electrician)

68. What is the purpose of pole shoe in D.C. generator?
(a) Reduce the air gap
(b) Increase the field strength
(c) Spread out flux uniformly in the air gap
(d) Minimize the magnetic losses

(I.T.I. August 2019 Electrician)

69. What is the function of split rings in D.C. generator?
(a) Maintains constant voltage
(b) Collects the current unidirectionally
(c) Reduces the voltage drop at brushes
(d) Increases the terminal voltage than rated

(I.T.I. August 2019 Electrician)

70. How does the magnetic circuit complete through the yoke and poles in a generator?
(a) Field coils
(b) Laminated pole core
(c) Armature core
(d) Winding conductors in armature

(I.T.I. August 2019 Electrician)

71. Which rule is used to find direction of dynamically induced e.m.f.?
(a) Cork screw rule
(b) Right hand palm rule
(c) Fleming's left hand rule
(d) Fleming's right hand rule

(I.T.I. August 2019 Electrician)

72. What is the name of the part of D.C. machine?

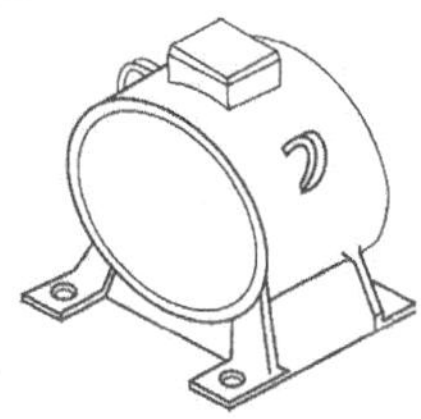

(a) Stator (b) Pole core
(c) Pole shoes (d) Yoke or frame

(I.T.I. August 2019 Electrician)

73. Name the part of D.C. generator?

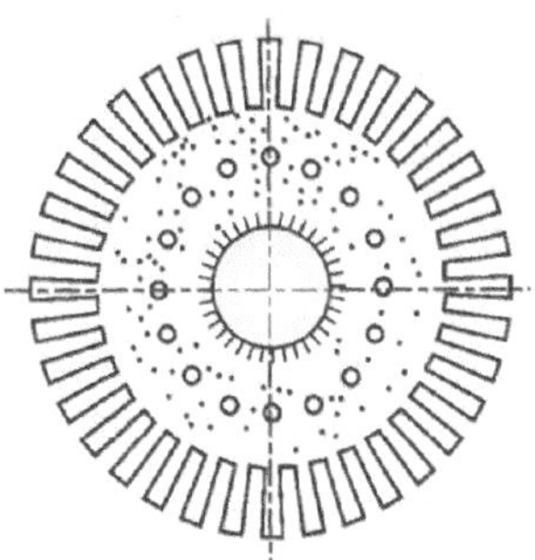

(a) Side end plates
(b) Pole shoe lamination
(c) Commutator segment
(d) Armature core lamination

(I.T.I. August 2019 Electrician)

74. Why D.C. generators are loosing their residual magnetism?
(a) Heavy short circuit in load
(b) Running without load continuously
(c) Change of direction of rotation very often
(d) Continuous running without break
(I.T.I. August 2019 Electrician)

75. What is the purpose of compensating in D.C. generator?
(a) Minimises rough commutation
(b) Maintain constant output voltage
(c) Neutralises the demagnetising effect
(d) Decreases the excitation current of field coils
(I.T.I. August 2019 Electrician)

76. Why the armature cores are laminated in D.C. generator?
(a) Reduce the friction loss
(b) Reduce the windage loss
(c) Reduce the hysteresis loss
(d) Reduce the eddy current loss
(I.T.I. August 2019 Electrician)

77. Why commutators are sparking heavily?
(a) Incorrect direction of rotation
(b) Incorrect field connection
(c) Incorrect brush position
(d) Incorrect armature connection
(I.T.I. August 2019 Electrician)

78. Why the terminal voltage decreases if load increases in D.C. shunt generator?
(a) Because of brush voltage drop decreases
(b) Due to increase in armature resistance
(c) Because of armature reaction effect
(d) Due to increase in shunt field inductance
(I.T.I. August 2019 Electrician)

79. Which method is used to improve the insulation resistance in D.C. generator?
(a) Blowing hot air into the machine during maintenance
(b) Heating the machine by running periodically
(c) Cleaning the commutator segments regularly
(d) Replacing the brushes frequently
(I.T.I. August 2019 Electrician)

80. Which type of D.C. generator is used for electroplating process?
(a) Shunt generator
(b) Series generator
(c) Differential compound generator
(d) Cumulative compound generator
(I.T.I. August 2019 Electrician)

81. Which type of D.C. generator is used for long distance distribution lines?
(a) Shunt generator
(b) Cumulative compound generator
(c) Differential compound generator
(d) Series generator
(I.T.I. August 2019 Electrician)

KEY TO MULTIPLE CHOICE QUESTIONS

1. (c)	**2.** (c)	**3.** (b)	**4.** (d)	**5.** (c)	**6.** (d)	**7.** (a)	**8.** (a)	**9.** (d)	**10.** (a)
11. (d)	**12.** (b)	**13.** (a)	**14.** (b)	**15.** (d)	**16.** (d)	**17.** (d)	**18.** (d)	**19.** (d)	**20.** (d)
21. (c)	**22.** (a)	**23.** (d)	**24.** (c)	**25.** (c)	**26.** (b)	**27.** (c)	**28.** (a)	**29.** (d)	**30.** (a)
31. (c)	**32.** (a)	**33.** (a)	**34.** (b)	**35.** (a)	**36.** (a)	**37.** (d)	**38.** (b)	**39.** (c)	**40.** (b)
41. (c)	**42.** (b)	**43.** (c)	**44.** (c)	**45.** (b)	**46.** (b)	**47.** (d)	**48.** (d)	**49.** (b)	**50.** (b)
51. (b)	**52.** (a)	**53.** (a)	**54.** (a)	**55.** (d)	**56.** (a)	**57.** (b)	**58.** (b)	**59.** (b)	**60.** (b)
61. (a)	**62.** (b)	**63.** (d)	**64.** (a)	**65.** (b)	**66.** (d)	**67.** (c)	**68.** (c)	**69.** (b)	**70.** (c)
71. (d)	**72.** (d)	**73.** (d)	**74.** (a)	**75.** (c)	**76.** (d)	**77.** (c)	**78.** (c)	**79.** (a)	**80.** (a)
81. (b)									

Module 2

D.C. Motor

Multiple Choice Questions

1. A force will act on a conductor placed in a magnetic field when :
   (a) a current is flowing in the conductor
   (b) the conductor is rotating
   (c) the conductor is placed in the magnetic field and it is not carrying a current
   (d) it is placed at right angle to the field
2. The magnitude of E_b in a d.c. motor in the start is equal to:
   (a) zero (b) applied e.m.f.
   (c) more than applied e.m.f.
   (d) maximum
3. If in a shunt motor the applied e.m.f. is kept constant then on increasing the load, the shunt-field strength will become ______ its normal value.
   (a) less than (b) more than
   (c) equal to (d) zero
4. In a series motor, the magnetic field-flux stands:
   (a) proportional to armature current
   (b) equal to armature current
   (c) inversely proportional to armature current
   (d) constant
5. A series motor is wound with :
   (a) fine wire winding
   (b) nichrome wire winding
   (c) thick wire winding
   (d) double wire winding
6. In a shunt motor, a starter is used for :
   (a) speed control
   (b) back e.m.f. control
   (c) applied e.m.f. control
   (d) current control in the start
7. The rotating part of a d.c. motor is known as :
   (a) pole (b) stator
   (c) rotor (d) armature
8. Which of the following methods is most economical for finding the no-load losses of a long shunt d.c. motor ?
   (a) Hopkinson's test
   (b) Swinburne's test
   (c) Retardation test
   (d) None of these
9. The rated speed of a given shunt motor is 1050 r.p.m. To run this motor at 1200 r.p.m., which of the following speed control scheme will be suitable ?
   (a) Armature current resistance control
   (b) Field resistance control
   (c) Ward-Leonard control
   (d) None of these
10. Which type of test is being carried out in the given circuit of a d.c. motor ?

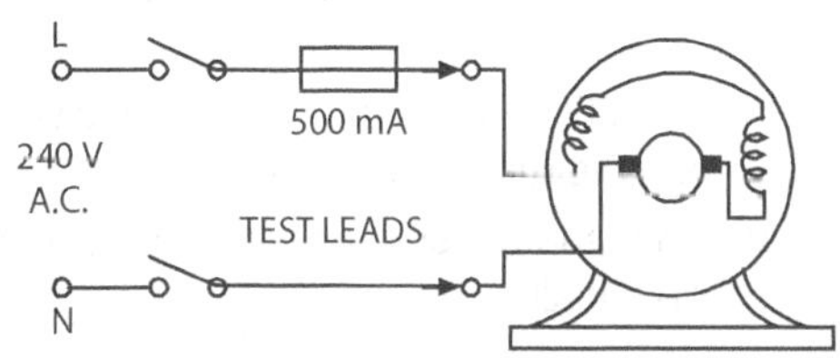

(a) Open-circuit test in d.c. shunt motor
(b) Test for armature short-circuit in d.c. series motor
(c) Open-circuit test in d.c. series motor
(d) Test for earthed field winding in compound motor.

11. During the normal operation of a DC series motor, if the field winding is suddenly open-circuited, what will be the effect on the motor ?
(a) It will run at slow speed
(b) It will run at high speed
(c) It will run at normal speed
(d) It will not run

12. The carbon brushes of DC machines become shorter by wearing out during the running time of the machine. At what length do you have to change the brushes?
(a) 1/2 of the original length
(b) 1/3 of the original length
(c) 3/4 of the original length
(d) 2/3 of the original length

(I.T.I. July 2010, Electrician)

13. How the HP, torque and speed are related together in a DC motor, if N is in 1/min (rpm) for DC motors?
(a) $HP = 2\pi NT$ (b) $HP = \frac{2pNT}{60}$
(c) $HP = \frac{2pNT}{60 \text{ ´ } 735}$ (d) $HP = \frac{2pNT}{735}$

(I.T.I. July 2012, Electrician)

14. The supply terminals of a D.C. shunt motor are reversed. What will be the effect on the motor ?
(a) It will run as a D.C. generator
(b) It will run in the reverse direction
(c) It will run in the same direction
(d) It will burn out

(I.T.I. February 2015, Electrician, Wiremen)

15. What will be the consequences if the field circuit of a D.C. shunt motor is opened when the motor is running at no load condition?
(a) The motor speed will be reduced drastically
(b) The torque of the motor will be increased
(c) The motor speed could become dangerously high
(d) Field current will become zero and hence motor stops.

(I.T.I. August 2014, Electrician)

16. Which type of D.C. motor is used for elevators?
(a) D.C. shunt motor
(b) D.C. series motor
(c) D.C. cumulative compound motor
(d) D.C. differential compound motor

(I.T.I. August 2015, Electrician)

17. Which type of DC motor is used in hoists and cranes?
(a) D.C. series motor
(b) D.C. shunt motor
(c) D.C. cumulative compound motor
(d) D.C. differential compound motor

(I.T.I February 2016, Electrician)

18. A customer requires a motor with the speed torque characteristic shown in the figure.

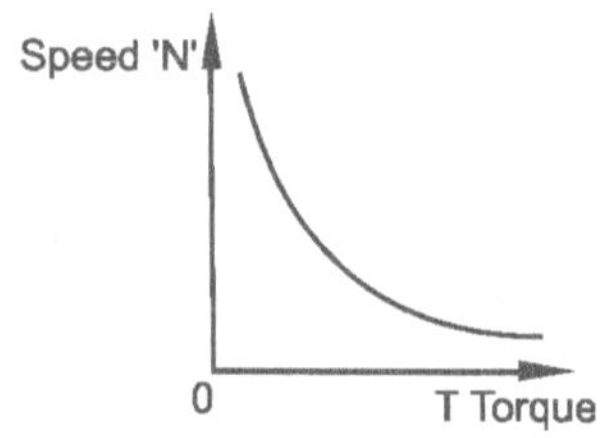

Which type of DC motor do you choose?
(a) DC series motor
(b) Long shunt compound motor
(c) Short shunt compound motor
(d) DC shunt motor

(I.T.I. August 2015, Electrician)

19. The series field diverter speed control method is mainly used in ....
(a) fan motors
(b) electric trains
(c) food mixers
(d) printing machines

(I.T.I. August 2015, Electrician)

20. Which method of speed control is mainly applied in electric trains?
(a) Series field tapping method
(b) Series field diverter method
(c) Series parallel method
(d) Supply voltage control method
(I.T.I. February 2016, Electrician)

21. The D.C. motor works on the principle that the ....
(a) current carrying conductor placed in a uniform magnetic field experiences a force on it.
(b) conductor moves when kept in a uniform magnetic field
(c) magnetic field set up by varying current which produces force in the conductor
(d) combined magnetic field set up by two current carrying conductors produces a force between them.
(I.T.I. February 2018, Electrician)

22. By which law, the direction of the induced e.m.f. will be identified?
(a) Faraday's first law of electromagnetic induction
(b) Faraday's second law of electromagnotic induction
(c) Lenz's law
(d) Fleming's left hand rule
(I.T.I. August 2014, Electrician)

23. The formula to calculate the back emf (E_b) of DC shunt motor is ....
(a) $E_b = V + I_{sh} R_{sh}$
(b) $E_b = V + I_a R_a$
(c) $E_b = V - I_a R_a$
(d) $E_b = V - I_{sh} R_{sh}$
(I.T.I. February 2016, Electrician)

24. The formula to find the armature current (I_a) in DC shunt motor is ....
(a) $I_a = \frac{V - E_b}{R_a}$ amp (b) $I_a = \frac{V + E_b}{R_a}$ amp
(c) $I_a = \frac{V}{R_a}$ amp (d) $I_a = \frac{E_b}{R_a}$ amp
(I.T.I. August 2015, Electrician)

25. The speed (N) of a DC motor is expressed as ....
(a) $N \propto E_b$ (b) $N \propto \frac{E_b}{f}$
(c) $N \propto \phi$ (d) $N \propto E_b \phi$
(I.T.I. February 2016, Wireman)

26. The turning (or) twisting moment of force about an axis is called as ....
(a) speed (b) momentum
(c) torque (d) velocity
(I.T.I. August 2015, Electrician)

27. To calculate the efficiency of motor is ....
(a) $\frac{\text{output}}{\text{input}}$ ´ 100
(b) output × input
(c) output + input/100
(d) output – input/100
(I.T.I. February 2016, Wireman)

28. The starting resistance of 3 point starter in D.C. motor is connected with ....
(a) parallel across the armature
(b) series with the armature
(c) parallel across the field winding
(d) series with the field winding
(I.T.I. February 2018, Electrician)

29. In a four point starter the protective resistor is connected in series with ....
(a) overload coil (b) no volt coil
(c) armature (d) shunt field
(I.T.I. February 2016, Wireman)

30. The starting current of a 220V, 10HP shunt motor having an armature resistance of 0.2 ohms, without starter is ....
(a) 11 amps (b) 110 amps
(c) 1100 amps (d) 11000 amps
(I.T.I. February 2019, Electrician)

31. A 220 V, D.C. shunt motor when runs at 500 rpm, the back e.m.f. (E_{bl}) is 210 volt and the torque is increased double time the back e.m.f. (E_{b2}) is 200 V, what is the new speed?
(a) 1000 rpm (b) 500 rpm
(c) 405 rpm (d) 476 rpm
(I.T.I. August 2015, Electrician)

32. The back e.m.f. E_b and supply voltage V_1 are connected in a D.C. motor as .....

(a) $E_b = V_1$ (b) $E_b > V_1$

(c) $E_b < V_1$ (d) Any one of these

(I.T.I. February 2017, Wireman)

33. The induced e.m.f. of D.C. machine is proporational to –

(a) Field flux only

(b) Speed of armature only

(c) Number of conductors

(d) All of these

(I.T.I. August 2016, Electrician)

34. Speed of motor is measured by –

(a) Ammeter (b) Voltmeter

(c) Megger (d) Tachometer

(I.T.I. February 2019, Electrician)

35. The direction of rotation of D.C. series motor can be reversed by inter changing connections to .....

(a) Supply (b) Field

(c) Armature (d) Either (b) or (c)

(I.T.I. February 2019, Wireman)

36. For a D.C. shunt motor:

(a) $I_A = I_L + I_F$ (b) $I_A = I_L - I_F$

(c) $I_F = I_L \times I_A$ (d) $I_L = I_F - I_A$

(I.T.I. February 2017, Wireman)

37. D.C. shunt motor is also called as ....

(a) Constant flux motor

(b) Variables voltage motor

(c) Constant voltage

(d) Constant current motor

(I.T.I. February 2018, Electrician)

38. D.C. shunt motors are commonly used in –

(a) Cranes (b) Electric traction

(c) Elevators (d) Lathe machines

(I.T.I. February 2018, Electrician)

39. The efficiency of D.C. machine is maximum when ....

(a) Iron losses are equal to mechanical losses

(b) Variable losses are equal to constant losses

(c) Field copper losses are equal to constant losses

(d) Stray losses are equal to copper losses

(I.T.I. August 2016, Electrician)

40. When the motor runs on no load, then –

(a) Back emf is almost equal to applied voltage

(b) Back emf will be greater than applied voltage

(c) Back emf will be less than applied voltage

(d) None of the above

(I.T.I. February 2019, Electrician)

41. In a constant power type load ......

(a) Torque is proportional to speed

(b) Torque is proportional to square of speed

(c) Torque is inversely proportional to speed

(d) Torque is independent of speed

(I.T.I. February 2018, Electrician)

42. A 4 Pole, 220 V, 10 H.P., D.C. shunt motor is rotating at a speed of 320 R.P.M. The developed useful torque will be .....

(a) 210. 647 newton metres

(b) 215. 547 newton metres

(c) 219.484 newton metres

(d) None of these

(I.T.I. February 2018, Electrician)

43. Armature resistance control method is also called as ....?

(a) Constant current control

(b) Constant power control

(c) Constant voltage control

(d) Constant torque control

(I.T.I. February 2019, Electrician)

44. The current drawn by the armature of D.C. motor is directly proportional to .....

(a) Speed of motor

(b) Torque required

(c) Voltage across the terminals
(d) Armature resistance

(I.T.I. February 2017, Electrician)

45. Which of the D.C. motor uses flywheel?
(a) Series
(b) Shunt
(c) Differentially compounded
(d) Cumulative compounded

(I.T.I. February 2017, Electrician)

46. D.C. series motor is best suited for electric traction as .....
(a) It develops high starting torque
(b) It is cheaper in cost
(c) It consumes less power
(d) None of these

(I.T.I. February 2019, Wireman)

47. In Ward Leonard method of speed control:
(a) 2 D.C. generator and 1 D.C. motor is used
(b) 1 D.C. generator and 2 D.C. motor is used
(c) 1 A.C. generators and 2 D.C. motors are used
(d) 2 A.C. generators and 1 D.C. motor is used

(I.T.I. February 2017, Electrician)

48. Ward-Lenard control is basically a ________ control method.
(a) Armature voltage
(b) Armature resistance
(c) Field
(d) Field diverter

(I.T.I. February 2019, Electrician)

49. If starter is not used with large D.C. motor for starting, it will draw a starting current which ....
(a) May damage the armature and commutator
(b) Is many times its rated full load current
(c) Will provide excessive voltage drop in line
(d) All of above

(I.T.I. August 2016, Electrician)

50. Direct-on-line (DOL) starters are not suitable for starting large DC motors, because –
(a) The motor may run away
(b) The starting torque becomes low
(c) The motor may not start
(d) The starting current will be enormously high

(I.T.I. August 2016, Electrician)

51. The starter used for D.C. shunt motor is ......
(a) Two point starter
(b) Three point starter
(c) Four point starter
(d) (b) and (c) of these

(I.T.I. August 2016, Wireman)

52. External connection points of a D.C. four points starter are :
(a) L, F, A (b) L, F_1, F_2, A
(c) L, F, A_1, A_2 (d) L +, L –, F, A

(I.T.I. February 2017, Wireman)

53. Following is TRUE about a four points starter:
(a) The coil of the no volt release is connected in series with the field circuit
(b) Coil of the no volt release does not carry the field current.
(c) By adjusting field rheostat, current through no volt release coil can be reduced
(d) All of these

(I.T.I. February 2017, Wireman)

54. Starting of a D.C. motor using 3 point starter and 4 point starter differs in connection of ......
(a) Overload release coil
(b) No volt release coil
(c) Armature supply
(d) Field supply

(I.T.I. February 2017, Wireman)

55. What Is the necessity of starter in D.C. motor?
(a) To improve the torque

(b) Reduce the field current
(c) To decrease the back e.m.f.
(d) Reduce the armature current

(I.T.I. August 2019 Electrician)

56. How NO VOLT coil is connected in a three point starter with D.C. shunt motor?
(a) Directly connected supply
(b) Connected in series with armature
(c) Connected in series with shunt field
(d) Connected in parallel with armature

(I.T.I. August 2019 Electrician)

57. Which material is used for starting resistance of D.C. starters?
(a) Eureka (b) Nichrome
(c) Manganin (d) Constantine

(I.T.I. August 2019 Electrician)

58. What is the name of the speed control method of DC motor?

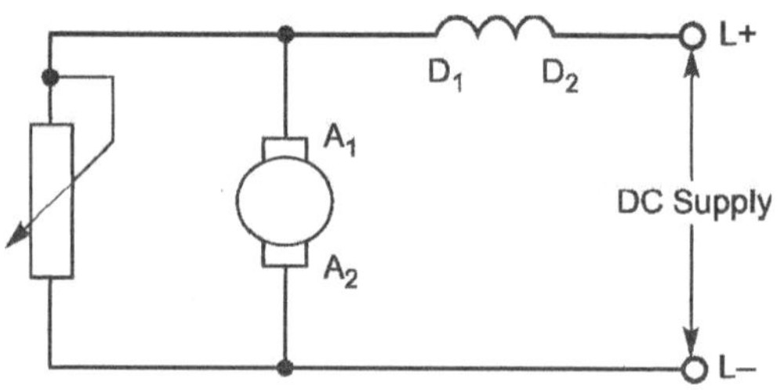

(a) Armature diverter method
(b) Field tapping method
(c) Voltage control method
(d) Field diverter method

(I.T.I. August 2019 Electrician)

59. What is the formula to calculate back e.m.f. in a D.C. motor?

(a) $E_b = \frac{XNP}{f.60A}$ (b) $E_b = \frac{NP}{Zf\ 60A}$

(c) $E_b = \frac{f\ ZNP}{60A}$ (d) $E_b = \frac{60Af}{ZNP}$

(I.T.I. August 2019 Electrician)

60. Which rule is applied to identify the direction of rotation of conductor in D.C. motor?
(a) Cork's screw rule
(b) Right hand grip rule
(c) Fleming's left hand rule
(d) Fleming's right hand rule

(I.T.I. August 2019 Electrician)

61. Which type of D.C. motor is used for sudden application of heavy loads?
(a) Shunt motor
(b) Series motor
(c) Differential compound motor
(d) Cumulative compound motor

(I.T.I. August 2019 Electrician)

62. Name the type of D.C. motor.

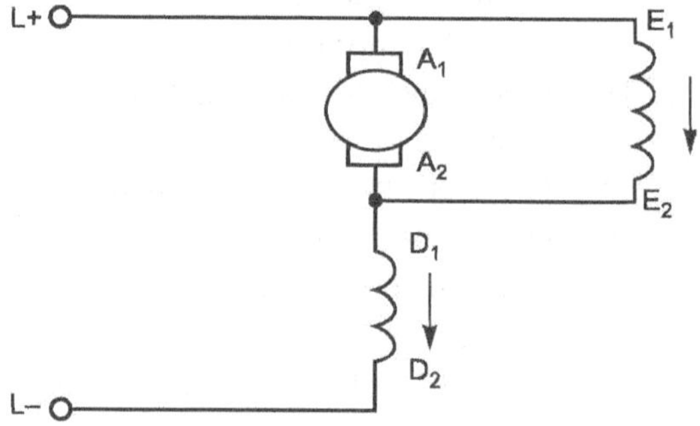

(a) Shunt motor (b) Series motor
(c) Long shunt compound motor
(d) Short shunt compound motor

(I.T.I. August 2019 Electrician)

63. Which D.C. compound motor is operated at constant speed under varying load?
(a) Cumulative long shunt
(b) Differential long shunt
(c) Differential short shunt
(d) Cumulative short shunt

(I.T.I. August 2019 Electrician)

64. Which speed control method is applied to obtain both below normal and above normal speed in D.C. motor?
(a) Field control method
(b) Armature control method
(c) Tapped field speed control
(d) Ward Leonard speed control

(I.T.I. August 2019 Electrician)

65. What is the best method to change the direction of rotation (D.O.R.) of a compound motor without change of its characteristics?
(a) Change armature current direction
(b) Change shunt field current direction
(c) Change series field current direction
(d) Change the current in armature and shunt field together

(I.T.I. August 2019 Electrician)

KEY TO MULTIPLE CHOICE QUESTIONS

1. (a)	**2.** (a)	**3.** (c)	**4.** (a)	**5.** (c)	**6.** (d)	**7.** (d)	**8.** (b)	**9.** (b)	**10.** (c)
11. (d)	**12.** (b)	**13.** (c)	**14.** (c)	**15.** (c)	**16.** (c)	**17.** (a)	**18.** (a)	**19.** (b)	**20.** (b)
21. (a)	**22.** (c)	**23.** (c)	**24.** (a)	**25.** (b)	**26.** (c)	**27.** (a)	**28.** (b)	**29.** (b)	**30.** (c)
31. (d)	**32.** (c)	**33.** (d)	**34.** (d)	**35.** (d)	**36.** (b)	**37.** (a)	**38.** (d)	**39.** (b)	**40.** (a)
41. (c)	**42.** (c)	**43.** (d)	**44.** (b)	**45.** (d)	**46.** (a)	**47.** (b)	**48.** (a)	**49.** (d)	**50.** (d)
51. (b)	**52.** (d)	**53.** (b)	**54.** (b)	**55.** (d)	**56.** (c)	**57.** (a)	**58.** (a)	**59.** (c)	**60.** (c)
61. (d)	**62.** (d)	**63.** (a)	**64.** (d)	**65.** (a)					

Module 3

Three Phase Motor and Windings

Multiple Choice Questions

Polyphase Induction Motors

1. The synchronous speed of a 2, 4 and 8 poles motor at 50 Hz can not be :

(a) 3000 R.P.M. (b) 1000 R.P.M.

(c) 1500 R.P.M. (d) 750 R.P.M.

2. The no. of poles of a 3-phase induction motor running at 500 R.P.M. will be : (supply frequency = 50 Hz)

(a) 12 (b) 76

(c) 4 (d) 2

3. The relation of the torque developed in an induction-motor and the supply voltage is given by the expression :

(a) $T \propto \sqrt{V}$ (b) $T \propto V$

(c) $T \propto V^2$ (d) $T \propto \frac{1}{V^2}$

4. The reason of drawing high current at the time of starting by an induction-motor is :

(a) high voltage (b) low power factor

(c) zero back e.m.f (d) none of these.

5. The principle of operation of an induction-motor is based on :

(a) Faraday's law

(b) Rotating magnetic field

(c) Fleming's right-hand rule

(d) Ohm's law.

6. The speed of rotor of an induction-motor is:

(a) the same as the synchronous speed

(b) higher than the synchronous speed

(c) lesser than the synchronous speed

(d) determined by the frequency and the number of poles.

7. To reverse the direction of rotation of a 3-phase induction-motor, it is necessary to:

(a) interchange all the three line connections

(b) reverse the polarity of the rotor's circuit

(c) increase the resistance of the rotor's circuit

(d) interchange any two out of the three line connections.

8. Squirrel-cage induction-motor differs from a slip-ring induction-motor as it has no :

(a) slip-rings (b) rotor windings

(c) stator windings (d) rotating part.

9. Squirrel-cage induction-motor has :

(a) zero starting torque

(b) very small starting torque

(c) very high starting torque

(d) highest starting torque amongst all types of induction motors.

10. It is advisable to avoid the use of a (d) O.L. starter for starting an induction-motor having 5 H.P. capacity because :

(a) the starting torque will be very low

(b) the motor will take five to seven times more starting current than its full-load current

(c) it will pick-up very high speed and may go out of step
(d) it will run in the reverse direction.

11. A thermal overload relay is provided in a starter to protect the motor against :
(a) short-circuit
(b) open-circuit
(c) excess current
(d) low voltage.

12. Total electrical degrees for a 4 pole (a) (c) machine is equal to :
(a) 180° (b) 360°
(c) 720° (d) 1080°

13. Fullloadlinecurrentofa3-phase,3H.P.,415V, 50 Hz, squirrel cage induction motor is approximately :
(a) 3 A (b) 4.5 A
(c) 6 A (d) 9 (a)

14. No-load current of an induction motor is approximately______ percent of full-load current.
(a) 10 (b) 20
(c) 40 (d) 75

15. The purpose of a back-up fuse is to provide protection against :
(a) overload (b) over voltage
(c) excess current (d) short-circuit.

16. A 4-pole, 3-phase, 50 Hz motor's speed is 95% of its synchronous speed then the actual speed of the motor is :
(a) 500 r.p.m. (b) 1300 r.p.m.
(c) 1400 r.p.m. (d) 1425 r.p.m.

17. Theinductionmotorofapumpisrunatasupply voltage 30% lesser than the normal voltage (415 V). The pump starts to work. After some time what possibly can happen ?
(a) The motor will become hot and will be damaged
(b) The motor will start after some time
(c) The motor will function normally
(d) None of the stated above will happen.

18. Regarding skewing of motor bars in a squirrel cage induction motor (SCIM), which statement is false?
(a) it prevents cogging
(b) it increases starting torque
(c) it produces more uniform torque
(d) it reduces motor "hum" duringt its operation

19. The effect of increasing the length of air gap in an induction motor will be to increase the
(a) power factor
(b) speed
(c) magnetising current
(d) air gap flux

20. The power factor of a squirrel cage induction motor is
(a) low at light loads only
(b) low at heavy loads only
(c) low at light and heavy loads both
(d) low at rated load only

21. The efficiency of a 3-phase induction motor is approximately proportional to
(a) $(1 - S)$ (b) S
(c) N (d) N_S

22. If stator voltage of a SCIM is reduced to 50 percent of its rated value, torque developed is reduced by ______ percent of its full load value.
(a) 50 (b) 25
(c) 75 (d) 57.7

23. The syncthronous speed of an induction motor refers to the speed of the ....
(a) rotor (b) armature
(c) rotating magnetic field
(d) value marked on the name plate

(I.T.I. February 2015, Electrician)

24. Calculate slip of an induction motor if it has 4 poles, 50Hz, rotor speed is 1400 rpm.
(a) 6.6% (b) 6.0%
(c) 5.5% (d) 5.0%

(I.T.I. February 2015, Wireman)

25. The motor suitable for loads where constant speed is required is ....
(a) induction motor
(b) synchronous motor
(c) capacitor motor
(d) universal motor

(I.T.I. February 2015, Electrician)

26. The magnetic field produced in the stator of a three phase induction motor travels at ....

(a) rotating speed.
(b) asynchronous speed
(c) synchronous speed
(d) slip speed

(I.T.I. February 2015, Electrician)

27. What is the possible cause, if the starter with single phasing preventer trips frequently during running of 3 phase induction motor?

(a) Improper function of relay contact
(b) Improper phase sequence
(c) Abnormal fluctuation in line voltage
(d) open in holding circuit

(I.T.I. February 2016, Electrician)

28. The formula to calculate the synchronous speed (N_S) of the 3 phase induction motor is ....

(a) $N_S = \frac{120f}{p}$ rpm (b) $N_S = \frac{60f}{p}$ rpm

(c) $N_S = \frac{pf}{60}$ rpm (d) $N_S = \frac{120}{pf}$ rpm

(I.T.I. February 2016, Electrician)

29. Calculate the synchronous speed N_S of 5 H.P., 6-Pole, 415 V, 50 Hz, 3 phase squirrel cage induction motor?

(a) 960 R.P.M (b) 1000 R.P.M
(c) 1400 R.P.M (d) 1500 R.P.M

(I.T.I. August 2015, Electrician)

30. A 3 phase induction motor is wound for 4 poles is supplied at 50 Hz frequency and running at the speed of 1440 rpm.

Calculate the percentage slip.

(a) 6% (b) 5%
(c) 4% (d) 3%

(I.T.I. February 2016, Electrician)

31. Calculate the number of poles of 3 phase induction motor running at 750 rpm, with 50 Hz frequency?

(a) 2 (b) 4
(c) 6 (d) 8

(I.T.I. February 2016, Electrician)

32. Formula of rotor frequency (f_r) is ....

(a) fractional slip × stator frequency
(b) $\frac{\text{fractional slip}}{\text{stator frequency}}$
(c) $\frac{\text{stator frequency}}{\text{fractional slip}}$
(d) stator frequency × percentage slip

(I.T.I. February 2016, Wireman)

33. The torque of the induction motor is directly proportional to ....

(a) rotor speed, fractional slip and rotor frequency
(b) stator flux, rotor current and rotor power factor
(c) stator current, synchronous speed and rotor power factor
(d) rotor frequency, stator flux and fractional slip

(I.T.I. February 2016, Electrician)

34. The most common type of A.C. motor is the .....

(a) Single phase induction motor
(b) Two phase induction motor
(c) Three phase induction motor
(d) Two phase squirrel cage motor

(I.T.I. February 2018, Electrician)

35. Most commonly used motor types in today's industry is ....

(a) D.C. motor
(b) Induction motor
(c) Synchronous motor
(d) All of these

(I.T.I. February 2017, Wireman)

36. The frequency of the induction e.m.f. in an induction motor at start is ....

(a) Greater than the supply frequency
(b) Lesser than the supply frequency
(c) Same as the supply frequency
(d) None of these

(I.T.I. February 2019, Electrician)

37. Induction motor runs at .....

(a) Synchronous speed
(b) Below synchronous speed

(c) Above synchronous speed
(d) None of these

(I.T.I. February 2016, Wireman)

38. The sequence of induction motor is RYB, then the direction of the induction motor can be changed by the following sequence:
(a) BYR (b) RBY
(c) BRY (d) None of these

(I.T.I. February 2017, Electrician)

39. The frame of induction motor and generator is usually made of :
(a) Silicon steel (b) Cast iron
(c) Aluminium (d) Mild steel

(I.T.I. February 2018, Electrician)

40. The shaft of an induction motor is made from ....
(a) High speed steel (b) Stainless steel
(c) Carbon steel (d) Cast iron

(I.T.I. February 2018, Electrician)

41. Slip rings are usually made of –
(a) Copper (b) Carbon
(c) Phosphor bronze (d) Aluminium

(I.T.I. February 2018, Electrician)

42. Slip of induction motor increases with –
(a) Increase in current and decrease in torque
(b) Increase in current and torque
(c) Decrease in current and torque
(d) Decrease in current and increase in torque

(I.T.I. February 2019, Electrician)

43. Which formula is used to calculate percentage slip of an AC 3 phase induction motor?
(a) $\frac{N_r - N_S}{N_S} \times 100$ (b) $\frac{N_S - N_r}{N_S} \times 100$
(c) $\frac{N_S - N_r}{N_r} \times 100$ (d) $\frac{N_r - N_S}{N_r} \times 100$

(I.T.I. August 2019, Electrician)

44. At standstill the value of slip of induction motor is ...
(a) 0 (b) 1
(c) Finite value (d) Infinite

(I.T.I. February 2017, Electrician)

45. In induction motors, parcentage slip is found in the range of ....
(a) 100% (b) 0%
(c) 5% (d) 40%

(I.T.I. August 2016, Electrician)

46. A 4 pole 50 Hz induction motor is running at 1470 R.P.M. What is the slip value?
(a) 0.2 (b) 0.004
(c) 0.02 (d) 0.4

(I.T.I. February 2017, Electrician)

47. 'N_S' is the synchronous speed and 'S' the slip, then actual running speed of an induction motor will be ...
(a) N . S (b) S . N
(c) $(1 - S)N_S$ (d) $(N_S - 1)S$

(I.T.I. August 2016, Electrician)

48. Find the number of poles when the frequency is 50 Hz and speed of the motor is 500 RPM?
(a) 5 (b) 10
(c) 12 (d) 24

(I.T.I. February 2017, Electrician)

49. 5 amperes, 3 phase, 440 V, 50 Hz induction motor has 4% slip. The freqnency of rotor e.m.f. will be....
(a) 200 Hz (b) 50 Hz
(c) 2 Hz (d) 0.2 Hz

(I.T.I. August 2016, Electrician)

Hint:

Frequency of Rotor e.m.f.
= slip × stator frequency
= 4% × 50 = 2Hz

50. Starting of induction mtoros "n" times full load current, where "*n*" ranges from .....
(a) 2 – 3 (b) 3 – 5
(c) 5 – 7 (d) 7 – 9

(I.T.I. February 2017, Wireman)

51. I.E. rule prohibits D.O.L. starting of 3-phase induction motor above ....
(a) 0.5 HP (b) 1 HP
(c) 3 HP (d) 5 HP

(I.T.I. February 2017, Wireman)

52. The shape of the torque/slip curve of induction motor is:

(a) Parabola
(b) Hyperbola
(c) Rectangular parabola
(d) Straight line

(I.T.I. August 2016, Electrician)

53. When maximum torque is obtained, the rotor power factor is ......

(a) Unity (b) Zero
(c) 0.707 lag (d) 0.5 lag

(I.T.I. February 2017, Electrician)

54. The condition for maximum running torque is .....

(a) R = X (b) R = S . X
(c) $R = \frac{X}{2}$ (d) $R = S\frac{X}{2}$

(I.T.I. February 2017, Electrician)

55. A three phase slip ring induction motor can be started using ......

(a) Direct on line starting
(b) Rotor resistance starting
(c) Auto transformer starting
(d) Star delta starting

(I.T.I. February 2017, Wireman)

56. A three phase 7.2 kW, 440 V, 50 Hz starter uses ....

(a) D.O.L starter
(b) Star delta starter
(c) Auto transformer starter
(d) Oil immersed starter

(I.T.I. August 2016, Wireman)

57. In star delta starting of three phase induction motor, the starting current of star connected stator reduces to ....

(a) One half of delta connected stator
(b) One third of the delta connected stator
(c) One seventh of the delta connected stator
(d) One eleventh of the delta connected stator

(I.T.I. February 2017, Wireman)

58. What is the function of timer in automatic star delta starter?

(a) Trip at overload
(b) Switch ON at present time
(c) Change from star to delta
(d) Switch OFF at preset time

(I.T.I. August 2019 Electrician)

59. What is the phase displacement between windings in three phase motor?

(a) 90° (b) 360°
(c) 180° (d) 120°

(I.T.I. August 2019 Electrician)

60. Which instrument is used to measure insulation resistance of a three phase induction motor?

(a) Multimeter
(b) Megger
(c) Shunt type Ohm meter
(d) Series type Ohm meter

(I.T.I. August 2019 Electrician)

61. Which loss is determined by blocked rotor test of three phase induction motor?

(a) Iron loss (b) Windage loss
(c) Friction loss (d) Copper loss

(I.T.I. August 2019 Electrician)

62. By which method of speed control two variable speeds only obtained in three phase motor?

(a) By changing the number of stator poles
(b) By changing applied frequency
(c) By changing the applied voltage
(d) By rotor rheostat control

(I.T.I. August 2019 Electrician)

63. What is the effect on three phase induction motor if one phase is cut off during running with load?

(a) Motor stops at once
(b) Motor will run at slow speed, but winding will burn out shortly
(c) Motor runs with humming noise with slow speed
(d) Motor will run normally

(I.T.I. August 2019 Electrician)

64. What is the defect in A.C. three phase induction motor that runs at low speed if loaded?

(a) Wrong motor connection

(b) Partially shorted in stator winding

(c) Open circuit in rotor winding

(d) Wrong starter connection

(I.T.I. August 2019 Electrician)

65. Why slip ring induction motor is fitted with wound rotor?

(a) To get high starting and running torque

(b) To control the speed

(c) To reduce the losses

(d) To reduce the slip

(I.T.I. August 2019 Electrician)

66. Why the rotor bars are mounted in a slightly skewed position in three phase motor?

(a) Generate maximum flux

(b) Reduce the stray losses

(c) Produce more uniform rotor field and torque

(d) Maintain the rotor speed constant

(I.T.I. August 2019 Electrician)

67. For which operation the following control circuit is used?

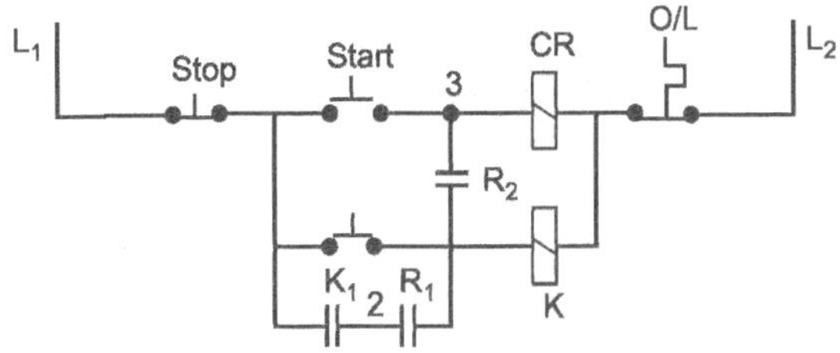

(a) Remote control

(b) Jogging/Inching

(c) Sequential control

(d) Forward and reverse

(I.T.I. August 2019 Electrician)

Winding

1. The 'maximum safe temperature' for class 'F' insulating materials is :

(a) 105°C (b) 120°C

(c) 130°C (d) 155°C

2. The type of slot used in small D.C. armature is :

(a) open type with tapered slot

(b) open type with dove tail

(c) semi-enclosed type

(d) closed type.

3. In a 8-pole duplex-lap winding, the total number of parallel paths will be

(a) 1 (b) 4

(c) 8 (d) 16.

4. Three types of slots are provided in armature cores which are :

(a) open, closed, semi-open

(b) open, closed, semi-enclosed

(c) round, sequare, triangular

(d) none of these.

5. Varnish coating is done on winding in order to:

(a) increase the insulation of the windings

(b) make the windings vibrationless

(c) provide an additional insulation layer on the windings

(d) all the three stated above.

6. Voltage drop test can be adopted :

(a) for short-circuit test only

(b) for open-circuit test only

(c) for 'earth' fault test only

(d) for all the three above stated tests.

7. In the open coil winding, the coil leads are kept:

(a) open-circuited

(b) short-circuited

(c) independent

(d) nothing of these.

8. The portion of a coil which remains out of the stator core's slot is called :

(a) active side (b) inactive side

(c) coil lead (d) unuseful side.

9. The three windings in a 3-phase induction-motor should be installed at a distance of :

(a) 90 electrical degrees

(b) 120 electrical degrees

(c) 360 electrical degrees

(d) 60 electrical degrees.

10. If the number of poles is 4 and the number of slots is 24 then the pole-pitch will be :

(a) 8 slots (b) 6 slots

(c) 4 slots (d) 2 slots

Hint : Pole-pitch = $\frac{\text{total number of slots}}{\text{number of poles}}$ ù ú û

11. In a single layer distributed winding if the number of coils is 24 then the number of slots will be :

(a) 8 (b) 16

(c) 24 (d) 48

12. To calculate the slot-distance in degrees for a stator

(a) $\frac{\text{total electrical degrees}}{\text{no. of slots}}$

(b) $\frac{\text{no. of slots}}{\text{total electrical degrees}}$

(c) $\frac{\text{no. of slots}}{90° \times \text{no. of slots}}$

(d) $\frac{90° \times \text{no. of poles}}{\text{no. of slots}}$

13. In distributed winding, all the coils will have the same :

(a) pitch (b) width

(c) length (d) thickness

14. The pitch factor K_p in alternator is expressed as $K_p = \cos\frac{a}{2}$ where angle "α" represents ....

(a) $\alpha = \frac{\text{number of stator slots}}{\text{number of poles}}$

(b) $\alpha = \frac{\text{difference in pitch}}{\text{full pitch}}$

(c) $\alpha = \frac{180°}{\text{number of slots per pole}}$

(d) $\alpha = \frac{180°}{\text{number of poles}}$

(I.T.I August 2014, Electrician)

15. Pole pitch is calculated by ....

(a) $\frac{\text{No. of poles}}{\text{Total No. of slots}}$

(b) $\frac{\text{Total No. of slots}}{\text{No. of poles}}$

(c) $\frac{\text{Total No. of coil}}{\text{No. of poles}}$

(d) $\frac{\text{Total No. of slots}}{\text{Total No. of coil}}$

(I.T.I February 2015, Electrician)

16. Total No. of coils per phase per pole is ....

(a) $\frac{\text{Total no. of coils}}{\text{No. of phases}}$

(b) $\frac{\text{Total no. of coils}}{\text{No. of phase} \times \text{No. of poles}}$

(c) $\frac{\text{Total no. of phases}}{\text{Total no. of coils}}$

(d) $\frac{\text{No of phases} \times \text{no. of poles}}{\text{Total no. of coils}}$

(I.T.I February 2015, Electrician)

17. Laying the coils clock wise, the connections to the commutator segments will proceed anti-clock wise, then this type of winding is called as ....

(a) lap winding.

(b) wave winding.

(c) progressive lap or wave winding.

(d) retrogressive lap or wave winding.

(I.T.I February 2015, Electrician)

18. Phase displacement in terms of slots in a stator of 3-phase is ....

(a) $\frac{120°}{\text{slot distance in degrees}}$

(b) $\frac{360°}{\text{slot distance in degrees}}$

(c) $\frac{\text{slot distance in degrees}}{360°}$

(d) $\frac{\text{slot distance in degrees}}{120°}$

(I.T.I February 2015, Electrician)

19. What are the two methods by which polarity of field (poles) of a winding can be checked?

(a) By internal glower and external glower
(b) Magnetic compass and search coil.
(c) Magnetic compass and internal glower
(d) Search coil and external glower

(I.T.I July 2009, Electrician)

20. Which type of armature winding is shown in the figure?

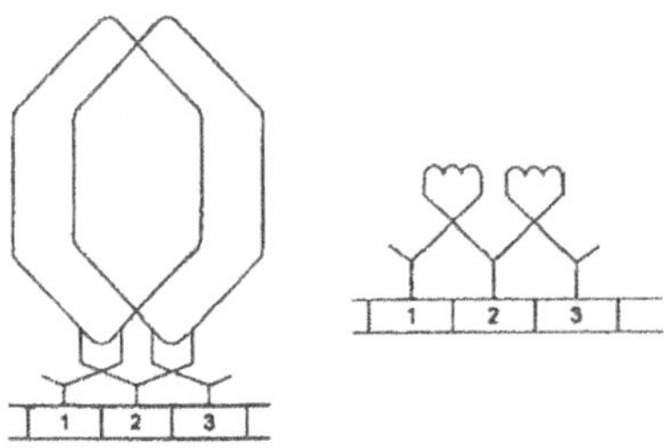

(a) Simplex lap winding
(b) Simplex wave winding
(c) Duplex progressive lap winding
(d) Simplex retrogressive wave winding

(I.T.I. February 2016, Electrician)

21. Which type of armature winding is shown in the figure?

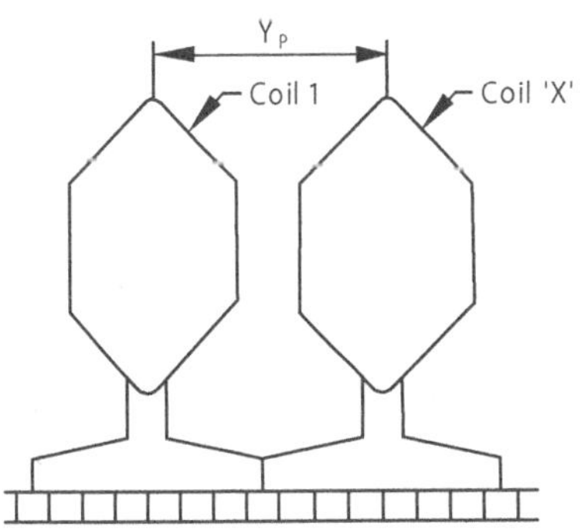

(a) Simplex wave winding
(b) Duplex wave winding
(c) Simplex lap winding
(d) Duplex lap winding

(I.T.I. August 2015, Electrician)

22. The electrical degree of 6 pole of AC motor is ....

(a) 1080° (b) 720°
(c) 540° (d) 360°

(I.T.I. February 2016, Electrician)

Hint: Total Number of electrical degrees = 180° × Number of poles

23. The number of parallel path 'A' for duplex lap wound 4 pole DC machine is ....

(a) 6 (b) 4
(c) 8 (d) 12

(I.T.I. August 2019, Electrician)

24. How many number of parallel paths are there for 4 pole duplex wave wound armature?

(a) 2 (b) 4
(c) 6 (d) 8

(I.T.I. February 2016, Electrician)

25. The distance between two adjacent poles in armature winding is called as ....

(a) resultant pitch
(b) commutator pitch
(c) pole pitch
(d) average pitch

(I.T.I. August 2015, Electrician)

26. The distance between the centre of two adjacent opposite poles is called ....

(a) coil throw (b) pole pitch
(c) coil span (d) coil pitch

(I.T.I. February 2016, Electrician)

27. How many number of coils are required for 36 slots, 4 pole, 3 phase single layer distributed winding?

(a) 36 coils (b) 28 coils
(c) 24 coils (d) 18 coils

(I.T.I. February 2016, Electrician)

28. If in any winding, the number of coils/pole/phase is one, then the winding is known as ....

(a) distributed winding
(b) skein winding
(c) concentrated winding
(d) concentric winding

(I.T.I. February 2016, Electrician)

29. The formula to calculate the number of coil groups in AC motor winding is ....

(a) $\frac{\text{No. of coils}}{\text{No. of phases}}$

(b) No. of coils × no. of poles

(c) No. of phases × no. of poles

(d) $\frac{\text{No. of coils}}{\text{No. of phases} \times \text{no. of poles}}$

(I.T.I. February 2016, Electrician)

30. Which one of the following insulating material belongs to class "E" insulation?

(a) cotton

(b) silicon elastomer

(c) Leatheroid paper

(d) glass fibre

(I.T.I. February 2016, Electrician)

31. In split phase motor the ratio of turns for starting winding to that of running winding is ....

(a) 2.0 (b) >1

(c) 1.0 (d) <1

(I.T.I. February 2017, Electrician)

32. If a slot consists of one coil then the winding is called ....

(a) Double layer winding

(b) Full pitch winding

(c) Single layer winding

(d) Short pitch winding

(I.T.I. February 2018, Electrician)

33. A winding having number of slots is greater than number of poles. Then this winding is called .....

(a) Concentrated winding

(b) Distributed winding

(c) Full pitch winding

(d) Integrated slot winding

(I.T.I. February 2018, Electrician)

34. What is the function of damper windings in synchronous motor?

(a) Maintain power factor

(b) Excite the field winding

(c) Start the synchronous motor

(d) Maintain constant speed

(I.T.I. August 2019, Electrician)

35. In a synchronous motor, damper winding is provided to –

(a) Stabilize rotor motion

(b) Suppress rotor oscillations

(c) Develop necessary starting torque

(d) Both (a) & (c)

(I.T.I. August 2018, Electrician)

36. The damping winding in a synchronous motor is generally used .....

(a) To provide starting torque only

(b) To reduce noise level

(c) To reduce eddy current

(d) The prevent hunting and proivde the starting torque

(I.T.I. February 2019, Electrician)

37. Calculate the pitch factor (K_p) for a winding having 36 stator slots, 4 poles with angle (θ) 30° in Alternator.

(a) 0.9429 (b) 0.9859

(c) 0.9789 (d) 0.9659

(I.T.I. August 2019 Electrician)

38. Which test in winding is illustrated?

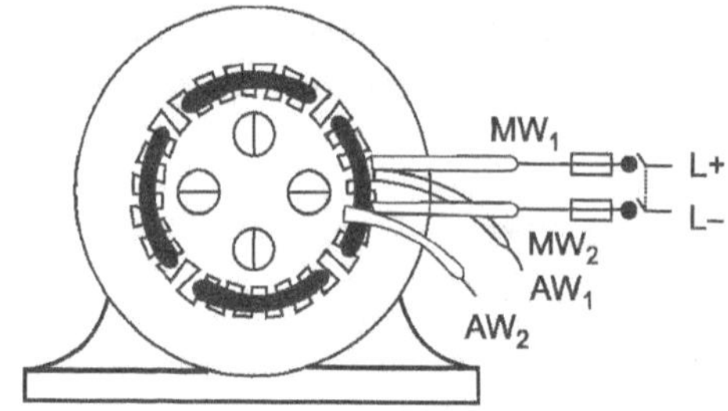

(a) Short circuit test (b) Ground test

(c) Countinuity test (d) Polarity test

(I.T.I. August 2019 Electrician)

39. Which type of armature winding is illustrated?

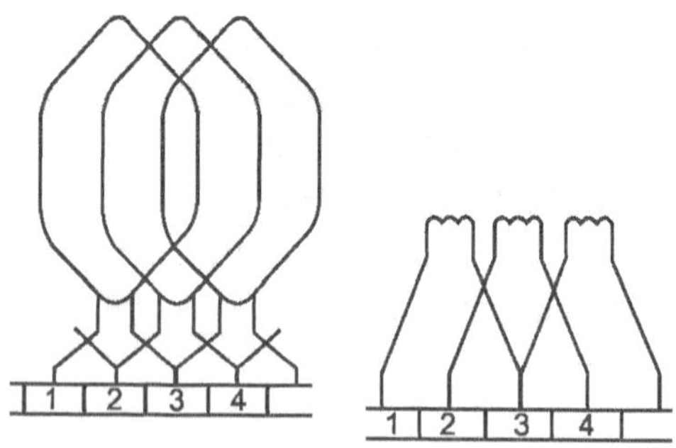

(a) Quadruplex lap winding
(b) Triplex lap winding
(c) Simplex lap winding
(d) Duplex lap winding

(I.T.I. August 2019 Electrician)

40. Which growler test for armature is illustrated?

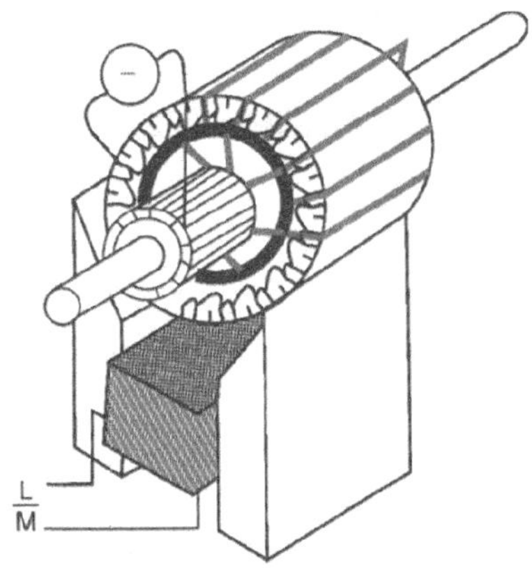

(a) Open coil test
(b) Grounded coil test
(c) Shorted coil test
(d) Shorted commutator test

(I.T.I. August 2019 Electrician)

41. What is the name of winding?

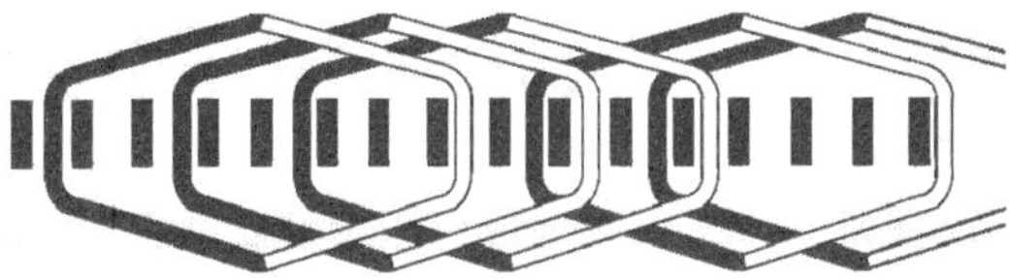

(a) Skew winding
(b) Diamond coil winding
(c) Concentric winding
(d) Basket winding

(I.T.I. August 2019 Electrician)

KEY TO MULTIPLE CHOICE QUESTIONS

Polyphase Induction Motors

1. (b)	**2.** (a)	**3.** (c)	**4.** (c)	**5.** (b)	**6.** (c)	**7.** (d)	**8.** (a)	**9.** (b)	**10.** (b)
11. (c)	**12.** (c)	**13.** (b)	**14.** (c)	**15.** (d)	**16.** (d)	**17.** (a)	**18.** (b)	**19.** (c)	**20.** (a)
21. (a)	**22.** (c)	**23.** (c)	**24.** (a)	**25.** (b)	**26.** (c)	**27.** (c)	**28.** (a)	**29.** (b)	**30.** (c)
31. (d)	**32.** (a)	**33.** (b)	**34.** (c)	**35.** (b)	**36.** (c)	**37.** (b)	**38.** (b)	**39.** (b)	**40.** (c)
41. (c)	**42.** (b)	**43.** (b)	**44.** (b)	**45.** (c)	**46.** (c)	**47.** (c)	**48.** (c)	**49.** (c)	**50.** (c)
51. (d)	**52.** (d)	**53.** (c)	**54.** (b)	**55.** (b)	**56.** (b)	**57.** (b)	**58.** (c)	**59.** (d)	**60.** (b)
61. (d)	**62.** (a)	**63.** (b)	**64.** (b)	**65.** (a)	**66.** (c)	**67.** (b)			

Winding

1. (d)	**2.** (c)	**3.** (d)	**4.** (b)	**5.** (d)	**6.** (d)	**7.** (c)	**8.** (b)	**9.** (b)	**10.** (b)
11. (d)	**12.** (a)	**13.** (a)	**14.** (b)	**15.** (b)	**16.** (b)	**17.** (d)	**18.** (a)	**19.** (b)	**20.** (d)
21. (a)	**22.** (a)	**23.** (c)	**24.** (b)	**25.** (c)	**26.** (b)	**27.** (d)	**28.** (c)	**29.** (c)	**30.** (c)
31. (d)	**32.** (a)	**33.** (b)	**34.** (c)	**35.** (d)	**36.** (d)	**37.** (d)	**38.** (d)	**39.** (d)	**40.** (b)
41. (d)									

Module 4

Single Phase Motors

Multiple Choice Questions

1. The type of the motor that does not have a commutator is :
   (a) repulsion motor (b) induction motor
   (c) d.c. shunt motor (d) a.c. series motor
2. The rotor of a split-phase motor receives its power from :
   (a) the line voltage
   (b) slip-rings
   (c) a centrifugal switch
   (d) magnetic induction
3. The capacitance of a small (less than 1 H.P.) single-phase motor will be in the order of :
   (a) hundreds of farads
   (b) tens of farads
   (c) tens and hundreds of pico farads
   (d) tens and hundreds micro farads
4. Which one of the following motors will have the lowest starting torque :
   (a) universal motor
   (b) shaded pole motor
   (c) capacitor-start capacitor-run motor
   (d) repulsion-induction motor
5. The purpose of the shaded coils in a shaded-pole motor is to :
   (a) produce rotating magnetic field
   (b) reduce friction loss
   (c) reduce rough commutation
   (d) reduce armature reaction
6. A D.O.L. starter is necessary for a single phase motor of more than one H.P. for :
   (a) starting the motor
   (b) stopping the motor
   (c) providing overload protection to the motor
   (d) all the above three reasons
7. Capacitor-start and capacitor-start-capacitor-run motors employ :
   (a) ceramic capacitors
   (b) mica capacitors
   (c) paper or non-polarised electrolytic capacitors
   (d) styroflex capacitors
8. If a winding makes electrical contact with the stator core, the winding is said to be :
   (a) short-circuited (b) open-circuited
   (c) closed (d) grounded
9. When a universal motor is operated on no load, its speed is limited by.....
   (a) Armature reaction
   (b) Armature weight
   (c) Windage and friction
   (d) Supply voltage frequency

   (I.T.I. February 2019, Electrician)
10. The direction of rotation of a universal motor is changed by :
    (a) interchanging the field connections only
    (b) interchanging the armature connections only

(c) interchanging either the field or armature connections
(d) none of the above

11. For the same rating the size of the single phase induction motor is about .... that of the corresponding three phase induction motor

(a) three times (b) the same as
(c) 1.5 times (d) one-third

12. Synchronous speed of an induction moto can be increased by

(a) reducing mechanical friction
(b) increasing supply voltage
(c) increasing number of poles
(d) increasing supply frequency

13. During plugging of an induction motor

(a) phase sequence is reversed
(b) a D.C. source is connected to stator
(c) one phase is open circuited
(d) power is fed back to mains

14. For toys, the preferred motor is

(a) shaded pole motor
(b) capacitor motor
(c) reluctance motor
(d) universal motor

15. The position of shaded ring with respect to main pole of the shaded pole motor determines ....

(a) speed of motor
(b) direction of rotation
(c) efficiency
(d) torque

(I.T.I. January 2014, Wireman)

16. Which type of AC single phase induction motor is used in wet grinders ?

(a) Capacitor start - Capacitor run motor
(b) Capacitor start - Induction run motor
(c) Permanent capacitor motor
(d) Universal motor

(I.T.I. February 2016, Electrician)

17. The capacitor start capacitor run motors have lesser efficiency and cannot take overloads. To avoid these problems capacitors of ....

(a) one large value used
(b) one low value used
(c) one medium value used
(d) one larger and one low value used

(I.T.I. August 2015, Wireman)

18. The permanent capacitor motor is most commonly used in ....

(a) compressors
(b) blowers
(c) ceiling fans
(d) washing machine

(I.T.I. February 2016, Electrician)

19. In a permanent capacitor motor, the capacitor is in series with ....

(a) starting winding
(b) running winding
(c) supply
(d) starting and running winding

(I.T.I. February 2016, Wireman)

20. The purpose of the capacitor used in single phases motor is to ....

(a) increase starting torque of the motor
(b) split the phases for producing the rotating magnetic field
(c) reduce the noise of the water
(d) protect the motor from overload

(I.T.I. February 2015, Electrician)

21. In single phase motors, the main winding and starting winding are placed apart from the degree of ....

(a) 180° (b) 120°
(c) 90° (d) 60°

(I.T.I. August 2015, Electrician)

22. The purpose of centrifugal switch provided in single phase induction motor is to ....

(a) connect the starting and running winding during running
(b) trip-off the motor, when the short circuit fault occurs
(c) disconnect the starting winding after obtaining 75% of speed
(d) control the speed of the motor

(I.T.I. February 2016, Electrician)

23. The full load torque develops at starting by universal motor is about ....

(a) 300% (b) 350%

(c) 400% (d) 450%

(I.T.I. February 2016, Wireman)

24. The stepper motor is a ....

(a) D.C. motor

(b) Single phase A.C. motor

(c) Two phase motor

(d) Multi phase motor

25. Which one of the following types of motors is most suitable for a computer printer drive?

(a) Reluctance motor

(b) Hysteresis motor

(c) Shaded pole motor

(d) Stepper motor

26. A stepper motor has four concentric coils on its 4 pole stator. It has an unwound variable reluctance motor with 5 teeth. The step angle for this motor is ....

(a) 36° (b) 18°

(c) 9° (d) 72°

27. A stepper motor having resolution of 300 steps per revolution and running at 2400 R.P.M. has a pulse ratio of .... p p s.

(a) 4000 (b) 6000

(c) 8000 (d) 10000

(I.T.I. February 2017, Electrician)

28. The rotor of stepper motor has no .....

(a) Windings (b) Commutator

(c) Brushes (d) None of these

(I.T.I. February 2017, Electrician)

29. Following type of motor is preferably used in a computer printer .....

(a) Universal motor

(b) Split phase motor

(c) Hysteresis motor

(d) Stepper motor

(I.T.I. February 2017, Wireman)

30. A stepper motor is .....

(a) A D.C. motor

(b) A single phase A.C. motor

(c) A multi phase motor

(d) A two phase motor

(I.T.I. August 2016, Electrician)

31. Single phase motors are commercially manufactured upto .....

(a) 1 H.P (b) 2 H.P

(c) 5 H.P (d) 10 H.P

(I.T.I. February 2019, Electrician)

32. Most of the single phase induction are motors ..... machines

(a) 2 pole (b) 6 pole

(c) 8 pole (d) 4 pole

(I.T.I. February 2017, Electrician)

33. A single phase induction motor ....

(a) is self starting

(b) operates at a fixed speed

(c) is less reliable than

(d) none of the above

(I.T.I. February 2018, Electrician)

34. In single phase induction motors, phase shift and torque can be provided by ....

(a) Capacitor only

(b) Shaded poles

(c) Inductors and resistors

(d) All of these

(I.T.I. February 2017, Wireman)

35. A single phase motor generally used for small air compressor is ....

(a) Capacitor start capacitor run

(b) Split phase

(c) Reluctance type

(d) Shaded pole

(I.T.I. February 2017, Electrician)

36. In a capacitor start capacitor run single phase motor, the ratio of starting torque T_{st} to full load torque T_{FL} is ....

(a) 1 (b) 2

(c) 3 (d) 4

(I.T.I. February 2017, Wireman)

37. A capcitor start, capacitor run single phase inductor motor is basically a ....

(a) AC series motor

(b) DC series motor

(c) 2 phase induction motor
(d) 3 phase induction motor
(I.T.I. February 2019, Electrician)

38. A capacitor start single phase induction motor will usually have a power factor of –
(a) Unity (b) 0.8 leading
(c) 0.6 leading (d) 0.6 lagging
(I.T.I. February 2018, Electrician)

39. The starting torque of a capacitor start motor is –
(a) Zero
(b) Low
(c) Same as rated toque
(d) More than rated torque
(I.T.I. February 2019, Electrician)

Hint: It has high starting torque i.e., 1.5 times of full load torque.

40. Which of the following motors will operate at high power factor?
(a) Universal motor
(b) Capacitor start motor
(c) Capacitor run motor
(d) Split phase motor
(I.T.I. February 2019, Electrician)

41. The starting capacitor of a single phase motor is –
(a) Electrolytic capacitor
(b) Ceramic capacitor
(c) Paper capacitor
(d) None of the above
(I.T.I. February 2019, Electrician)

42. Which of the following is non-reversible motor?
(a) Universal motor
(b) Capacitor start split phase motor
(c) Resistance start split phase motor
(d) Permanent split capacitor motor
(I.T.I. February 2019, Electrician)

43. The efficiency of shaded pole motor is in the range–
(a) 80 to 95 percent (b) 70 to 80 percent
(c) 50 to 70 percent (d) 5 to 35 percent
(I.T.I. February 2019, Electrician)

44. The electric motor used in toys which run on A.C. are usually .....
(a) Capacitor start motor
(b) Split phase motor
(c) Capacitor start capacitor run motor
(d) Shaded pole motor
(I.T.I. February 2017, Electrician)

45. Which of the following method is the most economical method of starting single phase motor?
(a) Resistance start method
(b) Capacitor start method
(c) Split phase method
(d) Inductance start method
(I.T.I. February 2017, Electrician)

46. A hysteresis motor works on the principle of ...
(a) Hysteresis loss
(b) Magnetisation of motor
(c) Eddy current loss
(d) Electromagnetic induction
(I.T.I. February 2016, Electrician)

47. A universal motor is one which ......
(a) Can run on large variations in A.C. supply voltage
(b) Has infinitely varying speeds
(c) Can operate on D.C. as well as A.C.
(d) Can work on single phase or three phase supply
(I.T.I. February 2017, Electrician)

48. Out of the following motors, which will give the highest starting torque?
(a) Universal motor
(b) Capacitor start motor
(c) Shaded pole motor
(d) All have zero starting torque
(I.T.I. February 2019, Electrician)

49. Which motor would you select for vacuum cleaners?
(a) Universal motor
(b) Repulsion motor
(c) Hysteresis motor
(d) Reluctance motor
(I.T.I. February 2018, Electrician)

50. The shortcoming of repulsion motor is –
(a) Variation of speed with load
(b) Low power factor
(c) Tendency to spark at brushes
(d) All of these

(I.T.I. February 2018, Electrician)

51. Reluctance motor ......
(a) is a self starting motor
(b) starts as an induction motor and runs as a synchronous motor
(c) needs no D.C. excitation
(d) All the above

52. A reluctance motor .....
(a) is self starting
(b) is constant speed motor
(c) needs no D.C. excitation
(d) All of these

(I.T.I. February 2016, Electrician)

53. Relutance motors are .....
(a) Double excited (b) Single excited
(c) Either (a) or (b) (d) None of these

(I.T.I. February 2019, Electrician)

54. Which single phase motor is fitted with wound rotor?
(a) Repulsion motors
(b) Shaded pole motors
(c) Permanent capacitor motors
(d) Capacitor start capacitor run motors

(I.T.I. August 2019 Electrician)

55. In which single phase motor, tapped field speed control method is employed?
(a) Capacitor start capacitor run motor
(b) Shaded pole motor
(c) Capacitor start induction run motor
(d) Universal motor

(I.T.I. August 2019 Electrician)

56. What is the relation between running winding and starting winding of a single phase induction motor with respect to resistance?
(a) Both resistances will be equal
(b) Running winding is more than starting winding less
(c) Running winding is less than starting winding more
(d) Running winding is less than starting winding infinity

(I.T.I. August 2019 Electrician)

57. How the direction of rotation of a capacitor start induction run motor is reversed?
(a) By changing the supply terminals
(b) By interchanging main winding terminals
(c) By changing the capacitor connections
(d) By interchanging both main and auxiliary winding terminals

(I.T.I. August 2019 Electrician)

58. How the direction of rotation of repulsion motors is to be reversed?
(a) By changing the compensating winding terminals
(b) By interchanging the supply terminals
(c) By changing the main winding terminals
(d) By shifting the brush axis

(I.T.I. August 2019 Electrician)

59. What is the effect if the centrifugal switch is not disconnected after the motor starts?
(a) Motor will run normally
(b) Motor will stop immediately
(c) Starting winding will burn out
(d) Motor will run at very slow speed

(I.T.I. August 2019 Electrician)

60. What is the function of the part marked 'X' in shaded pole motor?

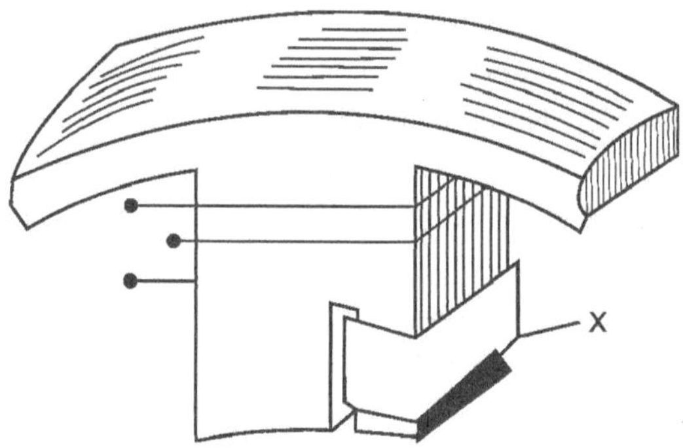

(a) Increase the efficiency
(b) Initiate the rotor movement
(c) Maintain constant speed
(d) Strengthen the magnetic field

(I.T.I. August 2019 Electrician)

KEY TO MULTIPLE CHOICE QUESTIONS

1. (b)	**2.** (d)	**3.** (d)	**4.** (b)	**5.** (a)	**6.** (d)	**7.** (c)	**8.** (d)	**9.** (c)	**10.** (c)
11. (c)	**12.** (d)	**13.** (a)	**14.** (a)	**15.** (b)	**16.** (b)	**17.** (a)	**18.** (c)	**19.** (a)	**20.** (b)
21. (c)	**22.** (c)	**23.** (d)	**24.** (d)	**25.** (d)	**26.** (b)	**27.** (b)	**28.** (d)	**29.** (d)	**30.** (c)
31. (b)	**32.** (d)	**33.** (d)	**34.** (d)	**35.** (a)	**36.** (c)	**37.** (c)	**38.** (d)	**39.** (d)	**40.** (c)
41. (a)	**42.** (c)	**43.** (d)	**44.** (d)	**45.** (a)	**46.** (a)	**47.** (c)	**48.** (a)	**49.** (a)	**50.** (d)
51. (d)	**52.** (d)	**53.** (b)	**54.** (a)	**55.** (d)	**56.** (c)	**57.** (b)	**58.** (d)	**59.** (c)	**60.** (b)

Module 5

Alternators

Multiple Choice Questions

1. A 4-pole, 50 Hz alternator will have a rotational speed of :

   (a) 750 r.p.m. (b) 1500 r.p.m.
   (c) 3000 r.p.m (d) 1440 r.p.m.

2. An exciter used for field excitation of an alternator will be a :

   (a) series generator (b) shunt generator
   (c) series motor (d) shunt motor

3. D.C. and A.C. generators are similar in one important respect, that is :

   (a) the generated e.m.f. is D. C.
   (b) the generated e.m.f. is pulsating
   (c) the generated e.m.f. is A. C.
   (d) the generated e.m.f. is oscillating

4. Alternators designed to be driven by steam-turbines are built for :

   (a) high speed operation
   (b) low speed operation
   (c) medium speed operation
   (d) both low and medium speed operations.

5. Sinlge phase portable generators are generally made in the range of :

   (a) 5000 kVA (b) 1000 kVA
   (c) 100 kVA (d) 5 kVA

6. An alternator supplies electric power to various kinds of loads of a factory. If the phase sequence of the alternator is changed, it will affect the :

   (a) 3-phase heating circuits
   (b) 3-phase motor circuits
   (c) single phase motor circuits
   (d) lighting circuits

7. The values of pitch factor and distribution factor for an alternator are 0.9 and 0.8 respectively. This will develop an e.m.f. lesser than an alternator having full-pitch coil and concentrated winding by :

   (a) 0.7 (b) 0.72
   (c) 0.88 (d) 0.92

8. The winding of a 4-pole alternator is having 36 slots and a coil span of 1 to 8. It is short-pitched by............ electrical degrees.

   (a) 80 (b) 40
   (c) 20 (d) 10

   (**Hint :** For full-pitch winding the coil span should be 36/4 = 9, but here actual pitch is 1 to 8, *i.e.*, 7.

   $$\therefore \text{angle} = \frac{(\text{full pitch} - \text{actual pitch}) \times 180}{\text{full pitch}}\,)$$

9. The number of slip-rings used in a 3-phase, star connected alternator having a rotating armature will be:

   (a) 2 (b) 3
   (c) 4 (d) 6

10. The pitch factor is defined as the ratio of the e.m.fs. of the :

    (a) full-pitch winding to short-pitch winding
    (b) short-pitch coil to full-pitch coil

(c) full-pitch winding to concentrated winding
(d) distributed winding to full-pitch winding

11. The rotor of a salient pole alternator has 12 poles. The number of cycles of e.m.f. in one revolution will be :

(a) 2 (b) 4
(c) 6 (d) 8

(**Hint :**Two-pole alternator will generate one cycle of e.m.f. in one rotation)

12. The number of electrical degrees which passes through in one revolution of a 6-pole alternator is :

(a) 2160° (b) 1080°
(c) 720° (d) 360°

13. The waveform of the e.m.f. generated by an alternator should be :

(a) pulse shaped wave
(b) square wave
(c) sine wave
(d) triangular wave

14. Salient pole type rotors as compared to the cylindrical pole type rotors are :

(a) smaller in diameter and larger in axial length
(b) larger in diameter and smaller in axial length
(c) larger in diameter and axial length
(d) smaller in diameter and axial length

15. The power factor of an alternator is determined by its

(a) speed (b) load
(c) excitation (d) prime mover

16. Skew of rotor bar eliminates

(a) the effect of space harmonics
(b) the entire effect of crawling
(c) magnetic noise
(d) variation due to unequal force developed on rotor

17. The alternators are rated in ....

(a) kW (b) MW
(c) kVA (d) kV

(I.T.I. August 2019, Electrician)

18. A 10 pole alternator rotates at a constant speed of 1200 rpm. The periodic time of voltage in seconds per cycle will be

(a) 0.01 (b) 0.02
(c) 0.04 (d) 0.08

(I.T.I. February 2015, Electrician)

19. The field circuit of an alternator is supplied with ....

(a) an oscillating current
(b) a direct current
(c) an alternating current
(d) a pulsating current

(I.T.I. August 2014, Electrician,
I.T.I. February 2015, Wireman)

20. The D.C. and A.C. generators are similar in one important function. That is the ....

(a) generated emf is a direct current
(b) generated emf is an alternating current
(c) generated emf is a pulsating current
(d) generated emf is an oscillating current

(I.T.I. February 2015, Electrician)

21. Adjusting the speed of the prime mover of an alternator is primary necessary. Which one of the following is the reason for this?

(a) To get constant frequency
(b) To get constant voltage
(c) To get increase polarity
(d) None of the above

22. Which one of the following is the reason for using dampers in large alternator work?

(a) To increase stability
(b) To increase output
(c) To decrease output
(d) To reduce voltage fluctuation

23. What is the name of the part of an alternator shown in the figure

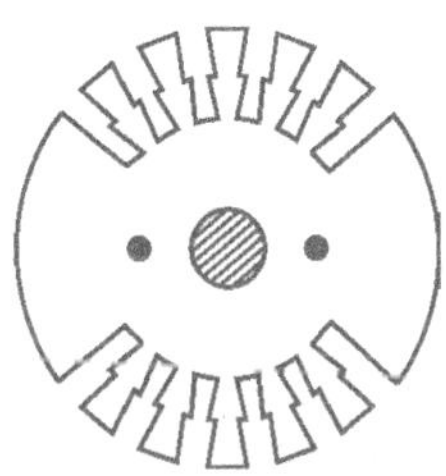

(a) salient pole type rotor
(b) semi enclosed slot type stator
(c) smooth cylindrical pole type rotar
(d) totally enclosed slot type stator

(I.T.I. February 2015, Electrician)

24. When all dark lamp method is used for synchronizing two 3-phase alternators, the lamps are not becoming dark at the same time but occurs at different timings. Bulbs used are of same wattage and voltage rating, connection of the lamps were checked and found OK. This is due to the difference in one of the factors stated below. Identify the one?

(a) Frequency (b) Speed
(c) Phase sequence (d) Voltage

(I.T.I. August 2015, Wireman)
(I.T.I. July 2009, Electrician)

25. What is the type of lamp method illustrated in the figure for synchronizing 3 phase alternators?

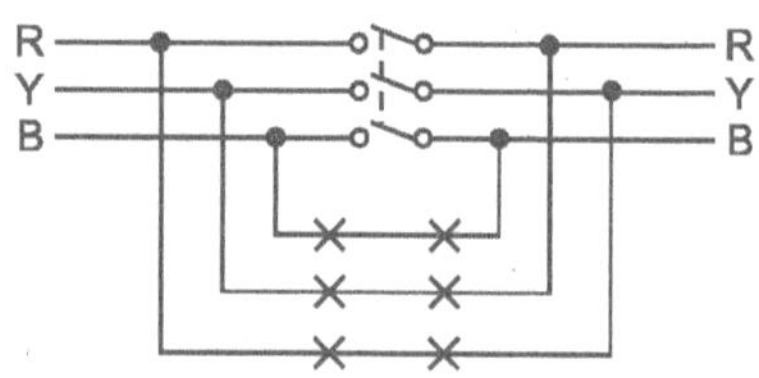

(a) All dark lamp method
(b) All bright lamp method
(c) Two bright and one dark lamp method
(d) Two dark and one bright lamp method

26. Which instrument is used to indicate the correct time for synchronization of two alternators ?

(a) phase sequence meter
(b) synchroscope
(c) power factor meter
(d) tachometer

(I.T.I. February 2016, Wireman)

27. The distribution factor (K_d) of an alternator is expressed as:

$$K_d = \frac{\sin\frac{m\,b}{2}}{m\,.\,\sin\frac{b}{2}}$$

The letter 'm' represents ....

(a) number of conductors per phase
(b) number of pair of poles per phase
(c) number of poles per phase
(d) number of slots per phase per pole

(I.T.I. August 2015, Electrician)

28. The emf equation of single phase alternator with considering pitch factor and distribution factor is ....

(a) $2.22\ \phi\ f\ T\ .\ K_d K_c$ volt
(b) $4.44\ .\ K_p K_d\ f\ \phi T$ volt
(c) $4.44\ .\ K_d\ f\phi T$ volt
(d) $2.22\ \phi\ f\ 2T$ volt

(I.T.I. February 2016, Electrician)

29. The magnitude of the induced emf (E_0) from the alternator is given by the formula "E_0 = BLV sin θ".
What does the letter "V" denotes in the formula ?

(a) Voltage produced by the alternator
(b) Number of magnetic lines of forces in weber
(c) Flux density in weber/square meter
(d) Velocity of the loop rotation in meter per second

(I.T.I. February 2016, Electrician)

30. Calculate the percentage voltage regulation of the alternator, when it's terminal voltage rises from 210 volt at full load to 220 volt at no load.

(a) 4.76 % (b) 4.65 %
(c) 4.54 % (d) 3.75 %

(I.T.I. February 2016, Electrician)

31. An alternator is also called –

(a) AC Generator
(b) Turbo generator
(c) Asynchronous generator
(d) Generator

(I.T.I. February 2019, Electrician)

32. The frequency of voltage generated by alternators having 8 poles and rotating at 250 R.P.M. is .....

(a) 60 Hz (b) 50 Hz
(c) 25 Hz (d) $16\frac{2}{3}$Hz

(I.T.I. February 2018, Electrician)

33. The power factor of an alternator depends on ....
(a) Load (b) Speed of rotor
(c) Core losses (d) Armature losses
(I.T.I. February 2018, Electrician)

34. Under short-circuit conditions, the power factor of an alternator is –
(a) Unity
(b) Almost zero Lagging
(c) Almost zero leading
(d) Same as normal condition
(I.T.I. February 2019, Electrician)

35. If the excitation is critical, the power factor of the alternator is ....
(a) Unity power factor
(b) Leading power factor
(c) Lagging power factor
(d) None of these
(I.T.I. February 2019, Electrician)

36. When excitation is increased (over excited) the synchronous generator operates at :
(a) Unity power factor
(b) Lagging power factor
(c) Leading power factor
(d) None of these
(I.T.I. February 2017, Electrician)

37. In an alternator, voltage drops occur in .....
(a) Armature resistance only
(b) Armature resistance and leakage reactance
(c) Armature resistance, leakage reactance and armature reaction
(d) Armature resistance, leakage reactance, armature reaction and earth connections
(I.T.I. February 2018, Electrician)

38. Active power of an alternator can be varied by ......
(a) Changing field excitation
(b) Changing of prime cover speed
(c) Changing the load
(d) None of these
(I.T.I. February 2019, Electrician)

39. In alternators fractional pitch coil are used to reduce the effect of ......
(a) Hunting
(b) Harmonics
(c) Armature reaction
(d) Power factor
(I.T.I. February 2019, Electrician)

40. In which of the following loads, armature reactance is magnetising in nature?
(a) Pure resistance load
(b) Pure inductive load
(c) Pure capacitive load
(d) R.L.C. load
(I.T.I. February 2017, Electrician)

41. In a synchronous machine if ϕ is the flux per pole and f is the frequency of induced e.m.f. then
(a) $E \propto \phi f$ (b) $E \propto \frac{f}{f}$
(c) $E \propto \frac{1}{f f}$ (d) $E \propto \frac{1}{f}$
(I.T.I. February 2017, Electrician)

42. The slots provided on the armature core of an alternator to house the armature coils are mainly of the following types :
(a) Open, wide open, closed
(b) Open, semiclosed, closed
(c) Open, wide open, semiclosed
(d) Wide open, semiclosed, closed
(I.T.I. February 2017, Wireman)

43. Following is true about a cylindrical rotor alternator:
(a) Poles are projecting outside
(b) Damper winding is required
(c) Noise free operation
(d) Suitable for slow speed hydro generators
(I.T.I. February 2017, Wireman)

44. Following statement (s) are true about an alternator:
(a) Field winding is placed on the rotor, while armature winding is placed on the stator
(b) Armature winding has more diameter than field winding
(c) Rotor is driven by a prime mover
(d) All of these
(I.T.I. February 2017, Wireman)

45. In a synchronous generator V curve is drawn between:
(a) Field current of the *X*-axis and armature current on the *Y*-axis
(b) Field current of the *Y*-axis and armature current on the *Y*-axis
(c) Field current of the *Y*-axis and power factor on the *X*-axis
(d) Field current of the *X*-axis and power factor on the *Y*-axis
(I.T.I. February 2017, Electrician)

46. Three phase alternator are generally .....
(a) Star connected (b) Delta connected
(c) Open circuited (d) Short circuited
(I.T.I. February 2019, Electrician)

47. What is/are the necessary conditions for parallel operation of alternators?
(a) Terminal voltage is same
(b) Frequency should be same
(c) Speed of alternators should be same
(d) Terminal voltage and frequency should be same
(I.T.I. February 2019, Electrician)

48. What are the advantages of parallel operation of alternators?
(a) Increase in reliability
(b) Increases in efficiency under light conditions
(c) Continuity in supply is maintained
(d) All of these
(I.T.I. February 2017, Electrician)

49. When two alternators are operating in parallel and at perfect synchronisation, the synchronising power will be....
(a) Negative (b) Infinite
(c) Positive (d) Zero
(I.T.I. February 2016, Electrician)

50. In the right lamp method fo synchronising two alternators incoming alternator is put on Bus, when .....
(a) One bulb is bright, two bulbs are dark
(b) Two bulbs are bright, one bulb is dark
(c) Two bulbs are bright, two bulbs are dark
(d) Three bulbs are bright
(I.T.I. February 2017 Wireman)

51. Which condition is to be satisfied before parallel operation of Alternators?
(a) Phase sequence must be same
(b) Rating must be same
(c) Rotor impedance must be same
(d) Stator impedance must be same
(I.T.I. August 2019 Electrician)

52. What is the cause for hunting effect in Alternators?
(a) Due to overload
(b) Due to continuous fluctuation in load
(c) Running with fluctuation of speed
(d) Running without load
(I.T.I. August 2019 Electrician)

53. Which formula is used to calculate the percentage voltage regulation in Alternator?
(a) $\frac{V_{FL} - VL}{V_{FL}}$ (b) $\frac{V_{NL} - V_{FL}}{V_{NL}}$
(c) $\frac{V_{NL} - V_{FL}}{V_{FL}}$ (d) $\frac{V_{FL} - V_{NL}}{V_{NL}}$
(I.T.I. August 2019 Electrician)

KEY TO MULTIPLE CHOICE QUESTIONS

1. (b)	**2.** (b)	**3.** (c)	**4.** (a)	**5.** (d)	**6.** (b)	**7.** (b)	**8.** (b)	**9.** (c)	**10.** (b)
11. (c)	**12.** (b)	**13.** (c)	**14.** (b)	**15.** (b)	**16.** (a)	**17.** (c)	**18.** (a)	**19.** (b)	**20.** (b)
21. (a)	**22.** (a)	**23.** (c)	**24.** (c)	**25.** (c)	**26.** (b)	**27.** (d)	**28.** (b)	**29.** (d)	**30.** (a)
31. (a)	**32.** (d)	**33.** (a)	**34.** (b)	**35.** (a)	**36.** (c)	**37.** (c)	**38.** (b)	**39.** (b)	**40.** (c)
41. (a)	**42.** (d)	**43.** (c)	**44.** (d)	**45.** (a)	**46.** (a)	**47.** (d)	**48.** (d)	**49.** (d)	**50.** (d)
51. (a)	**52.** (b)	**53.** (c)							

Module 6

Synchronous Motor

Multiple Choice Questions

1. An electric motor in which both the rotor and stator fields rotates with the same speed is called .... motor
   (a) D.C. (b) chrage
   (c) synchronous (d) universal
2. When load on a normally excited synchronous motor is increased, its power factor tends to
   (a) approach unity
   (b) become increasingly lagging
   (c) becomes increasingly leading
   (d) remain unchanged
3. If the field of a synchronous motor is under excited, the power factor will be
   (a) lagging (b) leading
   (c) unity (d) more than unity
4. In a synchronous motor, the rotor copper losses are met by
   (a) motor input (b) armature input
   (c) supply lines (d) D.C. source
5. Synchronous capacitor is
   (a) an ordinary static capacitor bank
   (b) an over-excited synchronous motor driving mechanical load
   (c) an over excited synchronous motor running without mechanical load
   (d) none of the above
6. The maximum valve of torque angle in a synchronous motor is .... degrees electrocal
   (a) 45
   (b) 90
   (c) between 45 and 90
   (d) below 60
7. The effect of increasing load on a synchronous motor running with normal excitation is to
   (a) increase both its I_a and p.f.
   (b) decrease I_a but increase p.f.
   (c) increase I_a but decrease p.f.
   (d) decrease both I_a and p.f.
8. The V-curves of a synchronous motor show relationship between :
   (a) excitation current and back e.m.f.
   (b) field current and power factor
   (c) D.C. field current and A.C. armature current
   (d) armature current and supply voltage
9. Synchronous motor can operate at –
   (a) Lagging power factor only
   (b) Leading power factor only
   (c) Unity power factor only
   (d) Lagging, leading and unity power factor only

   (I.T.I. February 2019, Electrician)
10. An unexcited single phase synchronous motor is .....
    (a) Reluctance motor
    (b) Repulsion motor

(c) Universal motor

(d) None of these

(I.T.I. February 2019, Electrician)

11. The maximum power developed in the synchronous motor will depend on .....

(a) Rotor excitation only

(b) Maximum value of coupling angle

(c) Supply voltage only

(d) Rotor excitation supply voltage and maximum value of coupling angle

(I.T.I. August 2016, Electrician)

12. The armature current of synchronous motor has large values of ....

(a) Low exciation only

(b) High excitation only

(c) Both low and high excitation

(d) Depends on other factors

(I.T.I. February 2019, Electrician)

13. A three phase synchronous motor will have .... slip rings

(a) No (b) 1

(c) 2 (d) 3

(I.T.I. February 2017, Electrician)

14. Oscillations of the motor about its equilibrium position due to sudden application or removal of load in synchronous motor is called .....

(a) Hunting (b) Swinging

(c) Galloping (d) Both (a) and (b)

(I.T.I. February 2017, Electrician)

15. The back emf set up in the stator of a synchronous motor will depend on ......

(a) Rotor speed only

(b) Rotor excitation only

(c) Rotor excitation and rotor speed

(d) Coupling angle, rotor speed and excitation

(I.T.I. February 2018, Electrician)

16. In a synchronous motor, the torque angle is ......

(a) The angle between the rotating stator flux and rotor poles

(b) The angle between magnetizing current and back emf

(c) The angle between the supply voltage and the back emf

(d) None of these

(I.T.I. February 2019, Electrician)

17. Why exciter is essential to run a synchronous motor?

(a) Carry more load in motor

(b) Improve the power factor

(c) Reduce the losses in motor

(d) Run the motor at synchronous speed

(I.T.I. August 2019 Electrician)

18. How the synchronous motor is used as a synchronous condenser?

(a) Varying the motor load

(b) Varying the rotor excitation

(c) Varying stator voltage in motor

(d) Varying stator current in motor

(I.T.I. August 2019 Electrician)

KEY TO MULTIPLE CHOICE QUESTIONS

1. (c)	**2.** (b)	**3.** (a)	**4.** (d)	**5.** (c)	**6.** (b)	**7.** (a)	**8.** (c)	**9.** (d)	**10.** (a)
11. (d)	**12.** (c)	**13.** (c)	**14.** (d)	**15.** (b)	**16.** (a)	**17.** (d)	**18.** (b)		

Module 7

Electronics Practice

Multiple Choice Questions

Basic Electronics

1. Which one of the following materials is used for 'doping' to prepare P-type semiconductor—
   (a) Arsenic (b) Antimony
   (c) Indium (d) Phosphorus
2. Which one of the following materials is used for 'doping' to prepare N-type semiconductor —
   (a) Indium (b) Gallium
   (c) Arsenic (d) Silver

   (I.T.I. February 2019, Electrician, Electroplator, Wireman)
3. The function of a diode can be compared with that of a—
   (a) fuse (b) relay
   (c) coil (d) switch

   (I.T.I. February 2019, Electrician, Electroplator, Wireman)
4. The potential difference across a diode is termed as —
   (a) reverse potential
   (b) PN potential
   (c) barrier potential
   (d) forward potential
5. What will be the voltage across the silicon diode shown in the given figure —

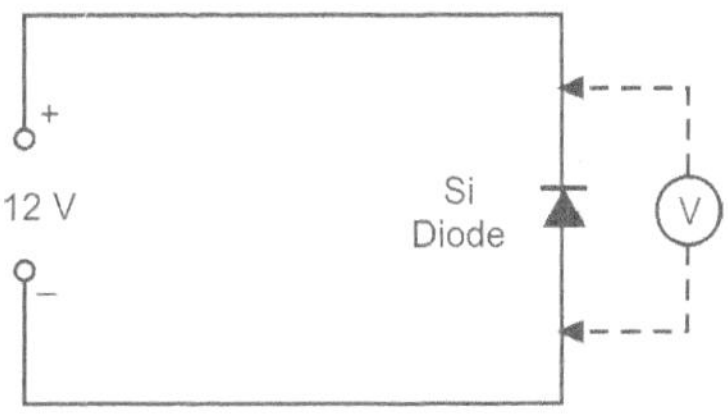

   (a) 0.3 V (b) 0.7 V
   (c) 1.0 V (d) 12.0 V
6. A semiconductor element in its pure form is called a/an ... semiconductor and in its 'doped' form is called a/an ... semiconductor —
   (a) crystal, compound
   (b) doped, undoped
   (c) intrinsic, extrinsic
   (d) normal, abnormal
7. What will be the voltage across the diode shown in the given figure —

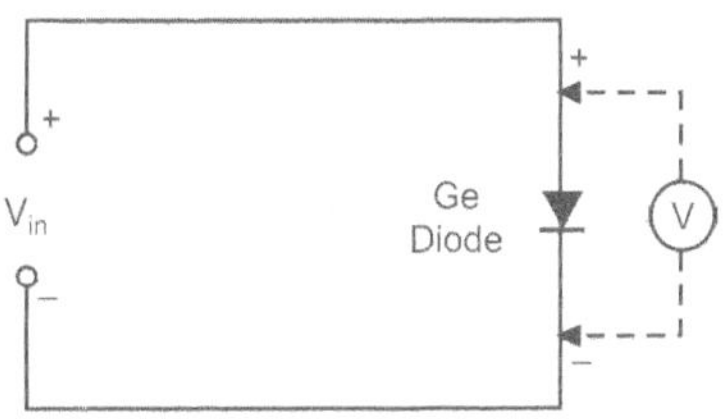

   (a) 0.3 V (b) 0.7 V
   (c) 1.0 V
   (d) equal to the applied voltage

8. If a diode shows very high resistane in one direction and a low resistance in the opposite direction then its condition is —

(a) good (b) short-circuited
(c) open-circuited (d) defective

(I.T.I. February 2019, Electrician, Electroplator, Wireman)

9. Silicon diodes are preferred over Germanium diodes because —

(a) their resistivity is lesser
(b) their barrier potential is higher
(c) their barrier potential is lower
(d) these are lesser temperature sensitive

(I.T.I. February 2019, Electrician, Electroplator, Wireman)

10. Below the breakdown region the leakage current of a silicon diode is of the order of —

(a) 1 μA (b) 100 μA
(c) 1 mA (d) 100 mA

11. The maximum reverse voltage that a diode can withstand is called —

(a) threshold voltage
(b) peak inverse voltage
(c) cut-in-voltage
(d) avalanche voltage

(I.T.I. February 2019, Electrician, Electroplator, Wireman)

12. PIV across each diode of a bridge rectifier is equal to—

(a) $2 \times V_s$ (rms) (b) $2 \times V_s$ (peak)
(c) V_s (rms) (d) V_s (peak)

13. The circuit shown in the given figure is a simple—

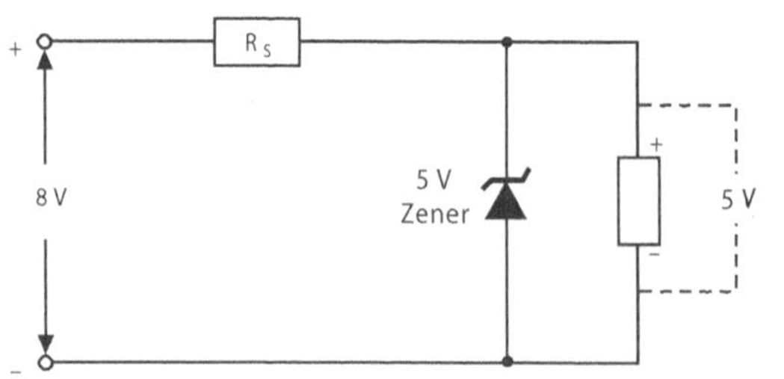

(a) inverter circuit
(b) H.W. rectifier circuit
(c) voltage regulator circuit
(d) zener controlled filter circuit

14. The phenomenon which causes reverse break down in a zener diode is called —

(a) Avalanche effect
(b) Hall effect
(c) breakdown effect
(d) Seeback effect

15. Which one is the symbol of zener diode out of the symbols shown in the given figure ?

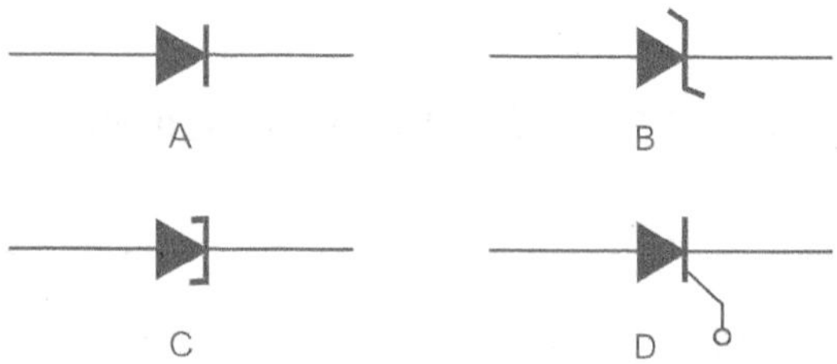

16. A filter circuit that allows signals above a particular frequency to pass through it is called a—

(a) low-pass filter (b) high-pass filter
(c) band-pass filter (d) band-stop filter

17. What type of filter circuits are used in graphic equalizers ?

(a) Low-pass filter (b) High-pass filter
(c) Band-pass filter (d) Band-stop filter

18. The filter circuit which attenuates or rejects a specific band of frequencies is called a —

(a) band-pass filter (b) band-stop filter
(c) high-pass filter (d) low-pass filter

19. Which type of filter is used to block the humming sound of 50 Hz produced by a.c. main ?

(a) Low-pass (b) High-pass
(c) Band-pass (d) Band-stop

20. The main component used in a filter circuit for reducing ripples is ....

(a) capacitor (b) resistor
(c) inductor (d) transformer

21. What will be the voltage across the diode in non-conducting state ....

(a) zero (b) 230 V
(c) 24 V (d) 0.7 V

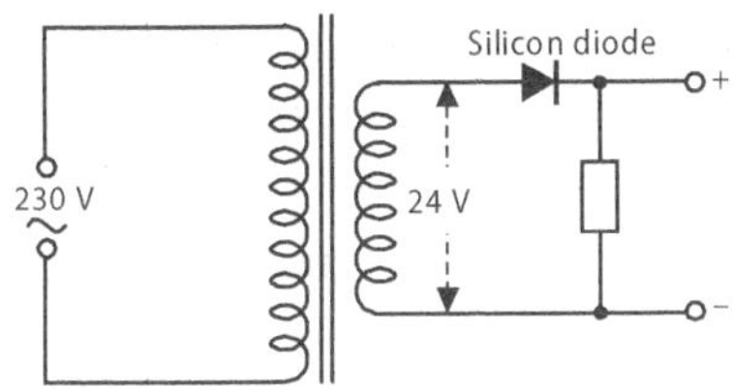

22. In P-type semiconductor the majority carriers are ....
(a) Holes (b) Free electrons
(c) Valence electrons (d) Neutrons
(I.T.I. August 2015, Electrician, Wireman)

23. Which one is Active component?
(a) Resistor (b) Capacitor
(c) Zener diode (d) Inductor
(I.T.I. August 2015, Electrician, Wireman)

24. The barrier potential of silicon semiconductor used in transistor is ....
(a) 0.3 volt (b) 0.5 volt
(c) 0.7 volt (d) 0.8 volt
(I.T.I. February 2016, Electrician, Wireman)

25. The barrier potential of germanium substance is ....
(a) 0.7 V (b) 0.5 V
(c) 0.3 V (d) 0.2 V
(I.T.I. February 2016, Electrician, Wireman)

26. In diode, the current which can allow to blow through it without getting damaged is called as ....
(a) maximum reverse current
(b) maximum average forward current
(c) maximum forward surge
(d) maximum load current
(I.T.I. August 2015, Electrician)

27. Diode is a ....
(a) Bi directional device
(b) Unidirectional device
(c) Multi directional device
(d) None of these
(I.T.I. February 2016, Electrician, Electroplator, Wireman)

28. A PN junction allows current flow when ......
(a) The P-type materials is more positive than the N-type material
(b) The N-type material is more positive than the P-type material
(c) Both the N-type and P-type materials have the same potential
(d) There is no potential on the N-type or P-type materials
(I.T.I. February 2016, Electrician, Electroplator, Wireman)

29. The conduction band is closest to the valence band in–
(a) Semiconductors
(b) Conductors
(c) Insulators
(d) The distance is the same for all of these
(I.T.I. August 2016, Electrician, Electroplator, Wireman)

30. The forbidden gap in Si is .... eV.
(a) 1.7 (b) 0.7
(c) 1.12 (d) 2.1
(I.T.I. February 2019, Electrician, Electroplator, Wireman)

31. Which one of the following is used as passive components in electronic circuits?
(a) Resistor (b) Diode
(c) Vacuum triode (d) Transistor
(I.T.I. February 2019, Electrician, Electroplator, Wireman)

32. Ohm's law is not applicable in .....
(a) D.C. circuit (b) High current
(c) Small resistor (d) Semiconductor
(I.T.I. August 2019 Electrician)

33. The most commonly used semiconductor is .....
(a) Germanium (b) Carbon
(c) Silicon (d) Sulphur
(I.T.I. August 2019 Electrician)

34. A semiconductor is formed by ...... bonds.
(a) Coordinate (b) Electrovalent
(c) Covalent (d) None of these
(I.T.I. August 2019 Electrician)

35. To make N type semiconductor we have to add .... valent impurity.
(a) 3 (b) 4
(c) 5 (d) 6
(I.T.I. August 2019 Electrician)

36. Diode is .... terminal device.
(a) 1 (b) 3
(c) 2 (d) 4
(I.T.I. August 2019 Electrician)

Oscillators

1. Feedback used in a RC oscillator is of —
(a) negative type
(b) positive type
(c) any type, positive or negative
(d) none of the stated above

2. In RC oscillator the phase of the feedback voltage is shifted to ... by employing a RC phase shifting network—
(a) 0º (b) 60º
(c) 90º (d) 180º

3. An oscillator which generates non sinusoidal waves is called —
(a) harmonic ocillator
(b) feedback oscillator
(c) relaxation oscillator
(d) none of the above stated

4. Barkhausen criterion for sustaining oscillation is expressed as —
(a) $A \cdot \beta = 0$ (b) $A \cdot \beta = 1$
(c) $A \cdot \beta > 1$ (d) $A \cdot \beta < 1$

5. The amplitude of damped waves —
(a) decreases with time
(b) increases with time
(c) is zero
(d) is constant

6. A Colpitt's Oscillator employs —
(a) inductive feedback
(b) capacitive feedback
(c) negative feedback
(d) stray capacitance feedback

7. The circuit shown in the given figure is of a —

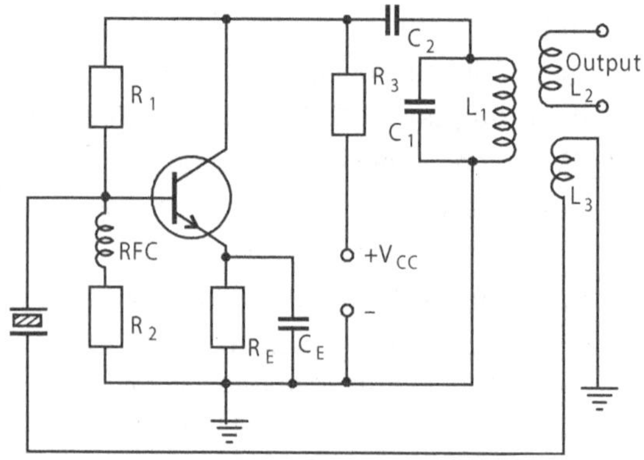

(a) Hartley sereis-fed oscillator
(b) Clapp oscillator
(c) Crystal socillator
(d) Colpitt's oscillator

8. Crystal oscillators are used in transmitters because of their —
(a) frequency stability property
(b) low cost
(c) rigidness
(d) simplicity

9. The circuit shwon in the given figure is of a ... and it is preferred for generating frequencies above1 MHz —

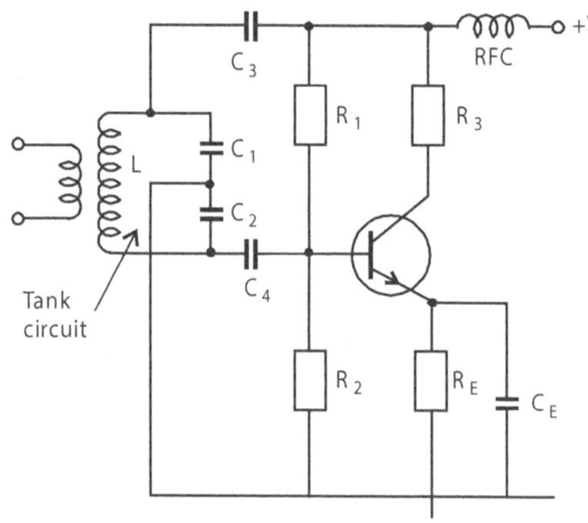

(a) Hartley parallel-fed oscillator
(b) Colpitt's oscillator

(c) Clapp oscillator
(d) Cyrstal oscillator

10. The frequency generated by a RC oscillator is given by the expression —

(a) $f = \frac{1}{2p \cdot R \cdot C}$ (b) $f = \frac{1}{2p\sqrt{R \cdot C}}$

(c) $f = \frac{1}{R \cdot C}$ (d) $f = 2\pi \cdot R \cdot C$

11. IC 555 is operated at —
(a) 3 to 10 volts d.c.
(b) 1 to 5 volts d.c.
(c) –5 to +5 volts d.c.
(d) +5 to +15 volts d.c.

12. An oscillator circuit which is meant for converting sine wave signal into square wave signal is called a —
(a) Schmmitt trigger
(b) blocking oscillator
(c) multivibrator
(d) Weinbridge oscillator

13. A sinusoidal oscillator is an amplifier with .... feedback.
(a) no (b) positive
(c) negative (d) zero

14. Crystal oscillator is used because
(a) it gives higher output voltage
(b) it has high efficiency
(c) the frequency of oscillations remains substantially constant.
(d) It requires very low D.C. supply voltage

15. Wein bridge oscillator is basically a
(a) pulse generator
(b) sine wave generator
(c) square wave generator
(d) triangular wave generator

16. In a Wein bridge oscillator with R = 51 kΩ and C = 0.001μ.F, frequency of oscillation will be
(a) 7.8 kilohertz
(b) 3.2 kilohertz
(c) 2.5 kilohertz
(d) 780 kilohertz

17. In an astable multivibrator, frequency of oscillation is given by
(a) 0.69 Rb . Cb (b) 1.38 Rb .Cb

(c) $\frac{1}{1.38\, R_b \,.\, C_b}$ (d) $\frac{1}{1.69\, R_b \,.\, C_b}$

18. A monostable multivibrator is widely used
(a) in counting circuits.
(b) in memory and timing circuits.
(c) for regeneration of distorted waves.
(d) for generating square waveforms.

19. Schmitt trigger is basically
(a) an astable multivibrator
(b) a bistable multivibrator.
(c) a monostable multivibrator
(d) an oscillator.

20. Trigger pulses are essential for the operation of a bistable multivibrator because
(a) there are no RC timing circuits
(b) the circuit cannot switch between its two states automatically.
(c) transistor operation is controlled by the triggering pulses.
(d) all of these

21. The frequency of oscillation in case of an astable multivibrator depends mainly on the
(a) value of transistor b.
(b) RC values of the circuit
(c) value of Vcc power supply.
(d) Collector resistors.

22. In a piezo electric crystal oscillator, the oscillation or turning frequency is linearly proportional to the
(a) mass of the crystal
(b) square root of the mass of the crystal
(c) square of the mass of the crystal
(d) inverse of the square root of the mass of the crystal

23. Circuit used for production of delay is
(a) an astable multivibrator
(b) a bistable multivibrator
(c) a monostable multivibrator
(d) none of the above

24. A 555 timer can be used as
(a) an astable multivibrator only
(b) a monostable multevibrator only
(c) a frequency divider only
(d) an astable multivibrator or a monostable multivibrator or a frequency divider

25. Square wave generator is a special type of
(a) variable frequency oscillator
(b) sweep frequency generator.
(c) function generator and pulse generator.
(d) none of the above.

26. IC 555 is available as a ....
(a) 4 pin DIL package
(b) 8 pin DIL package
(c) 14 pin DIL package
(d) 16 pin DIL package

(I.T.I. February 2016, Electronic Mechanic)

27. In IC 555, the pin No. 3 denotes ....
(a) VC.C (b) Output
(c) ground (d) trigger

(I.T.I. February 2016, Electronic Mechanic)

28. In function generator, the function selector is used to select the desired ....
(a) voltage (b) frequency
(c) wave form (d) amplitude

(I.T.I. February 2016, Electronic Mechanic)

29. The output waveform of the harmonic oscillator is ....
(a) square waveform
(b) triangular waveform
(c) sinusoidal waveform
(d) saw tooth waveform

(I.T.I. February 2016, Electrician)

30. Which type of oscillators are used in signal generator?
(a) Crystal oscillator
(b) Colpitt's oscillator
(c) Hartley oscillator
(d) Relaxation oscillator

(I.T.I. February 2016, Electrician)

31. Which electronic circuit produces signal waves or pulses without an input?
(a) Detector (b) Amplifier
(c) Oscillator (d) Modulator

(I.T.I. August 2019, Electrician)

32. Which type of oscillator is used in broadcast transmitters?
(a) Hartley oscillator
(b) Crystal oscillator
(c) R.C. phase shift oscillator
(d) Colpitt's oscillator

(I.T.I. August 2015, Electrician)

33. Which type of oscillator is used for generating frequency above 1 MHz?
(a) Colpitt's oscillator
(b) Hartley oscillator
(c) Crystal oscillator
(d) RC-phase shift oscillator

(I.T.I. February 2016, Electrician)

34. Which oscillator uses split capacitor in the tank circuit?
(a) Wein bridge oscillator
(b) RC phase shift oscillator
(c) Colpitt's oscillator
(d) Hartley oscillator

(I.T.I. February 2016, Electronic Mechanic)

35. The free running multivibrator is also known as ....
(a) bistable multivibrator
(b) monostable multivibrator
(c) astable multivibrator
(d) mono shot

(I.T.I. August 2015, Electrician)

36. The astable multivibrator can be used as a ....
(a) voltage controlled oscillator
(b) diode
(c) ripple filter
(d) power amplifier

(I.T.I. February 2016, Electrician)

37. Application of Astable multivibrator is ....
(a) electronic piano

(b) counter
(c) timer
(d) memory element

(I.T.I. February 2016, Electronic Mechanic)

38. Calculate the frequency produced by astable multivibrator if each stage is ON or OFF for time 0.5 milli second.

(a) 500 Hz (b) 1000 Hz
(c) 2000 Hz (d) 3000 Hz

(I.T.I. August 2014, Electronic Mechanic)

39. How many stable states has a bi-stable multivibrator?

(a) One (b) Two
(c) Three (d) Four

(I.T.I. February 2016, Electronic Mechanic)

40. The relaxation oscillator is one of the applications of ....

(a) Field Effect Transistor (FET)
(b) Bipolar transistor
(c) Uni–Junction Transistor (UJT)
(d) Silicon Controlled Rectifier (SCR)

(I.T.I. February 2016, Electrician)

41. An R.C. oscillator uses .....

(a) one R.C. combination
(b) two R.C. combination
(c) At least three R.C. combination
(d) Either (a) or (b)

(I.T.I. February 2018, Electrician)

42. A.R.C. oscillator uses :

(a) Positive feedback
(b) Negative feedback
(c) No feedback
(d) Both positive and negative feedbacks

43. Which circuit is essential to maintain oscillations or waves in an oscillator circuit?

(a) Rectifier with circuit
(b) Voltage multiplier
(c) Positive feedback
(d) Negative feedback

(I.T.I. August 2019, Electrician)

44. Which of the following oscillators is suitable for frequencies in the range of mega Hertz

(a) R.C. phase shift (b) Wein bridge
(c) Hartley (d) Both (a) and (c)

(I.T.I. February 2018, Electrician)

45. What is the formula to calculate the resonance frequency in an oscillator circuit?

(a) $F_r = \frac{1}{2pLC}$ (b) $F_r = \frac{1}{\sqrt{2pLC}}$

(c) $F_r = \frac{1}{2p\sqrt{LC}}$ (d) $F_r = \frac{1}{LC\sqrt{2p}}$

(I.T.I. August 2019 Electrician)

46. What is the average value of saturation factor in a practical circuit?

(a) 3 (b) 0
(c) 2 (d) 1

(I.T.I. August 2019 Electrician)
(I.T.I. February 2019, Electrician)

Semi-Conductor Devices

1. In amplifiers, transistors are used for —

(a) enlarging small signals
(b) reducing a signal
(c) converting an a.c. signal into d.c. signal
(d) all the above stated purposes

2. An amplifier in which the shape of applied input signal does not change is called —

(a) pulse amplifier
(b) linear amplifier
(c) non-linear amplifier
(d) none of these

3. The emitter bypass capacitor used in a CE type amplifier affects —

(a) the input impedance
(b) the gain of the circuit
(c) the frequency of the amplified signal
(d) all the above stated factors

4. In a *CE* amplifier which one of the following is an advantage of using an emitter bypass capacitor (C_E)?
(a) Increased gain
(b) Reduced input impedance
(c) Voltage gain changes throughout the input cycle
(d) All the above stated facts

5. Small signal amplifiers are also known as —
(a) large power amplifiers
(b) medium power amplifiers
(c) low power amplifiers
(d) both medium and low power amplifiers

6. Class 'A' amplifier conducts during —
(a) positive half cycle of the input only
(b) negative half cycle of the input only
(c) entire cycle of the input
(d) less than positive half cycle of the input

7. If a portion of output voltage is fed back to the input of an amplifier at a phase difference of 180º, the phenomenon is called —
(a) negative feedback
(b) positive feedback
(b) voltage feedback
(d) current feedback

8. A power amplifier is essentially —
(a) an impedance matching device
(b) a noise amplifier
(c) a current amplifier
(d) a voltage amplifier

9. The efficiency of a class 'B' amplifier is around—
(a) 10%-30% (b) 30%-50%
(c) 50%-60% (d) 100%

10. In class 'B' amplifier, the output current flows for—
(a) less than half input cycle
(b) more than half input cycle
(c) half input cycle
(d) entire input cycle

11. An Op-Amp. is designed to amplify —
(a) a.c. signals only
(b) d.c. signals only
(c) both a.c. and d.c. signals
(d) pulse signals only

12. A summing amplifier is basically a/an —
(a) voltage follower amplifier
(b) impedance matching amplifier
(c) inverting amplifier with more than one input
(d) none of the above stated

13. In an Op-Amp. the use of negative feedback increases —
(a) the frequency response
(b) stability of amplification
(c) gain
(d) none of the above stated

14. Which element is used as a semiconductor material?
(a) Copper (b) Plastic
(c) Silicon (d) Tungsten

15. Which is the element added to intrinsic semi-conductor material as an impurity to produce P-type semiconductor?
(a) Antimony (b) Phosphorus
(c) Bismuth (d) Indium

16. Name the element used as an impurity to produce N-type semiconductor.
(a) Antimony (b) Platinum
(c) Gallium (d) Indium

17. A Triac has three terminals viz ....
(a) Drain, source, gate
(b) Two main terminal and a gate teminal
(c) Cathode, anode, gate
(d) None of these

(I.T.I. February 2019, Electrician)

18. The given figure shows the digital multimeter display.
Which segments are to be connected to the power supply to show the number 5 at the display ?
(a) ABCDEF (b) ABCDG
(c) AFGCD (d) ABDEG

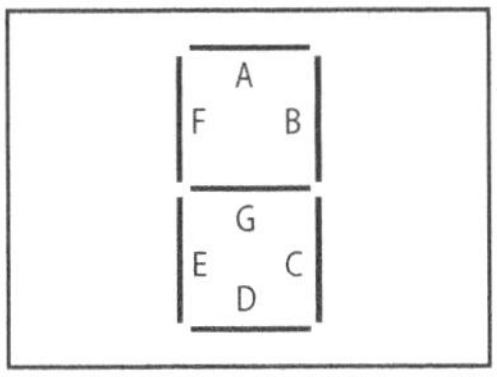

19. Which one of the symbols given below represents a FET ?

(a) 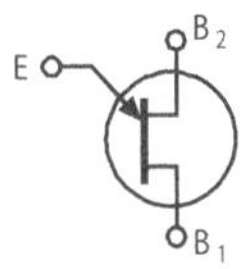

(b) 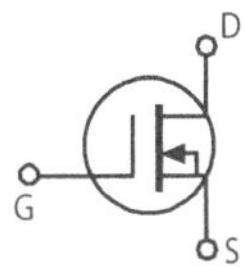

(c) 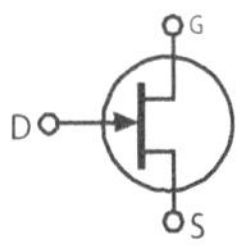

(d) 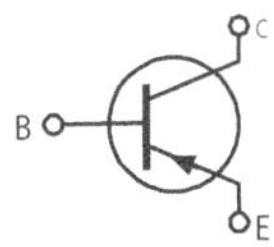

20. What will be the value of V_D in the given figure ?

(a) 0 V (b) 5 V

(c) 4.3 V (d) 0.7 V

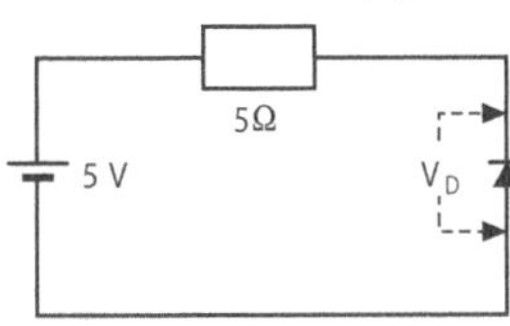

21. The time base of a CRO is adjusted to 20 ms/div. What is the frequency of the waveform shown in the figure.

(a) 25 Hz (b) 10 Hz

(c) 50 Hz (d) 25 kHz

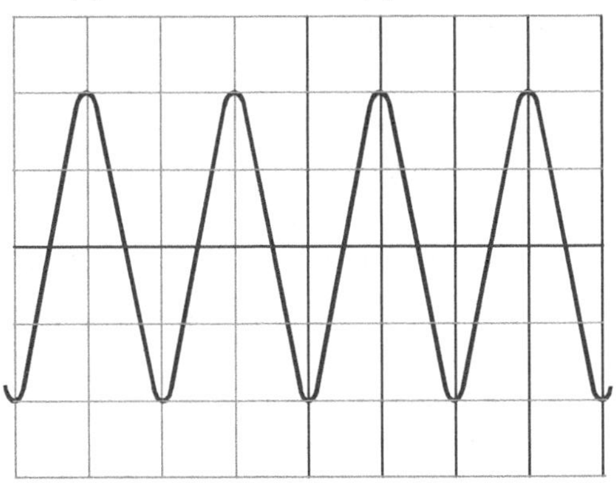

(**Hint :** The time of one cycle is 40 ms because setting is 20 ms / div.

f = d = d = 25 Hz)

22. The function of a zener diode is to maintain a constant:

(a) input voltage

(b) input current

(c) output voltage

(d) output current irrespective of load resistance.

23. Which part is called as heart of C.R.O.?

(a) Sweep generator

(b) C.R.T.

(c) Trigger circuit

(d) Amplifier

(I.T.I. February 2019, Electrician)

24. A triac is a

(a) unidirectional diode

(b) bidirectional thyristor

(c) a combination of two P.N. diodes

(d) none of these

25. A diac is equivalent to

(a) pair of SCRs

(b) triac with two gates

(c) pair of four layer SCRs

(d) diode with two transistors

26. Thyristor is

(a) P-N-P-N device (b) P-P-P-P device

(c) N-N-N-N device (d) none of these

27. A thyristor (SCR) is a

(a) currrent controlled device

(b) bilateral device

(c) voltage controlled device

(d) none of these

28. Thyristors are not made using germanium because

(a) it is a stable semiconductor

(b) group V elements cannot be doped with it

(c) its current handling capacity is less than silicon

(d) it has high leakage current.

29. Latching currrent of an SCR is .... of rated current.

(a) below 10% (b) 10% to 20%

(c) 20% to 25% (d) 25% to 30%

30. Suppose the anode current of a conducting SCR is 50 A. If its gate current is reduced to $\frac{1}{5}$ th, its anode current will become

(a) 10 A (b) 50 A
(c) 25 A (d) zero

31. In a class A amplifier, there will be a worst condition

(a) when signal input is zero
(b) when signal input is maximum
(c) when load resistance is high
(d) when there is a transformer coupling

32. In a class-A amplifier, conduction extends over 360°, because Q-point is

(a) located on load line
(b) located at or near cut-off point
(c) located near saturation point
(d) centred on load line

33. Transformer coupling is used in a class-A amplifier so as to make it

(a) more efficient (b) less bulky
(c) less costly (d) distortion free

34. The output of a class-B amplifier

(a) is distortion free
(b) comprises short duration current pulses.
(c) consists of positive half cycles only.
(d) is similar to that of a full wave rectifier.

35. A class-B push pull amplifier has the main advantage of being free from

(a) unwanted noise
(b) even order harmonic distortion
(c) amplitude distortion
(d) circuit imbalance

36. The type of power amplifier which exhibits cross over distortion in its output is

(a) class A (b) class B
(c) class AB (d) class C

37. Negative feedback in amplifier increases

(a) its gain
(b) its stability
(c) its distortion
(d) its gain or distortion

38. Which one is not the advantage of push pull amplifier?

(a) More output.
(b) Improved linearity.
(c) All harmonics (even or odd) removed.
(d) Balanced out effect of ripple in power supply.

39. Heat sink is used in the power amplifier circuit

(a) to increase the output power.
(b) to reduce the heat loss in a transistor.
(c) to increase the voltage gain of the power amplifier.
(d) to increase the collector dissipation ratings. of the transistor.

40. The negative feedback in amplifier

(a) reduces the voltage gain.
(b) increases the voltage gain.
(c) increases the gain bandwidth products.
(d) reduces the input impedance.

41. The voltage gain of an amplifier is 100. On applying negative feedback with $\beta = 0.04$, its gain will reduce to

(a) 4 (b) 20
(c) 60 (d) 99.96

42. An amplifier has gain without feedback as 10. To make it oscillate β must be

(a) 0.2 (b) 0.1
(c) 0.5 (d) 0.8

43. Feedback in amplifier always helps in

(a) controlling its output.
(b) increasing its gain.
(c) reducing its input impedance.
(d) stabilizes its gain.

44. An emitter follower is mainly used as

(a) an impedance matching device.
(b) a power amplifier.
(c) a follower of base signal.
(d) none of the above.

45. Emitter follower is an example of .... feedback

(a) negative current

(b) negative voltage
(c) positive current
(d) positive voltage

46. The ratio of latching current to holding current is
(a) 0.5 (b) 1.0
(c) 1.5 (d) 2.5

47. In a three stage amplifier the first stage gives a gain of 50, the second stage gives 10 and the third stage gives a gain of 5. Which one of the following is the overall gain of the amplifier ?
(a) 65 (b) 250
(c) 500 (d) 2500

48. A transistor is designated as "2N 3055". What does "2N" indicate?
(a) Indicates as diode
(b) Indicates the number of terminals
(c) Indicates the field effect transistor
(d) Indicates the two junction device (transistor)

(I.T.I. August 2015 Electrician)

49. The circuit shown in the given figure is a simple—

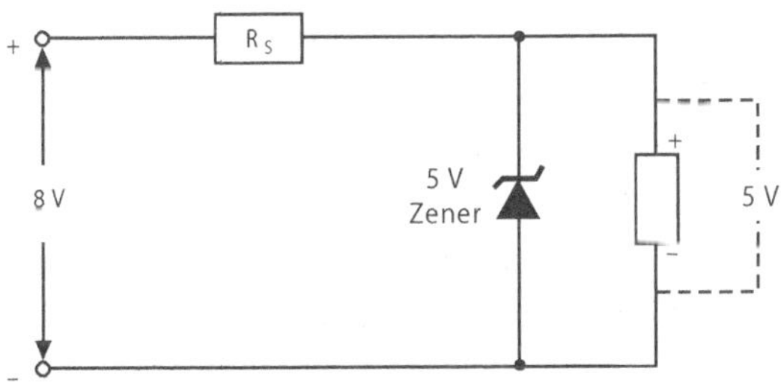

(a) inverter circuit
(b) H.W. rectifier circuit
(c) voltage regulator circuit
(d) zener controlled filter circuit

50. The power dissipation of a zener diode —
(a) is minimum when load current is zero
(b) is maximum when load current is zero
(c) is always zero
(d) remains constant irrespective of load current variations

51. The voltage drop across the series resistor (V_S) in zener voltage regulator is equal to —
(a) $V_{out} - V_z$ (b) $V_z - V_{out}$
(c) $V_{in} - V_z$ (d) $V_z - V_{in}$

52. The magnitude of current through a zener diode in a zener voltage regulator under full-load condition—
(a) is minimum
(b) is maximum
(c) remains constant irrespective of load
(d) is zero

53. Zener diodes are used as ....
(a) rectifiers (b) filters
(c) regulators (d) amplifier

(I.T.I. February 2016, Electronic Mechanic)

54. Which diode works in the breakdown region?
(a) Tunnel diode (b) Rectifier diode
(c) Zener diode (d) Pin diode

(I.T.I. February 2016, Electronic Mechanic)

55. Which of the following diodes is used in reverse biased condition for its normal working?
(a) Signal diode
(b) Rectifier diode
(c) Light emitting diode
(d) Zener diode

(I.T.I. February 2016, Wireman)

56. In which type of clipper, the positive half cycle of the input voltage is removed?
(a) Positive clipper
(b) Negative clipper
(c) Biased clipper
(d) Combination clipper

(I.T.I. February 2016, Electronic Mechanic)

57. Basic components required for a clamping circuits are ....
(a) diode, capacitor and resistor
(b) inductor, capacitor and transistor
(c) transistor, inductor, resistor
(d) diode, transformer and transistor

(I.T.I. February 2016, Electronic Mechanic)

58. Most of the transistors are made by using silicon compared to germanium semiconductor because, it has ....

(a) higher barrier voltage
(b) higher resistance range
(c) higher thermal stability
(d) higher current carrying capacity

(I.T.I. August 2015, Electrician)

59. When transistors are used in digital circuits, they usually operate in the ....
(a) active region
(b) breakdown region
(c) saturation and cut off region
(d) linear region

(I.T.I. August 2014, Instrument Mechanic)

60. Which one is the application of low power transistor?
(a) Loud speaker
(b) TV receiver
(c) Second stage amplifier
(d) Microphone

(I.T.I. February 2016, Electrician, Wireman)

61. The application of the medium power transistor is....
(a) microphone (b) audio amplifier
(c) transducer (d) tape head

(I.T.I. February 2016, Electrician)

62. Which one is the application of high power transistor?
(a) Microphone (b) Loud speaker
(c) Tape recorder (d) Transducers

(I.T.I. August 2015, Electrician)

63. Which type of transistor biasing is commonly used in digital circuits?
(a) Self bias
(b) Base bias
(c) Voltage divider bias
(d) Emitter bias

(I.T.I. August 2015, Electrician)

64. Which type of transistor biasing compensates for the variations in temperature in transistors?
(a) Base bias
(b) Self bias
(c) Voltage divider bias
(d) Emitter bias

(I.T.I. February 2016, Electrician)

65. Photo transistor is a combination of ......
(a) photodiode and bipolar transistor
(b) resistor and bipolar transistor
(c) photodiode and uni-polar transistor
(d) resistor and diode

(I.T.I. February 2016, Electronic Mechanic)

66. How many PN junctions are in PNP transistors?
(a) 4 (b) 3
(c) 2 (d) 1

(I.T.I. February 2016, Electrician)

67. How many layers are there in SCR?
(a) 2 (b) 3
(c) 4 (d) 5

(I.T.I. February 2016, Electronic Mechanic)

68. In NPN transistor, current carriers come from......
(a) emitter (b) collector
(c) base (d) V_{CC}

(I.T.I. February 2016, Electronic Mechanic)

69. The unit of input impedance of a transistor is ....
(a) ampere (b) ohm
(c) mho (d) volt

(I.T.I. February 2016, Electronic Mechanic)

70. Which layer of transistor is lightly doped?
(a) Vcc-terminal (b) Emitter region
(c) Base region (d) Collection region

(I.T.I. February 2016, Electronic Mechanic)

71. The NPN transistors are prepared more than PNP transistor, because ....
(a) it has higher switching speed
(b) it has lower switching speed
(c) it dissipates the heat quickly
(d) it consumes less power

(I.T.I. February 2016, Electrician)

72. If the transistor in the figure is in saturation, then .....

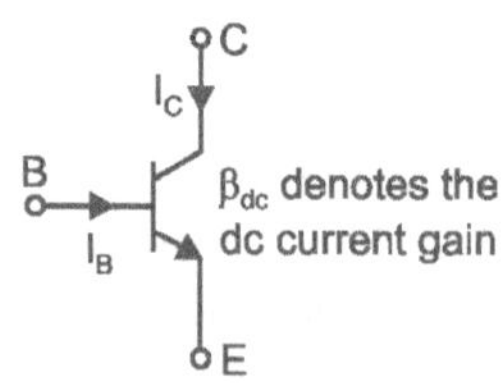

(a) I_C is always equal $\beta_{dc} I_B$.
(b) I_C is always equal to $-\beta_{dc} I_B$.
(c) I_C is greater than or equal to $\beta_{dc} I_B$.
(d) I_C is less than or equal to $\beta_{dc} I_B$.

73. The terminals of bipolar junction transistors are.....
(a) gate (G) collector (C) source (S)
(b) source (S) collector (C) emitter (E)
(c) base (B) collector (C) emitter (E)
(d) drain (D) emitter (E) base (B)
(I.T.I. Auguest 2015, Electrician)

74. In an NPN bipolar Junction transistor, the connections are arranged as follows:
(a) Base to P type, Emitter to N type material, Collector to N type
(b) Base to N type, Emitter to N type material, Collector to P type
(c) Base to N type, Emitter to P type material, Collector to N type
(d) None of the above
(I.T.I. February 2015, Mech. Consumer Electronics)

75. Bipolar transistors are .... than field effect transistors.
(a) Less sensitive and slower
(b) More sensitive and slower
(c) More sensitive and faster
(d) Less sensitive and faster
(I.T.I. February 2015, Mech. Consumer Electronics)

76. FET (Field Effect Transistor) is also known as ....
(a) bipolar transistor
(b) unipolar transistor
(c) Current controlled device
(d) power controlled device
(I.T.I. August 2015, Electrician)

77. FET is a ....
(a) current controlled device
(b) voltage controlled device
(c) controlled device
(d) None of the above
(I.T.I. February 2015, Mech. Consumer Electronics)

78. Which of the following is a voltage controlled device?
(a) SCR (b) FET
(c) BJT (d) UJT
(I.T.I. February 2016, Electronic Mechanic)

79. The application of UJT is ....
(a) fitter circuits
(b) power amplifier
(c) relaxation oscillator
(d) sinusoidal oscillator
(I.T.I. August 2015, Electrician)

80. Application of UJT is ....
(a) generation of sine wave
(b) generation of square wave
(c) generation of saw tooth wave
(d) generation of rectangular wave
(I.T.I. February 2016, Electronic Mechanic)

81. Which of the following is NOT considered as the characteristic of a UJT?
(a) Peak point current
(b) Valley current
(c) Valley voltage
(d) Maximum permitted peak voltage
(I.T.I. February 2016, Electronic Mechanic)

82. Consider the following statements :
1. BJT is a current controlled device with a high input impedance and high gain bandwidth.
2. FET is a voltage controlled device with high input impedance and low gain bandwidth.
3. UJT is a negative resistance device and can be used as an oscillator.
4. BJT, FET and UJT can all be used for amplification

Which of the statements given above are correct ?
(a) 1 and 2 (b) 2 and 3
(c) 3 and 4 (d) 1 and 4

83. Mobility of electrons in N-channel JEFT and mobility of holes in P-cannel JEFT are ....
(a) large, poor (b) poor, large
(c) large, large (d) poor, poor

84. IGBT stands for ....
- (a) Integrated Gate Bipolar Transistor
- (b) Intrinsic Gallium Bipolar Thyristor
- (c) Insulated Gate Bipolar Transistor
- (d) Integrated Gain Bipolar Thyristor

(I.T.I. August 2015, Electrician)

85. IGBT stands for ....
- (a) Integrated Gate Bi-directional Transistor
- (b) Integrated Gate Bipolar Transistor
- (c) Integrated Gate Blocky Transistor
- (d) Insulated Gate Bipolar Transistor

(I.T.I. February 2016, Electrician, Wireman)

86. IGBT is a ....
- (a) current controlled device
- (b) phase controlled device
- (c) voltage controlled device
- (d) frequency controlled device

(I.T.I. February 2016, Electronic Mechanic)

87. The BIS symbol shown in the figure represents .....

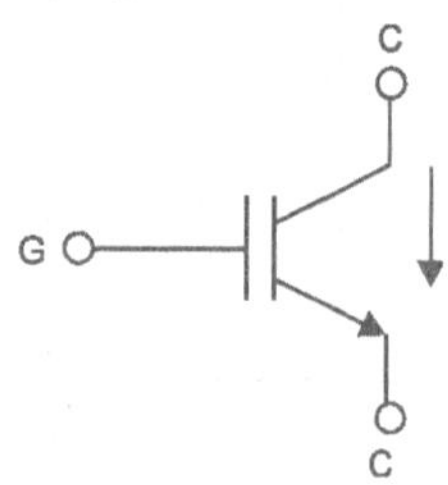

- (a) GTO (b) SCR
- (c) FET (d) IGBT

(I.T.I. February 2016, Electrician)

88. GTO stands for ....
- (a) Gated Tuned Off Thyristor
- (b) Gate Turn Off Thyristor
- (c) Gain Triggered Off Transistor
- (d) Gate Triggering Off Thyristor

(I.T.I. August 2015, Electrician)

89. One of the application of GTO is ....
- (a) high power amplifier
- (b) Single diodes
- (c) induction heating
- (d) oscillators

(I.T.I. August 2015, Electrician)

90. Input impedance of MOSFET is ....
- (a) Less than of FET but more than BJT
- (b) More than that of FET and BJT
- (c) More than that of FET but less than BJT
- (d) Less than that of FET and BJT

(I.T.I. February 2015, Mech. Consumer Electronics)

91. In MOSFET, the insulating layer is generally made of ....
- (a) gallium dioxide (b) silicon dioxide
- (c) germanium plate (d) bismuth plate

(I.T.I. August 2015, Electrician)

92. Which one of the following is an application of SCR ?
- (a) Voltage amplifier
- (b) Speed control of motors
- (c) Oscillators
- (d) Signal generator

(I.T.I. February 2016, Electrician)

93. The speed of the A.C. motors using SCR can be controlled by varying ....
- (a) the holding current of SCR
- (b) the firing angle of the SCR
- (c) the load current
- (d) the gate current

(I.T.I. August 2015, Electrician)

94. The function of snubber circuit connected across the SCR is to –
- (a) Suppress dv/dt (b) Increase dv/dt
- (c) Decrease dv/dt (d) Decrease di/dt

(I.T.I. February 2015, Mech. Consumer Electronics)

95. The formula to find A.C. current gain (A_i) of CE amplifier is given as ....
- (a) $A_i = \frac{i_b}{i_c}$ (b) $A_i = \frac{i_c}{i_b}$
- (c) $A_i = i_c \times i_b$ (d) $A_i = i_c + i_b$

(I.T.I. August 2015, Electrician)

96. In common emitter amplifier the output characteristic curves can be obtained by plotting ....
- (a) I_C on the "Y" axis *VS* V_{CE} on the "X" axis

(b) I_B on the "Y" axis *VS* V_{BE} on the "X" axis
(c) V_{BE} on the "Y" axis *VS* I_B on the "X" axis
(d) V_{CE} on the "Y" axis *VS* I_C on the "X" axis

(I.T.I. August 2015, Electician)

97. The cascode amplifier is a multistage configuration of –

(a) CC-CB (b) CE-CB
(c) CB-CC (d) CE-CC

(I.T.I. February 2015, Mech. Consumer Electronics)

98. What is the effect of cascading the amplifier stages ?

(a) To increase the voltage gain and increase the bandwidth
(b) To increase the voltage gain and reduce the bandwidth
(c) To decrease the voltage gain and increase the bandwidth
(d) To decrease the voltage gain and reduce the bandwidth

99. The power gain (A_P) of transistor amplifier is expressed as A_P = ...

(a) $A_P = \frac{A_V}{A_i}$ (b) $A_P = A_v . A_i$

(c) $A_P = \frac{A_i}{A_V}$ (d) $A_P = A_i + A_v$

(I.T.I. February 2016, Electrician, Wireman)

100. Which type of amplifier has the greatest power gain?

(a) Common Emitter Amplifier
(b) Common Collector Amplifier
(c) Common Base Amplifier
(d) Voltage Series Feedback Amplifier

(I.T.I. August 2015, Electrician)

101. Which one of the following is the power amplifier?

(a) Common Base Amplifier
(b) Common Collector Amplifier
(c) Operational Amplifier
(d) Class B Amplifier

(I.T.I. February 2016, Electrician)

102. Which coupling is also called as impedance coupling?

(a) RC coupling
(b) LC coupling
(c) Direct coupling
(d) Transformer coupling

(I.T.I. February 2016, Electronic Mechanic)

103. Which type of power amplifier is used as the first stage amplifier?

(a) Class A power amplifier
(b) Class B power amplifier
(c) Class C power amplifier
(d) Class AB power amplifier

(I.T.I. February 2016, Electrician)

104. The power gain of the amplifiers is usually expressed in ....

(a) percentage (%) (b) number
(c) decibel (dB) (d) watt

(I.T.I. February 2016, Electrician)

105. The gain of an op. Amp is specified at different frequencies is termed as ....

(a) frequency gain
(b) gain band width product
(c) band width gain
(d) op. Amp voltage gain

(I.T.I. August 2015, Electrician)

106. The Op – Amp circuit that contains feedback capacitor is ....

(a) integrator
(b) differentiator
(c) comparator
(d) summing amplifier

(I.T.I. February 2016, Electronic Mechanic)

107. An operational amplifier is designed to amplify ....

(a) AC voltage (b) DC voltage
(c) Both AC and DC voltage
(d) Pulse signals only

(I.T.I. February 2016, Electronic Mechanic)

108. An ideal operational amplifier has ....

(a) Infinite output impedance
(b) Zero input impedance
(c) Infinite band width
(d) All of these

(I.T.I. August 2014, and I.T. & E.S.M.)

109. Which type of amplifiers provide good noise rejection?
(a) Class A amplifier
(b) Class C amplifier
(c) Push Pull amplifier
(d) Direct coupled amplifier
(I.T.I. February 2016, Electrician)

110. Generally, the gain of transistor amplifier falls at high frequencies due to the ....
(a) Internal capacitances of the device
(b) coupling capacitor at the input
(c) skin effect
(d) coupling capacitor at the output
(I.T.I. February 2015 Mech. Consumer Electronic)

111. The BJT amplifier which offers highest input impedance and least voltage gain is ....
(a) CE (b) CB
(c) CC
(d) Cascode amplifier

112. One of the disadvantages of the class B push pull amplifier is ....
(a) the single ended input and output
(b) the positive half cycle only get amplified
(c) having a cross over distortion
(d) having centre tapped drive transformer
(I.T.I. August 2015, Electrician)

113. A summing amplifier is basically ....
(a) an inverting amplifier with more than one input
(b) non inverting amplifier with more than one input
(c) an impedance matching circuit
(d) a voltage follower
(I.T.I. February 2016, Electronic Mechanic)

114. The emitter follower configuration is frequently used as ....
(a) an amplifier circuits
(b) switching circuits
(c) negative feedback circuits
(d) an impendance matching circuits
(I.T.I. August 2015, Electrician)

115. The maximum output voltage of IC 7812 regulator is ....
(a) 12 V (b) 10 V
(c) 5 V (d) 2 V
(I.T.I. February 2016, Electronic Mechanic)

116. In Large Scale integration (LSI) IC, the number of transistors are built in it with the range of ....
(a) 1 to 10 transistors
(b) 10 to 100 transistors
(c) 100 to 1000 transistors
(d) 1000 and above
(I.T.I. February 2016, Electrician)

117. What is the other name of Printed Circuit Boards (PCB)
(a) Copper laminate boards
(b) Tag boards
(c) Lug boards
(d) Bread boards
(I.T.I. February 2016, Wireman)

118. The process of removing portions of a metal foil on PCB is known as ....
(a) lamination (b) etching
(c) desoldering (d) masking
(I.T.I. February 2016, Wireman)

119. Which one of the following chemical is not used for etching a PCB —
(a) Cupric chloride
(b) Ferric chloride
(c) Carbon tetra chloride
(d) Alkaline ammonia

120. Etching of a PCB's circuit side requires about —
(a) 5 to 20 minutes (b) 20 to 40 minutes
(c) 40 to 60 minutes (d) 60 to 80 minutes

121. The circular rings of copper made on PCBs for providing solder connections of components are called—
(a) patches (b) islands
(c) discs (d) pads

122. The side of a PCB on which the components are mounted is called —
(a) copper side (b) solder side
(c) track side (d) components side

123. Which one is NOT an advantage of LED ?
(a) LED have no filaments
(b) LED lasts much longer
(c) LED terminals get rusted
(d) LED are always cool
(I.T.I. August 2015, Wireman)

124. The forward voltage drop of green colour LED is ....
(a) 2.2 volt (b) 2.1 volt
(c) 2.0 volt (d) 1.8 volt
(I.T.I. August 2016, Electrician)

125. Which colour LED has the lowest forward voltage drop?
(a) Red (b) Orange
(c) Green (d) Yellow
(I.T.I. February 2016, Electronic Mechanic)

126. The forward voltage drop of red colour LED is ....
(a) 1.8 volt (b) 2.0 volt
(c) 2.1 volt (d) 2.2 volt
(I.T.I. February 2016, Wireman)

127. Which one of the following LED emitts an invisible light –
(a) Laser LED (b) Infra-red LED
(c) Bicolour LED (d) Tricolour LED

128. Terminals of a DIAC are ....
(a) MT_1 and MT_2
(b) anode and cathode
(c) base 1 and base 2
(d) anode 1 and anode 2
(I.T.I. February 2016, Electronic Mechanic)

129. What are the terminal markings in TRIAC?
(a) Source (S), Gate (G), Drain (D)
(b) Main terminal one (MT_1), Main terminal two (MT_2), Gate (G)
(c) Base (B), Collector (C), Emitter (E)
(d) Anode one, Anode Two
(I.T.I. August 2015, Electrician)

130. When TRIAC is 'ON', the current flowing between MT_1 and MT_2 is known as ....
(a) principal current (b) latching current
(c) holding current (d) trigger current
(I.T.I. February 2016, Electronic Mechanic)

131. Which is the common terminal both in TRIAC and SCR?
(a) Gate (b) Anode
(c) Cathode (d) MT_1 and MT_2
(I.T.I. February 2016, Electronic Mechanic)

132. The expansion of D.S.O. is ....
(a) Digital storage oscilloscope
(b) Dual storage oscilloscope
(c) Digital storage oscillator
(d) Dual storage oscillator
(I.T.I. February 2016, Electronic Mechanic)

133. The vertical amplitude of any wave form in C.R.O. indicates ....
(a) R.M.S. values (b) average values
(c) instantaneous values
(d) peak to peak values
(I.T.I. August 2015, Electrician)

134. Which control switch is used to control the sharpness of the trace in CRO?
(a) Intensity control switch
(b) Time base switch
(c) CAL - switch
(d) Alternator switch
(I.T.I. February 2016, Electrician)

135. If the secondary voltage of a transformer is fed to a CRO then which one of the following value will appear on the screen –
(a) Average voltage
(b) R.M.S. voltage
(c) Peak voltage
(d) Peak to peak voltage

136. Principally C.R.O. is a ......
(a) Ammeter (b) Watt hour mater
(c) Wattmeter (d) Voltmeter
(I.T.I. February 2019, Electrician)

137. CRO gives that visual representation of time varying signals. The display of the signal is .....
(a) One dimensional
(b) Four dimensional
(c) Three dimensional
(d) Two dimensional
(I.T.I. February 2019, Electrician)

138. An oscilloscope cannot be used to incidate –

(a) Frequency
(b) Peak signal voltage
(c) Energy
(d) Wave shape

(I.T.I. August 2016, Electrician)

139. The base of a transistor is .... doped.

(a) heavily (b) moderately
(c) lightly (d) None of these

(I.T.I. February 2019, Electrician)

140. The collector of a transistor is .... doped

(a) heavily (b) very heavily
(c) lightly (d) moderately

(I.T.I. February 2019, Electrician)

141. A transistor is a .... operated device.

(a) Both voltage and current
(b) Voltage
(c) Current
(d) None of these

(I.T.I. February 2019, Electrician)

142. The element that has the biggest size in a transistors is ....

(a) Collector-Base- junction
(b) Base
(c) Emitter
(d) Collector

(I.T.I. February 2019, Electrician)

143. In a CE amplifier circuit the A.C. voltage between emitter and ground ......

(a) is very high
(b) may be high or low
(c) is zero
(d) depends upon circuit cofiguration

(I.T.I. February 2018, Electrician)

144. Photo diode is used for detection of ......

(a) Visible light (b) Invisible light
(c) No light (d) Both (a) & (b)

(I.T.I. February 2018, Electrician)

145. An LED emits colour due to ......

(a) Transition of an electron from valence band to conduction band
(b) Transition of a hole from valance band to conduction band
(c) Transition of an electron from inner shell to valence shell
(d) Transistion of an electron from higher energy band to lower energy band

(I.T.I. February 2017, Wireman)

146. Colour of light emitted by LED depends on .....

(a) Its forward bias
(b) Its reverse bias
(c) Forward current
(d) Semiconductor material

(I.T.I. February 2018, Electrician)

147. One that is based on forward biased PN junction is ....

(a) Photo diode (b) LED
(c) Photo voltaic cell (d) Both (a) & (b)

(I.T.I. February 2018, Electrician)

148. How will electrons flow through a n-channel JFET?

(a) From source to gate
(b) From drain to gate
(c) From source to drain
(d) From drain to source

(I.T.I. February 2017, Electrician)

149. An NPN transistor conducts load current with which of the following biasing :

(a) Base diode reverse biased, collector diode reverse biased
(b) Base diode forward biased, collector diode forward biased
(c) Base diode reverse biased, collector diode forward biased
(d) Base diode forward biased, collector diode reverse biased

(I.T.I. February 2017, Wireman)

150. In a BJT circuit, a PNP transistor is replaced by NPN transistor. To analyse the new circuit.....

(a) All calculations done earlier have to be repeated
(b) Replace all calculated voltages by reverse values

(c) Replace all calculated currents by reverse values
(d) Replace all calculated voltages and currents by reverse values

(I.T.I. February 2018, Electrician)

151. A JFET –
(a) Is a current-controlled device
(b) Has a low input resistance
(c) Is a voltage-controlled device
(d) Is always forward-biased

(I.T.I. August 2016, Electrician)

152. The three terminals of the IGBT are –
(a) Base, emitter & collector
(b) Gate, source & drain
(c) Gate, emitter & collector
(d) Base, source & drain

(I.T.I. August 2016, Electrician)

153. The three terminals of power MOSFET –
(a) Collector, Emitter, base
(b) Drain, source, base
(c) Drain, source, gate
(d) Collector, emitter, gate

(I.T.I. August 2016, Electrician)

154. The depletion-mode MOSFET –
(a) Can operate with only positive gate voltages
(b) Can operate with only negative gate voltages
(c) Cannot operate in the ohmic region
(d) Can operate with positive as well as negative gate voltages

(I.T.I. August 2016, Electrician)

155. The number of PN junction in a SCR is/are ....
(a) 1 (b) 2
(c) 3 (d) 4

(I.T.I. February 2017, Electrician)

156. Circuit turn off time of an SCR is defined as the time –
(a) Taken by the SCR turn to be off
(b) Required for the SCR current to become zero
(c) For which the SCR is reverse biased by the commutation circuit
(d) For which the SCR is reverse biased to reduce its current

(I.T.I. February 2019, Electrician)

157. An SCR is considered to be a semi-controlled device because–
(a) It can be turned OFF but not ON with a gate pulse
(b) It conducts only during one half cycle of an alternating current wave
(c) It can be turned ON but not OFF with a gate pulse
(d) It can be turned ON only during one half-cycle of alternating voltage

(I.T.I. February 2019, Electrician)

158. Gate triggering is done when SCR is in –
(a) Reverse biased condition
(b) Foward biased condition
(c) Either forward or reverse
(d) None of these

(I.T.I. February 2019, Electrician)

159. Which triggering is the most reliable?
(a) Forward voltage triggering
(b) Thermal triggering
(c) dv / dt triggering
(d) Gate triggering

(I.T.I. February 2019, Electrician)

160. Holding current is –
(a) Minimum anode current required above which thyristor gets turn on
(b) Minimum current below which anode current must fall for turning on
(c) Minimum anode current above which thyristor gets turn off
(d) Minimum current below which anode current must fall for turning off

(I.T.I. February 2019, Electrician)

161. Which device is made up of using the methods of point contact, grown, diffusion and alloy junctions?
(a) Inductor (b) Resistor
(c) Capacitor (d) Transistor

(I.T.I. August 2019 Electrician)

162. Why the collector region is physically made larger than emitter region in a transistor?
(a) It has to dissipate more heat
(b) Output taken from collector terminal
(c) Base collector region is reverse biased
(d) Collector region always operate with high voltage
(I.T.I. August 2019 Electrician)

163. What is the function of a transistor if emitter to base and collector to base are forward biased?
(a) Acts as an Amplifier
(b) Acts as an oscillator
(c) Acts as an open circuit
(d) Acts as a closed switch
(I.T.I. August 2019 Electrician)

164. Transistor is used as ...
(a) Resistor (b) Rectifier
(c) Amplifier (d) All of these
(I.T.I. August 2019 Electrician)

165. In an NPN transistor, the arrow indicates ....
(a) the emitter
(b) the base
(c) depends on the configuration
(d) the collector
(I.T.I. August 2019 Electrician)

166. In the operation of an NPN transistor, the electrons cross which region?
(a) Emitter region
(b) The region where there is high depletion
(c) "P" type base region
(d) The region where there is low depletion
(I.T.I. August 2019 Electrician)

167. A thyristor equivalent of a thyratron tube is ....
(a) Diac
(b) Triac
(c) MOSFET
(d) Silicon controlled rectifier
(I.T.I. August 2019 Electrician)

168. In a thyristor the ratio of latching current to holding current is ....
(a) 0.5 (b) 1
(c) 5 (d) 2.7
(I.T.I. August 2019 Electrician)

169. A thyristor is basically .....
(a) a set of S.C.R., diac and a triac
(b) a combination of diac and triac
(c) a set of S.C.Rs
(d) PNPN device
(I.T.I. August 2019 Electrician)

170. In series connected thyristors ....
(a) L is used for tuning out junction capacitance
(b) R, C is called snubber circuit
(c) L and C is used for filtering out the ripple
(d) L is intended to increase $\frac{di}{dt}$ at switch on
(I.T.I. August 2019 Electrician)

171. LED stands for ....
(a) Listing device
(b) Line emitting device
(c) Light emitting device
(d) Limen detector
(I.T.I. August 2019 Electrician)

172. What is the work of a driver in case of an LED?
(a) To convert A.C. to D.C.
(b) To convert D.C. to A.C.
(c) Both (a) and (b)
(d) None of these
(I.T.I. August 2019 Electrician)

173. How many junctions does UJT have?
(a) 2 (b) 1
(c) 3 (d) 4
(I.T.I. August 2019 Electrician)

174. What is the main application of unijunction transistor?
(a) Rectification
(b) Amplification
(c) Regulator circuits
(d) Triggering circuits
(I.T.I. August 2019 Electrician)

175. In a BJT, the base current (I_B) is about .... of emitter current (I_E).
(a) 5% (b) 20%
(c) 25% (d) 50%
(I.T.I. August 2019 Electrician)

176. Which among the following indicates early effect in BJT?
(a) Zener break down
(b) Thermal break down
(c) Avalanche break down
(d) Base narrowing
(I.T.I. August 2019 Electrician)

177. Which device has very high input impedance, low noise output, good linearity and low inter electrode capacity?
(a) NPN transistor
(b) PNP transistor
(c) Field effect transistor
(d) Unijunction transistor
(I.T.I. August 2019 Electrician)

178. What is the difference in current control of MOSFET compared to JFETs?
(a) Insulating layer instead of junction
(b) Using N material instead of P material
(c) Using P material instead of N material
(d) Using N material gate instead of P material
(I.T.I. August 2019 Electrician)

179. What is the cause for the fault if the output voltage of UPS is higher than normal?
(a) Battery got short circuited
(b) Defective feed back circuit
(c) Input voltage is very high
(d) Relay points are joined together
(I.T.I. August 2019 Electrician)

180. What is the value of input voltage, if AB = 1 and input voltage is 5 volts?
(a) 0 (b) 5
(c) 2 (d) 10
(I.T.I. August 2019 Electrician)

181. What is the main advantage of a class A amplifier?
(a) Minimum distortion
(b) Maximum current gain
(c) Maximum voltage gain
(d) Minimum signal to noise ratio losses
(I.T.I. August 2019 Electrician)

182. What is the name of amplifier?

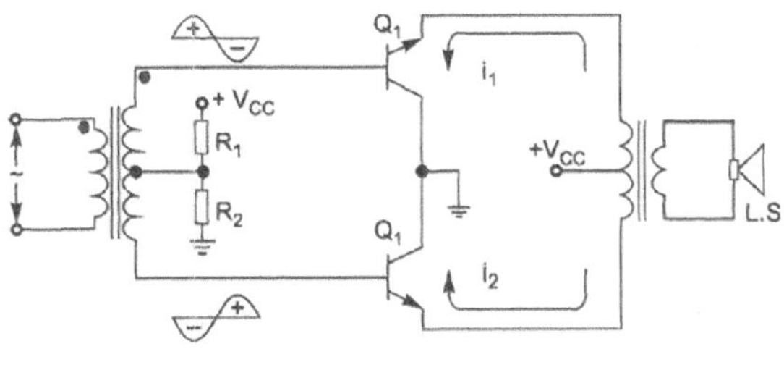

(a) Common emitter amplifier
(b) Class B push pull amplifier
(c) Common collector amplifier
(d) Class AB push pull amplifier
(I.T.I. August 2019 Electrician)

183. What is the name of amplifier circuit?

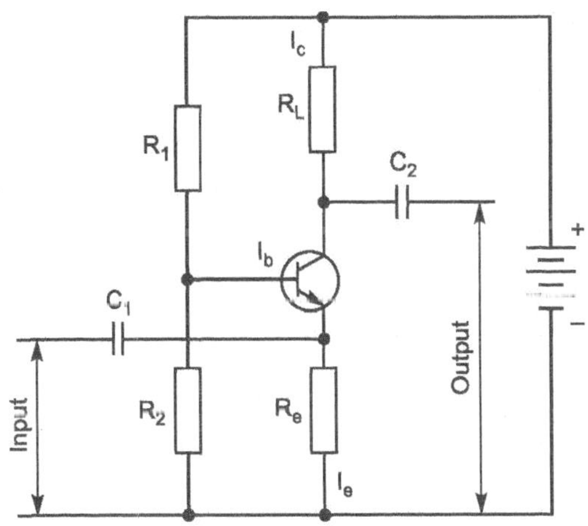

(a) Common base amplifier
(b) Common emitter amplifier
(c) Class B push pull amplifier
(d) Common collector amplifier
(I.T.I. August 2019 Electrician)

184. Front panel of a C.R.O. has .... sections.
(a) 15 (b) 8
(c) 4 (d) 20
(I.T.I. August 2019 Electrician)

185. What is the role of the focus section in a C.R.O.?

(a) Increases the focus
(b) Decreases the focus
(c) Controls sharpness
(d) Maintains the focus zero

(I.T.I. August 2019 Electrician)

186. Which instrument provides a visual representation of measure or tested quantities?
(a) Voltage stabiliser
(b) Function generator
(c) Cathode ray oscilloscope
(d) Radio frequency generator

(I.T.I. August 2019 Electrician)

187. The advantage of S.C.S. over S.C.R. is ...
(a) Slow switching time and large V_H
(b) Slow switching time and smaller V_H
(c) Faster switching time and large V_H
(d) Faster switching time and smaller V_H

(I.T.I. August 2019 Electrician)

Boolean Algebra and Logic Gates

1. A logic is an electronic circuit which......
(a) operates on binary algebra
(b) performs arithmetic and logic functions
(c) allows flow of electrons only in one direction
(d) alternates between 0 and 1 values

2. The voltage levels for a negative logic system
(a) must necessarily be negative
(b) could be negative or positive
(c) must necessarily be positive
(d) must necessarily be either zero or –5 V

3. The following truth table is for which gate?

A	B	Y
0	0	0
0	1	1
1	0	1
1	1	1

(a) OR (b) AND
(c) NAND (d) Ex. OR

(I.T.I. August 2015, Electronics Mechanic Trade Theory)

4. If negative logic is used, the diode gate shown in the given figure will represent......

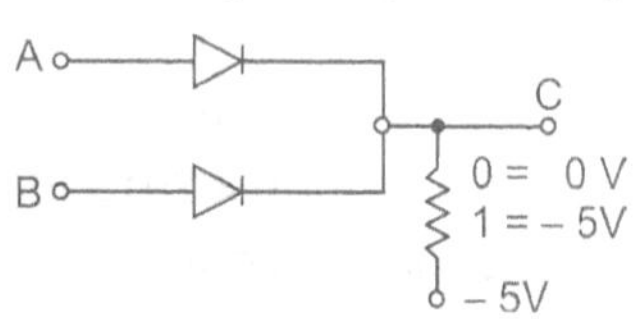

(a) OR gate (b) AND gate
(c) NOR gate (d) NAND gate

5. The circuit shown in the given figure is......

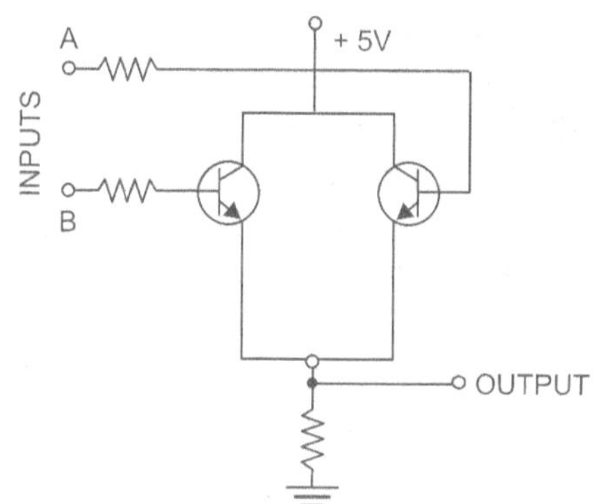

(a) an AND gate (b) an OR gate
(c) a XOR gate (d) a NAND gate

6. An AND gate......
(a) implements logic addition
(b) gives high output only when all inputs are low
(c) is equivalent to a series switching circuit
(d) is equivalent to a parallel switching circuit

7. Identify the type of gate below from the equation:

$$X = A \oplus B = \overline{A} \cdot \overline{B} + A . \overline{B}$$

(a) Ex. NOR gate (b) OR gate
(c) Ex. OR gate (d) NAND gate

(I.T.I. August 2015, I.T. & E.S.M.)

8. According to De-Morgan's second theorem ......

(a) A NAND gate is always complementary to an AND gate.

(b) An AND gate is equivalent to a bubbled NAND gate.

(c) A NAND gate is equivalent to a bubbled AND gate.

(d) A NAND gate is equivalent to a bubbled OR gate.

9. The complete set of only those Logic Gates designated as Universal Gates is......

(a) NOT, OR and AND gates

(b) XNOT, NOR and NAND gates

(c) NOR and NAND gates

(d) XOR, NOR and NAND gates

10. Which of the following expressions is in the sum of products (SOP) form?

(a) (A + B) . (C + D) (b) (A) . B . (C . D)

(c) A . B (C . D) (d) A . B + C . D

(I.T.I. August 2014, Instrument Mechanic)

11. The output of a logic gate is '1' when all its inputs are at logic "0". The gate is either......

(a) a NAND or an EX-OR gate

(b) a NOR or an EX-OR gate

(c) an AND or an EX-NOR gate

(d) a NOR or an EX-NOR gate

12. The Boolean expression C + C . D is equal to ....

(a) C (b) C . D

(c) C + D (d) 1

(I.T.I. August 2014, Instrument Mechanic)

13. The Boolean expression C + $\overline{C}$. D is equal to ....

(a) C (b) C . D

(c) C + D (d) 1

(I.T.I. August 2014, Instrument Mechanic)

14. Derive the Boolean expression for the logic circuit shown below:

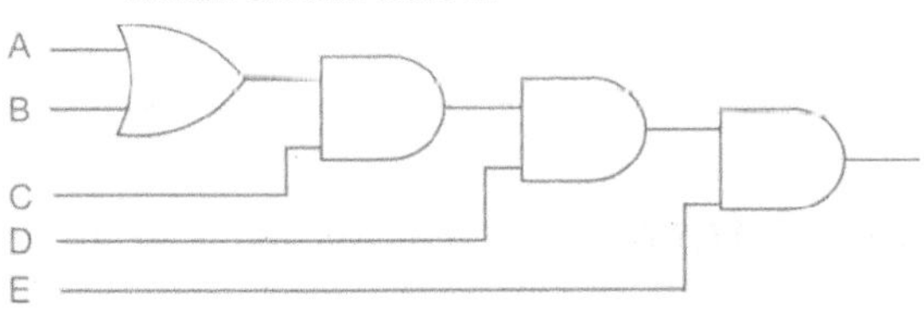

(a) C · (A + B) · D · E

(b) C · (A + B) · D · E

(c) C · (A + B) · D · $\overline{E}$

(d) A · B · C · D · E

(I.T.I. August 2014, Instrument Mechanic)

(I.T.I. August 2015, I.T. & E.S.M.)

15. The BIS symbol shown in the figure represents ....

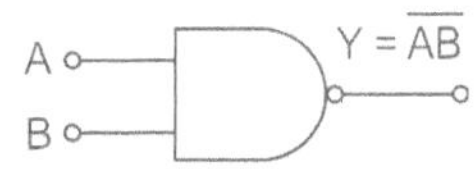

(a) OR gate (b) NAND gate

(c) NOR gate (d) AND gate

(I.T.I. February 2016, Electrician Trade Theory)

16. The following truth table is for which gate?

A	Y
0	1
1	0

(a) OR gate (b) AND gate

(c) NOT gate (d) EX-OR gate

(I.T.I. February 2016, Electronics Mechanic, Mech. Consumer Electronics, Mechanic Consumer Electronics Appliances Trade Theory)

17. Determine the values of A, B, C and D that make the product term $\overline{A} \cdot B \cdot \overline{C} \cdot D$ equal to 1.

(a) A = 0; B = 1; C = 0; D = 1

(b) A = 1; B = 1; C = 0; D = 1

(c) A = 0; B = 1; C = 1; D = 1

(d) A = 0; B = 1; C = 0; D = 0

(I.T.I. August 2014, Instrument Mechanic)

18. The given figure shows electrical equivalent of ... gate —

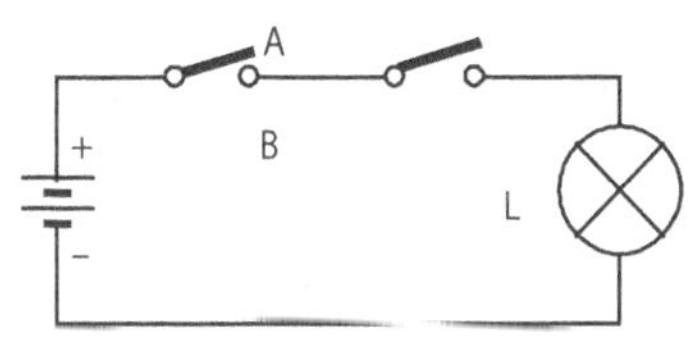

(a) OR (b) NOR

(c) NOT (d) AND

19. The logic gate which has high output when its inputs are different (low-high or high-low) is —

(a) Ex-OR (b) Ex-NOR
(c) NAND (d) NOR

20. If the inputs of the "NAND" gate are "A & B" then the output (Y) = ...

(a) Y = A . B (b) $Y = \overline{A \cdot B}$
(c) Y = A + B (d) $Y = \overline{A + B}$

(I.T.I. August 2015, Electrician)

21. Full adder is a —

(a) Sequential circuit
(b) Combinational circuit
(c) Combinational circuit with feedback
(d) None of these

(I.T.I. August 2016, Electrician)

22. The output of an AND gate is LOW —

(*a*) All the time
(*b*) When any input is LOW
(*c*) When any input is HIGH
(*d*) When all inputs are HIGH

(I.T.I. August 2016, Electrician)

23. The output of an exclusive-NOR gate is 1. Which input combination is correct?

(a) A = 1, B = 0 (b) A = 0, B = 1
(c) A = 0, B = 0 (d) None of these

(I.T.I. August 2016, Electrician)

24. Which of the following comes under the category of combinational circuits?

(*i*) Encoder (*ii*) Counter
(*iii*) Adder (*iv*) Register

(a) i & iii (b) i & iv
(c) ii & iii (d) ii & iv

(I.T.I. February 2017, Electrician)

KEY TO MULTIPLE CHOICE QUESTIONS

Basic Electronics

1. (c)	**2.** (c)	**3.** (d)	**4.** (c)	**5.** (d)	**6.** (c)	**7.** (a)	**8.** (a)	**9.** (d)	**10.** (a)
11. (b)	**12.** (d)	**13.** (c)	**14.** (a)	**15.** (b)	**16.** (b)	**17.** (c)	**18.** (b)	**19.** (a)	**20.** (a)
21. (c)	**22.** (a)	**23.** (c)	**24.** (c)	**25.** (c)	**26.** (b)	**27.** (b)	**28.** (a)	**29.** (b)	**30.** (c)
31. (a)	**32.** (d)	**33.** (c)	**34.** (c)	**35.** (c)	**36.** (d)				

Oscillators

1. (b)	**2.** (d)	**3.** (c)	**4.** (b)	**5.** (a)	**6.** (b)	**7.** (c)	**8.** (a)	**9.** (b)	**10.** (a)
11. (d)	**12.** (a)	**13.** (b)	**14.** (c)	**15.** (b)	**16.** (b)	**17.** (c)	**18.** (c)	**19.** (b)	**20.** (d)
21. (b)	**22.** (d)	**23.** (c)	**24.** (d)	**25.** (c)	**26.** (b)	**27.** (b)	**28.** (c)	**29.** (c)	**30.** (a)
31. (c)	**32.** (b)	**33.** (a)	**34.** (c)	**35.** (c)	**36.** (a)	**37.** (a)	**38.** (c)	**39.** (b)	**40.** (c)
41. (c)	**42.** (a)	**43.** (c)	**44.** (c)	**45.** (c)	**46.** (d)				

Semi-Conductor Devices

1. (a)	**2.** (b)	**3.** (a & b)	**4.** (d)	**5.** (c)	**6.** (c)	**7.** (a)	**8.** (c)	**9.** (c)	**10.** (c)
11. (c)	**12.** (c)	**13.** (b)	**14.** (c)	**15.** (d)	**16.** (a)	**17.** (b)	**18.** (c)	**19.** (c)	**20.** (b)

21. (a)	**22.** (c)	**23.** (b)	**24.** (b)	**25.** (c)	**26.** (a)	**27.** (a)	**28.** (d)	**29.** (a)	**30.** (b)
31. (b)	**32.** (d)	**33.** (a)	**34.** (c)	**35.** (b)	**36.** (b)	**37.** (b)	**38.** (c)	**39.** (d)	**40.** (a)
41. (b)	**42.** (b)	**43.** (a)	**44.** (a)	**45.** (a)	**46.** (d)	**47.** (d)	**48.** (d)	**49.** (c)	**50.** (b)
51. (c)	**52.** (a)	**53.** (c)	**54.** (c)	**55.** (d)	**56.** (a)	**57.** (a)	**58.** (c)	**59.** (c)	**60.** (d)
61. (b)	**62.** (b)	**63.** (b)	**64.** (d)	**65.** (a)	**66.** (c)	**67.** (c)	**68.** (a)	**69.** (b)	**70.** (c)
71. (a)	**72.** (d)	**73.** (c)	**74.** (a)	**75.** (a)	**76.** (b)	**77.** (b)	**78.** (b)	**79.** (c)	**80.** (c)
81. (d)	**82.** (b)	**83.** (c)	**84.** (c)	**85.** (d)	**86.** (c)	**87.** (d)	**88.** (b)	**89.** (c)	**90.** (b)
91. (b)	**92.** (b)	**93.** (b)	**94.** (a)	**95.** (b)	**96.** (a)	**97.** (b)	**98.** (b)	**99.** (b)	**100.** (a)
101. (d)	**102.** (b)	**103.** (a)	**104.** (c)	**105.** (b)	**106.** (b)	**107.** (c)	**108.** (d)	**109.** (c)	**110.** (a)
111. (c)	**112.** (c)	**113.** (a)	**114.** (d)	**115.** (a)	**116.** (d)	**117.** (a)	**118.** (b)	**119.** (c)	**120.** (a)
121. (d)	**122.** (d)	**123.** (c)	**124.** (a)	**125.** (a)	**126.** (a)	**127.** (b)	**128.** (a)	**129.** (b)	**130.** (a)
131. (a)	**132.** (a)	**133.** (d)	**134.** (a)	**135.** (d)	**136.** (d)	**137.** (d)	**138.** (c)	**139.** (c)	**140.** (d)
141. (c)	**142.** (d)	**143.** (c)	**144.** (d)	**145.** (d)	**146.** (d)	**147.** (b)	**148.** (c)	**149.** (d)	**150.** (d)
151. (c)	**152.** (c)	**153.** (c)	**154.** (d)	**155.** (c)	**156.** (c)	**157.** (c)	**158.** (b)	**159.** (d)	**160.** (d)
161. (d)	**162.** (a)	**163.** (d)	**164.** (c)	**165.** (a)	**166.** (c)	**167.** (d)	**168.** (d)	**169.** (d)	**170.** (b)
171. (c)	**172.** (a)	**173.** (b)	**174.** (d)	**175.** (a)	**176.** (d)	**177.** (c)	**178.** (a)	**179.** (b)	**180.** (b)
181. (a)	**182.** (b)	**183.** (d)	**184.** (c)	**185.** (c)	**186.** (c)	**187.** (d)			

Boolean Algebra and Logic Gates

1. (b)	**2.** (d)	**3.** (a)	**4.** (b)	**5.** (b)	**6.** (c)	**7.** (c)	**8.** (d)	**9.** (c)	**10.** (d)
11. (d)	**12.** (a)	**13.** (c)	**14.** (a)	**15.** (b)	**16.** (c)	**17.** (a)	**18.** (d)	**19.** (a)	**20.** (b)
21. (b)	**22.** (b)	**23.** (c)	**24.** (a)						

Module 8

Control Panel Wiring

Multiple Choice Questions

1. What is the purpose of control transformer used in control panel wiring?
   (a) To maintain constant terminal voltage
   (b) To supply the power to the auxiliary circuits
   (c) To control the supply voltage to the contactor
   (d) To protect the control elements from over voltage fault
   (I.T.I. August 2019 Electrician)
2. What is the name of the accessory used in control panel wiring?

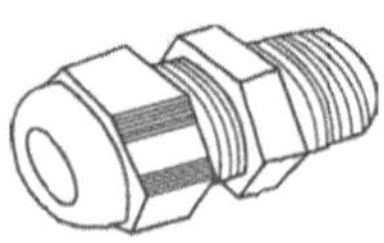

   (a) Lugs
   (b) Thimble
   (c) Grommet
   (d) Terminal connector
   (I.T.I. August 2019 Electrician)
3. What is the purpose of DIN-rail used in control panel wiring?
   (a) It provides a pathway for electrical wiring
   (b) Install the high powered circuit accessories
   (c) Mounting the double deck terminal connectors
   (d) Mounting the control accessories without screws
   (I.T.I. August 2019 Electrician)
4. Which device controls the operations in sequential control systems?
   (a) Timer (b) Relays
   (c) Contactor
   (d) Control transformer
   (I.T.I. August 2019 Electrician)
5. Which device operates with an actuator by the motion of a machine part or an object?
   (a) Contactor (b) Limit switch
   (c) Push button switch
   (d) Miniature Circuit Breaker (M.C.B.)
   (I.T.I. August 2019 Electrician)
6. Which D.C. load is represented by the $D.C_4$ standard duty cycle of contactors?
   (a) Resistive loads except motor loads
   (b) Starting and stopping of shunt motor
   (c) Starting and stopping of series motor
   (d) Starting and stopping within charge and braking
   (I.T.I. August 2019 Electrician)
7. Why the motor is not changing the direction, if reverse push button is pressed in forward and reverse control star delta starter?
   (a) No volt coil is not energised
   (b) Fault in forward contactor
   (c) Due to interlock in reverse contactor
   (d) No voltage exists in reverse contactor
   (I.T.I. August 2019 Electrician)

KEY TO MULTIPLE CHOICE QUESTIONS

1. (b)	2. (c)	3. (d)	4. (a)	5. (b)	6. (c)	7. (c)

Module 9

Electrical Drives

Multiple Choice Questions

1. The basic elements of an electric drive are:
   (a) Electric motor, and the transmission system.
   (b) Electric motor, the transmission and control system.
   (c) The transmission and control system
   (d) Electric motor and conversion equipment.
2. Electric drive is becoming more and more popular because ....
   (a) It is simple, clean, compact and reliable.
   (b) It provides easy and smooth control, flexibility in layout, easy starting and facility for remote control.
   (c) It is cheaper in initial as well as in maintenance cost.
   (d) All of the above
3. The type of drive used for a paper mill requiring constant speed operation and flexibility of control is:
   (a) group drive
   (b) multi-motor drive
   (c) individual or multi-motor drive
   (d) individual drive
4. In individual drive, ...
   (a) each operator has complete control of his machine.
   (b) no load-losses are eliminated.
   (c) the initial cost is less than that of group drive.
   (d) both (a) and (b)
5. Speed control by variation of field flux results in:
   (a) constant power drive
   (b) variable power drive
   (c) constant torque drive
   (d) variable torque drive
6. The least expensive drive is:
   (a) belt drive (b) rope drive
   (c) chain drive (d) gear drive
7. The range of rating of electric motor used for rolling mills is of the order of:
   (a) 10 to 25 kW (b) 25 to 85 kW
   (c) 85 to 400 kW (d) 400 to 1,000 kW
8. The least significant electrical characteristic in selection of electric motor for a flour mill is:
   (a) starting characteristics
   (b) braking
   (c) running characteristics
   (d) efficiency
9. Variable speed operation is preferred in:
   (a) water pump (b) ceiling fan
   (c) exhaust fan (d) refrigerator
10. The application in which the motor is to start with high acceleration is:
    (a) lifts and hoists
    (b) centrifugal pump
    (c) oil expeller
    (d) flour mill

11. DC motors are not so widely used as AC ones because:

(a) DC motors are usually more expensive than AC motors for similar operating conditions.

(b) additional equipment is required for converting existing AC supply into DC supply.

(c) DC motors have commutators that subject to trouble resulting from sparking, brush wear, arc-over and presence of moisture and destructive fumes in the surrounding air.

(d) All of the above.

12. In variable speed DC motor drive the armature voltage control leads to:

(a) constant power operation

(b) constant torque operation

(c) variable torque operation

(d) randomly varying power operation

13. A DC shunt motor runs at rated speed. If its field circuit gets open circuited, then soon after this the motor speed would tend to:

(a) decrease

(b) remain unchanged

(c) increase

(d) fluctuate around its previous speed

14. A DC series motor is running at rated speed without any additional resistance in series. If an additional resistance is placed in series, the speed of motor:

(a) increases

(b) decreases

(c) remains unchanged

(d) oscillates around the rated speed

15. In case of centrifugal pumps, the starting torque is usually:

(a) less than running torque

(b) same as running torque

(c) slightly higher than running torque

(d) double of running torque

16. Which motor should not be used for centrifugal pumps?

(a) Shunt (b) Series

(c) Cumulatively compound

(d) Differentially compound

17. A flywheel is normally fitted to:

(a) DC series motor driving a constant torque load

(b) separately excited DC motor driving pulsed torque load

(c) cumulatively compound motor driving pulsed torque load

(d) differentially compound motor driving pulsed torque load.

18. A flywheel is generally used in:

(a) a cement mill drive

(b) a paper mill drive

(c) a rolling mill drive

(d) a sugar centrifuge drive

19. For quick speed reversal, the motor preferred is:

(a) DC motor

(b) squirrel cage induction motor.

(c) slip ring induction motor

(d) synchronous motor

20. .... DC motor has the best speed regulation.

(a) Series (b) Shunt

(c) Cumulatively compound

(d) Differentially compound

21. The motor having wider range of speed control is:

(a) DC shunt motor

(b) squirrel cage induction motor

(c) synchronous motor

(d) double squirrel cage induction motor

22. Dynamic braking is very effective if the DC motor

(a) is series excited (b) is shunt excited

(c) is separately excited

(d) has cumulative compound excitation

23. It is preferable to employ ...... braking for controlling the speed of motor during deceleration.

(a) regenerative (b) plugging

(c) DC dynamic (d) mechanical

24. The most economical method of electric braking is:
(a) plugging
(b) dynamic braking with separate excitation
(c) dynamic braking with self excitation
(d) regenerative braking

25. Regenerative braking ....
(a) can be easily applied to DC shunt motors.
(b) can be easily applied to DC series motors
(c) can be used for stopping the motor
(d) cannot be used when the load on motor has overhauling characteristics.

26. Thyristors can be used for control of DC:
(a) separately excited motor
(b) series motor
(c) shunt motor
(d) all of the above

27. Squirrel cage induction motors have the disadvantages of:
(a) higher starting current and poor starting torque
(b) high sensitivity to fluctuations in supply voltage and low power factor at light load.
(c) no possibility of speed regulation
(d) all of the above

28. Slip-ring motors have the disadvantages of:
(a) low efficiency and low power factor in comparison to squirrel cage induction motors.
(b) low power factor at light load.
(c) sensitivity to fluctuations in supply voltage
(d) all of the above

29. The induction motor has lagging power factor during:
(a) starting only
(b) operation only
(c) both starting and operation
(d) none of the above

30. When the supply voltage to an induction motor is reduced by 10%, the maximuym torque will decrease by approximately:
(a) 5% (b) 10%
(c) 20% (d) 40%

31. Which of the following motors has the lowest ratio of starting torque to full load torque?
(a) DC series motor
(b) Wound rotor induction motor
(c) DC cumulative compound motor
(d) Squirrel cage induction motor

32. Very large three phase induction motors are started:
(a) direct on-line
(b) by star-delta starter
(c) by auto transformer starter
(d) none of these

33. In AC motor control $\frac{V}{f}$ is kept constant to:
(a) make maximum use of the magnetic circuit
(b) make minimum use of the magnetic circuit
(c) maximize the current drawn
(d) make the power constant

34. In a variable speed, induction motor drive $\frac{V}{f}$ is kept constant over a wide range of its frequency variation. The motor operates in the following mode:
(a) constant power mode
(b) constant torque mode
(c) variable power and variable torque mode
(d) constant slip mode

35. What is the advantage of A.C. drive compared to D.C. drive?
(a) Requires less space
(b) Installation and running cost is less
(c) Fast response and wide speed range of control
(d) Power circuit and control circuits are simple

(I.T.I. August 2019 Electrician)

36. What is the purpose LCD in Basic Operator Panel (BOP) in AC drive?
(a) Indicate the status of drive
(b) Monitor the parameter of AC drive
(c) Indicate the display error in reading
(d) Indicate the incorrect operation of BOP

(I.T.I. August 2019 Electrician)

37. Which is the correct sequence operation of key button in BOP of AC drive to change the direction of rotation?
(a) Press ON → REV → ON
(b) Press OFF → REV → ON
(c) Press ON → OFF → REV → ON
(d) Press ON → REV → OFF → ON

(I.T.I. August 2019 Electrician)

38. What is the purpose of PROG / DATA button in BOP of AC drive?
(a) To change the parameter setting
(b) To store the entered data and factory stored data
(c) To display the data direction of rotation forward / REV
(d) To display the data status of frequency and current

(I.T.I. August 2019 Electrician)

39. What is the name of the component marked 'X' in the block diagram of AC drive?

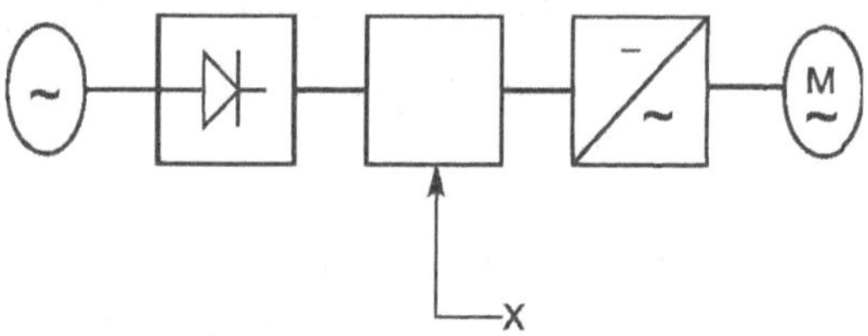

(a) Rectifier (b) D.C. bus
(c) Inverter (d) A.C. motor

(I.T.I. August 2019 Electrician)

40. What is the disadvantage of D.C. drive?
(a) Not suitable for high speed operation
(b) More complex with a single power conversion
(c) More expensive than A.C. drive for high capacity motor
(d) Installation of D.C. drives is more complicated

(I.T.I. August 2019 Electrician)

41. What is the function of field supply unit (FSU) in D.C. drive?
(a) Produces required firing current to the firing circuit
(b) Provides variable voltage to the field winding of motor
(c) Provides a constant voltage to the field winding of motor
(d) Provides a constant voltage to the armature of motor

(I.T.I. August 2019 Electrician)

42. What is the necessity of maintenance?
(a) Increase the total production
(b) Improve the attitude of workers
(c) Reduce the electric power consumption
(d) Increase the life and efficiency of the machines

(I.T.I. August 2019 Electrician)

43. Which maintenance of the machine to be carried out after machine gets fault / repair?
(a) Routine maintenance
(b) Periodical maintenance
(c) Preventive maintenance
(d) Brake down maintenance

(I.T.I. August 2019 Electrician)

44. Which maintenance is carried out on machines and controls at regular intervals before machines develop trouble?
(a) Routine maintenance
(b) Periodical maintenance
(c) Preventive maintenance
(d) Brake down maintenance

(I.T.I. August 2019 Electrician)

45. V.V.V.F stands for..........
(a) Variac variable voltage of Fan
(b) Valid voltage variable frequency
(c) Variable voltage and variable frequency
(d) None of these

(I.T.I. August 2019 Electrician)

46. Which is the application of single quadrant loads operating in first quadrant in drives?
(a) Hoists
(b) Elevators
(c) Conveyors
(d) Centrifugal pumps
(I.T.I. August 2019 Electrician)

KEY TO MULTIPLE CHOICE QUESTIONS

1. (b)	**2.** (d)	**3.** (c)	**4.** (d)	**5.** (a)	**6.** (a)	**7.** (c)	**8.** (b)	**9.** (b)	**10.** (a)
11. (d)	**12.** (b)	**13.** (c)	**14.** (b)	**15.** (a)	**16.** (b)	**17.** (c)	**18.** (c)	**19.** (a)	**20.** (b)
21. (a)	**22.** (c)	**23.** (c)	**24.** (d)	**25.** (a)	**26.** (d)	**27.** (d)	**28.** (d)	**29.** (c)	**30.** (c)
31. (d)	**32.** (c)	**33.** (a)	**34.** (b)	**35.** (b)	**36.** (b)	**37.** (c)	**38.** (b)	**39.** (b)	**40.** (a)
41. (c)	**42.** (d)	**43.** (d)	**44.** (c)	**45.** (c)	**46.** (d)				

Module 10

Converters, Relays and Voltage Regulators

Multiple Choice Questions

Converter and Rectifier

1. Which type of motor is used in large sized motor-generator sets ?
   (a) Squirrel-cage induction motor
   (b) Wound-rotor induction motor
   (c) Synchronous motor
   (d) A.C. commutator motor.
2. The equipment used to convert D.C. into A.C. is called :
   (a) a converter (b) an inverter
   (c) a rectifier (d) a regulator.
3. A single machine which can generate A.C. or D.C. as well as A.C. and D.C. both is called :
   (a) M.G. set
   (b) synchronous motor
   (c) inverter
   (d) rotary converter.
4. The main advantage of a M.G. set is this that :
   (a) it does not require any maintenance
   (b) the generated D.C. remains completely isolated from the A.C
   (c) its efficiency is very high
   (d) none of these.
5. A rotary converter can be started by :
   (a) d.c. supply
   (b) a.c. supply
   (c) mechanical power
   (d) either of the above three.
6. Which of the following is known as a metal rectifier?
   (a) Gas tube diode
   (b) Silicon diode
   (c) Copper oxide rectifier
   (d) All the above stated.
7. A three phase halfwave rectifier has an AC input of 415 V 50 Hz. The ripple frequency of DC output in pulses/sec is ....
   (a) 50 (b) 100
   (c) 150 (d) 300
   (I.T.I. July 2009, Electrician)
8. A 3-phase full wave rectifier has an AC input of 415 V, 50 Hz. The ripple frequency of D.C output in pulses per second is ....
   (a) 50 (b) 100
   (c) 150 (d) 300
   (I.T.I. January 2013, Electrician)
9. If the input A.C. voltage is 30 volts, then output D.C. voltage of the half wave rectifier will be ....
   (a) 30.0 volt (b) 27.0 volt
   (c) 13.5 volt (d) 10.8 volt
   (I.T.I. August 2015, Electrician)

   Hint:

   In half wave rectifier $V_{d.c.} = 0.45 \times$ input A.C. voltage

10. Bridge rectifiers are preferred because ....

(a) they require small transformer

(b) they have less peak inverse voltage

(c) they need small transformer and also have less peak inverse voltage

(d) they have low ripple factor

11. If in a mercury arc rectifier, connections of anode and cathode are interchanged ....

(a) the rectifier will get damaged

(b) the rectifier will operate at reduced output

(c) the rectifier will not operate

(d) none of the above

12. When a large power is required, which conversion method can be used?

(a) Metal Rectifiers

(b) Motor Generator set

(c) Mercury Arc Rectifier

(d) Rotary converter

(I.T.I. August 2015, Electrician)

13. The PIV across the diode of a H.W. rectifier is equal to –

(a) half the peak secondary voltage

(b) peak secondary voltage

(c) peak-to-peak secondary voltage

(d) peak-to-peak primary voltage

14. The disadvantage of a F.W. rectifier as compared with H.W. rectifier is –

(a) its ripple frequency is higher

(b) its D.C. output level is higher

(c) each diode carries half load current

(d) the need of a bulky transformer having a centre tapped secondary winding

15. The PIV (Peak Inverse Voltage) rating of diodes in a full-wave bridge rectifier should be at least:

(a) 2 × peak A.C. voltage

(b) peak A.C. voltage

(c) 4 × peak A.C. voltge

(d) 6 × peak A.C. voltage

16. The frequency of rectified A.C. of a full-wave rectifier is:

(a) the same as the supply frequency

(b) twice the supply frequency

(c) zero

(d) half of the supply freqnecy

17. The ripple present in the output of a rectifier is defined as :

(a) varying direct current

(b) another name of alternating current

(c) the varying positive and negative components

(d) filtered direct current

18. In full wave rectification, if the input frequency is 50 Hz, then frequency at the output of filter is ....

(a) 50 Hz (b) 0 Hz

(c) 100 Hz (d) 75 Hz

(I.T.I. February 2019, Electrician, Electroplator, Wireman)

19. In a full-wave rectifier, the current in each of the diodes flows for –

(a) The complete cycle of the input signal

(b) Half cycle of the input signal

(c) Less than half cycle of the input signal

(d) Zero time

(I.T.I. February 2019, Electrician, Electroplator, Wireman)

20. The efficiency of a half wave rectifier is ....

(a) About 50% (b) About 40%

(c) About 80% (d) About 100%

(I.T.I. February 2019, Electrician, Electroplator, Wireman)

21. In a half-wave rectifier, the load current flows for–

(a) The complete cycle of the input signal

(b) Only for the positive half-cycle of the input signal

(c) Less than half cycle of the input signal

(d) More than half cycle but less than the complete cycle of the input signal

(I.T.I. February 2019, Electrician, Electroplator, Wireman)

22. How many diodes are required for bridge rectifier?

(a) 3 (b) 4
(c) 2 (d) 1

(I.T.I. August 2019 Electrician)

23. Which converter is having high efficiency?

(a) Mercury arc rectifier
(b) Rotary converter
(c) Motor generator set
(d) S.C.R. converter

(I.T.I. August 2019 Electrician)

Relays and Voltage Regulators

1. What will happen when 12 V, 50 Hz A.C. relay is connected to a 12 V D.C. supply?

(a) The switching speed will be increased because of more current drawn by relay
(b) The switching capacity of the relay will be decreased very strongly
(c) The relay draws more current in the absence of inductive reactance, which bruns out the coil
(d) D.C. operation causes the core and armature to become partly permanent magnet and fails to function after a few operations

2. Which type of relay is used to protect the damage of equipment during under and over voltage fault?

(a) current sensing relay
(b) under current relay
(c) voltage sensing relay
(d) latching relay

(I.T.I. February 2015, Electrician)

3. The protective relay is specified as '4PDT' contact. What is the full form of it ?

(a) 4 Point Double Time Delay Contact
(b) 4 Pole Double Throw Contact
(c) 4 Package Double Throw Contact
(d) 4 Pole Double Time Delay Contact

(I.T.I. August 2015, Electrician)

4. In I.C. type voltage regulator, it is designated as "7805". What does the number "78" indicate ?

(a) The positive output regulator
(b) The negative output regulator
(c) The input positive voltage
(d) The output negative voltage

(I.T.I. August 2015, Electrician)

5. In the protection scheme, Relay functions as .....

(a) Switching device (b) Sensing device
(c) Breaking device (d) Isolating device

(I.T.I. August 2016, Electrician)

6. Which component ensures the safety of the line from damage?

(a) Relay
(b) Circuit breaker
(c) Bus bars
(d) Current transformer

(I.T.I. February 2017, Electrician)

7. Which device sends the information to the circuit breaker in case of fault?

(a) Relay
(b) Switch
(c) Circuit breaker itself
(d) Fuse

(I.T.I. February 2017, Electrician)

8. What is the actuating quantity for the relays?

(a) Magnitude (b) Frequency
(c) Phase angle (d) All of these

(I.T.I. February 2017, Electrician)

9. Protective relays can be designed to respond to .....

(a) Fault current
(b) Temperature
(c) Resistance, reactance or impedance
(d) All of these

(I.T.I. August 2016, Wireman)

10. Regulator I.C. series starts with ....
(a) 65xx (b) 54xx
(c) 78xx (d) 55xx
(I.T.I. August 2019 Electrician)

11. Application of regulator I.C. is ....
(a) Current stability (b) Voltage stability
(c) Power stability (d) None of these
(I.T.I. August 2019 Electrician)

12. Which type of regulator is considered more efficient?
(a) Special regulator
(b) Switching regulator
(c) Spike regulator
(d) All of these
(I.T.I. August 2019 Electrician)

13. On which principle the constant voltage transformer works?
(a) Self induction principle
(b) Fall in potential principle
(c) Ferro resonant principle
(d) Mutual induction principle
(I.T.I. August 2019 Electrician)

14. Which transformer is used in servo voltage stabiliser?
(a) Step up transformer
(b) Step down transformer
(c) Toroidal auto transformer
(d) Constant voltage transformer
(I.T.I. August 2019 Electrician)

15. What is the name of the relay?

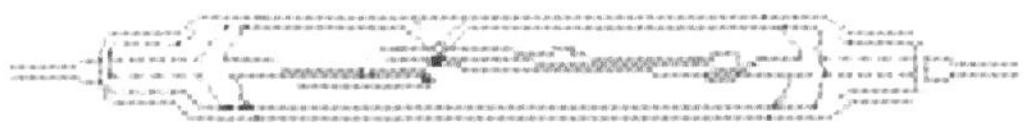

(a) Impulse relay
(b) Electromagnetic relay
(c) Dry reed relay
(d) Mercury wetted contact relay
(I.T.I. August 2019 Electrician)

KEY TO MULTIPLE CHOICE QUESTIONS

Converter and Rectifier

1. (c)	**2.** (b)	**3.** (d)	**4.** (b)	**5.** (d)	**6.** (c)	**7.** (c)	**8.** (d)	**9.** (c)	**10.** (c)
11. (c)	**12.** (c)	**13.** (b)	**14.** (d)	**15.** (b)	**16.** (b)	**17.** (a)	**18.** (b)	**19.** (b)	**20.** (b)
21. (b)	**22.** (b)	**23.** (d)							

Relays and Voltage Regulators

1. (c)	**2.** (c)	**3.** (b)	**4.** (a)	**5.** (b)	**6.** (a)	**7.** (a)	**8.** (d)	**9.** (d)	**10.** (c)
11. (b)	**12.** (b)	**13.** (c)	**14.** (c)	**15.** (c)					

Module 11

Generation of Electrical Energy

Multiple Choice Questions

1. The largest size hydroelectric unit in India is
   (a) 350 MW (b) 500 MW
   (c) 165 MW (d) 90 MW
2. In hydroelectric power plants
   (a) operating cost is low and initial cost is high
   (b) operating cost is high and inital cost is low
   (c) both operating and initial costs are high
   (d) both operating and initial costs are low
3. The function of a surge tank is to
   (a) supply water at constant pressure
   (b) relieve water hammer pressures in the penstock pipe
   (c) produce surge in the pipeline
   (d) none of the above
4. One kilogram of natural uranium gives energy equivalent to about
   (a) 100 kg of coal
   (b) 1000 kg of coal
   (c) 5000 kg of coal
   (d) 27.5 lakh kg of coal
5. Graphite is used in nuclear power plant as a
   (a) fuel (b) coolant
   (c) moderator (d) electrode
6. Which material is used in controlling chain reaction in a nuclear reactor?
   (a) Thorium (b) Heavy water
   (c) Boron (d) Beryllium
7. In a nuclear reactor, chain reaction is controlled by introducing
   (a) iron rods (b) cadmium rods
   (c) graphite rods (d) brass rods
8. Reflectors of a nuclear reactor are made of
   (a) cast iron (b) beryllium
   (c) steel (d) boron
9. Which one of the following fuels is used by the slow thermal nuclear reactors for power generation?
   (a) U^{235} (b) U^{238}
   (c) Th^{232} (d) Pu^{239}
10. In a nuclear reactor thermal energy is obtained from
    (a) fission of radioactive materials
    (b) fusion of radioactive materials
    (c) burning of the fuel rods in oxygen
    (d) none of these
11. The function of moderator in a nuclear reactor is
    (a) to absorb the excess neutrons
    (b) to increases the energy of the neutrons
    (c) to slow down the neutrons
    (d) none of these
12. The main sources of production of biogas are:
    (a) wet cow dung
    (b) human waste
    (c) wet livestock waste
    (d) all of the above

13. Biogas consists
(a) only methane
(b) methane and carbon dioxide with some impurities
(c) only ethane
(d) a special organic gas

14. Tidal energy utilizes
(a) kinetic energy of water
(b) potential energy of water
(c) both potential as well as kinetic energy of water
(d) none of the above

15. The energy radiated by sun on a bright sunny day is about
(a) 2.5 kW/m^2 (b) 1.0 kW/m^2
(c) 500 W/m^2 (d) 200 W/m^2

16. In solar power plants, the solar heat is transferred to
(a) molten salts (b) liquid metals
(c) water steam (d) any of the above

17. Which of the following area is preffered for solar plants?
(a) Coastal areas
(b) Hot arid zones
(c) Mountain tops
(d) High rainfall zones

18. Reflector mirrors employed for exploiting solar energy are called the
(a) mantle (b) heliostats
(c) diffusers (d) ponds

19. A pyrometer can be used for measurement of
(a) diffuse radiations only
(b) direct radiation only
(c) both direct and diffuse radiations
(d) none of the above

20. Temperature attained by cylindrical parabolic collector is of the order of
(a) 50–100°C
(b) 100–150°C
(c) 150–200°C
(d) 200–300°C and above

21. The flat plate collector gives a
(a) temperature of about 90°C with an efficiency of 30–35%
(b) temperature of about 120°C with an efficiency of 45%
(c) temperature of about 150°C with an efficiency of 52%
(d) none of the above

22. In a solar collector, the transparent cover is provided to
(a) protect the collector from dust
(b) reduce the heat losses from collector beneath to atmosphere
(c) transmit solar radiation only
(d) all of the above

23. The efficiency of solar cell is about
(a) 25% (b) 15%
(c) 40% (d) 60%

24. Solar cells are made of
(a) aluminium (b) germanium
(c) silicon (d) cadmium

25. The output of a solar cell is of the order of
(a) 0.5 – 1 V (b) 1 – 2 V
(c) 2 – 3 V (d) 4 – 5 V

26. The nature of the current developed in MHD generator is
(a) A.C. (b) D.C.
(c) A.C. or D.C. (d) none of these

27. In an MHD generator, the conductor is made of
(a) copper or aluminium
(b) liquid metal
(c) gas
(d) liquid metal or gas

28. Which of the following liquid metal is not used as an MHD working fluid?
(a) Potassium (b) Sodium
(c) Lithium (d) all of the above

29. Seeding material commonly used in MHD generator is
(a) sodium (b) aluminium
(c) potassium (d) cesium

30. The concept of MHD generation depends much more on the .... of the gas in the duct
(a) permeability (b) permittivity
(c) conductivity (d) none of these

31. The advantages of an MHD system are
(a) low pollution and good efficiency
(b) conservation of fuel
(c) absence of moving parts
(d) all of the above

32. Which of the following power plants is the least reliable?
(a) Wind (b) Tidal
(c) Geothermal (d) Solar

33. What is the approximate total power installed capacity in India?
(a) 445 GW (b) 175 GW
(c) 20 GW (d) 50 GW

34. Which of the following power plants can generate power at unpredictable or uncontrollable time?
(a) Tidal power plant
(b) Wind power plant
(c) Solar power plant
(d) Any of the above

35. Which of the following plants will take least time in starting from cold conditions to full load operation?
(a) Nuclear power plant
(b) Steam power plant
(c) Hydroelectric plant
(d) Gas turbine plant

36. The commercial sources of energy are
(a) fossil fuels, water and radioactive substances
(b) solar, wind, biomass
(c) wood, animal wastes and agricultural wastes
(d) none of the above

37. Major share of power generated in India is through which means –
(a) Hydroelectric power plants
(b) Nuclear power plants
(c) Thermal power plants
(d) Gas turbine power plants

(I.T.I. February 2018, Electrician)

38. The source of energy of the sun is –
(a) Nuclear fission
(b) Chemical reaction
(c) Nuclear fusion
(d) Photoelectric effect

(I.T.I. February 2018, Electrician)

39. The efficiency of the process of conversions of chemical energy from coal into electrical energy is limited to a maximum of ......
(a) 30% (b) 40%
(c) 50% (d) 60%

(I.T.I. February 2017, Electrician)

40. At which of the following voltage level ranges is power generated in a power station?
(a) 1.1 kV to 6.6 kV
(b) 11 kV to 30 kV
(c) 3.3 kV to 66 kV
(d) 230 volts to 415 volts

(I.T.I. February 2017, Electrician)

41. In a super thermal power plant in India, what is the unit capacity in use?
(a) 300 MW (b) 400 MW
(c) 500 MW (d) 600 MW

(I.T.I. February 2017, Electrician)

42. In a hydro electric plant a conduct system for taking water from the intake works to the turbine is known as .....
(a) Dam (b) Reservoir
(c) Pen stock (d) Surge tank

(I.T.I. February 2018, Electrician)

43. For which of the following reasons are hydrogenerating stations sited in mountain regions?
(a) Large land area is required
(b) Safety
(c) Earthquake free region
(d) Availability of the required head of water

(I.T.I. February 2017, Electrician)

44. The efficiency of a thermal power plant is about ......
(a) 100% (b) 85%
(c) 80% (d) 35%

(I.T.I. August 2016, Electrician)

45. Which of the following is a conventional source of energy?
(a) Coal (b) Solar
(c) Small Hydro (d) Wind
(I.T.I. August 2016, Electrician)

46. Coal used in thermal power plant has ash content of ....
(a) About 5% (b) About 10%
(c) About 20% (d) About 0%
(I.T.I. August 2016, Electrician)

47. In a thermal power plant, cooling towers are used to ....
(a) Condense low pressure steam
(b) Cool condensed steam
(c) Cool water used in condenser for condensing steam
(d) Cool feed water of boiler
(I.T.I. August 2016, Electrician)

48. Which of the following is considered to be superior quality of coal?
(a) Bituminous (b) Coke
(c) Lignite (d) Peat
(I.T.I. August 2016, Wireman)

49. Pulverized coal is .....
(a) Coal free from ash
(b) Non-smoking coal
(c) Coal which burns for the long time
(d) Coal broken down into fine particles
(I.T.I. August 2017, Electrician)

50. Coal used in power plants is also known as .....
(a) Soft coal (b) Steam coal
(c) Coke (d) Charcoal
(I.T.I. February 2017, Electrician)

51. Economiser is used to heat ......
(a) Air (b) Flue gases
(c) Feed water (d) All of these
(I.T.I. February 2017, Electrician)

52. Which of the following has the highest grid connected installed capacity ....
(a) Solar (b) Biomass
(c) Wind (d) Small Hydro
(I.T.I. August 2016, Electrician)

53. Which of the following is an advantage of wind as a prime source of energy?
(a) Unpredictable (b) Unsteady
(c) Renewable (d) None of these
(I.T.I. February 2017, Electrician)

54. The first nuclear power plant installed in India is ...
(a) Rajasthan (Kota)
(b) Tarapur (Maharashtra)
(c) Kalpakam (T.N.)
(d) Narora (U.P.)
(I.T.I. August 2016, Wireman)

55. The disadvantage of renewable source of energy is .....
(a) Requires backup
(b) Availability in low energy densities
(c) Intermittency
(d) All of the above
(I.T.I. August 2016, Eectrician)

56. Both power and manure is provided by ......
(a) Nuclear plants
(b) Thermal plants
(c) Biogas plants
(d) Hydroelectric plant
(I.T.I. August 2016, Electrician)

57. Which of the following is a form of biomass?
(a) Woody (b) Crop residues
(c) Animal dung (d) All of these
(I.T.I. August 2016, Electrician)

58. Geothermal energy is ......
(a) A renewable energy resource
(b) Alternative energy source
(c) Inexhaustible energy source
(d) Any of the above
(I.T.I. August 2016, Electrician)

59. Which of the following is used to generate power in geothermal station?
(a) Heat in the atmosphere
(b) Heat in the ionosphere
(c) Heat inside the earth
(d) Heat of the sun
(I.T.I. February 2017, Electrician)

60. Which of the following is the prime source of energy in a gas fired power station?
(a) L.P.G. (b) Natural gas
(c) Hotwater springs (d) None of these
(I.T.I. February 2017, Electrician)

61. Which of the following describes co-generation?
(a) Using coal fired and hydro generating stations in combination
(b) Using coal fired and diesel oil fired generating stations in combination
(c) Using large capacity thermal units
(d) Using electricity, steam or hot water for industrial purposes.
(I.T.I. February 2017, Electrician)

62. Bagasse is ....
(a) A variety of coal
(b) A fuel consisting of wood etc.
(c) Fibrous portion of sugarcane left after extracting the juice
(d) A king of rice straw
(I.T.I. February 2017, Electrician)

63. Photovoltaic solar energy conversion system makes use of ......
(a) Fuel cell (b) Solar cell
(c) Solar pond (d) Solar collectors
(I.T.I. August 2016, Electrician)

64. In a fuel cell electrical energy is produced by ....
(a) Thermionic action
(b) Reaction of hydrogen with oxygen
(c) Combustion of fuel in the absence of oxygen
(d) All of these
(I.T.I. August 2019 Electrician)

65. Which is the natural source of energy?
(a) Sun (b) Heat
(c) Coal (d) Biogas
(I.T.I. August 2019 Electrician)

66. What is the advantage of non conventional energy source?
(a) More reliable
(b) Low initial cost
(c) Efficiency is high
(d) Green house effect is avoided
(I.T.I. August 2019 Electrician)

67. Which is the disadvantage of non conventional power generation over conventional power generation?
(a) Increased pollution
(b) Security risk is more
(c) Required more maintenance
(d) Cannot be used for base load demand
(I.T.I. August 2019 Electrician)

68. In a MHD generator, the conductor is made of ....
(a) Copper
(b) Liquid metal or gas
(c) Gas
(d) Liquid metal
(I.T.I. August 2019 Electrician)

69. Power output per unit volume of a MHD generator is porportional to ....
(a) Specific electrical conductivity of gas
(b) Square of the fluid velocity
(c) Square of the magnetic field strength
(d) All of these
(I.T.I. August 2019 Electrician)

70. What is the function of governing system in diesel power plant?
(a) Start the engine from cold by supplying the compressed air
(b) Maintain the speed of the engine constant
(c) Control the heat from the engine cylinder to keep safe temperature
(d) Reduce the friction of moving parts in engine
(I.T.I. August 2019 Electrician)

71. What is the advantage of pressurised water reactor (PWR)?
(a) No heat loss
(b) High thermal efficiency
(c) It has high power density
(d) Metal surcace temperature is lower
(I.T.I. August 2019 Electrician)

72. Which power generation plant is having more reliability in operation?

(a) Hydro power plant
(b) Diesel power plant
(c) Nuclear power plant
(d) Thermal power plant

(I.T.I. August 2019 Electrician)

73. In a solar collector, why is the transparent cover provided for?

(a) Reduce the heat losses from collector beneath to atmosphere
(b) Transmit solar radiation only
(c) Physical safety
(d) All of these

(I.T.I. August 2019 Electrician)

74. Tidal energy utilises ....

(a) Kinetic energy of water
(b) Potential energy of water
(c) Both (a) and (b)
(d) None of these

(I.T.I. August 2019 Electrician)

75. What is the function of turbine used in tidal power generation?

(a) Prevents water flow to other parts of dam
(b) Converts potential energy into kinetic energy
(c) Keeps the water flow from low to higher level
(d) Converts kinetic energy into potential energy

(I.T.I. August 2019 Electrician)

76. What is the major disadvantage of wind power generation?

(a) Pollution effect is more
(b) Requires high technology
(c) Plant installation more complicated
(d) Wind power is not constant and steady

(I.T.I. August 2019 Electrician)

77. Which component in a steam power plant is used to heat the feed water from the flue gas?

(a) Boiler
(b) Economiser
(c) Super heater
(d) Air preheater

(I.T.I. August 2019 Electrician)

78. Which is the main constituent of the biogas mixture?

(a) Methane
(b) Nitrogen
(c) Hydrogen
(d) Carbon dioxide

(I.T.I. August 2019 Electrician)

79. Name the constituent marked 'X' of the schematic arrangement of hydro electric plant.

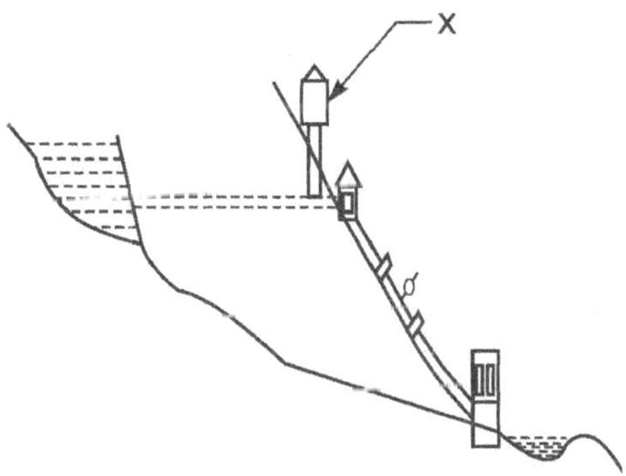

(a) Penstock (b) Surge tank
(c) Valve house (d) Power house

(I.T.I. August 2019 Electrician)

KEY TO MULTIPLE CHOICE QUESTIONS

1. (c)	**2.** (a)	**3.** (b)	**4.** (d)	**5.** (c)	**6.** (c)	**7.** (b)	**8.** (b)	**9.** (a)	**10.** (a)
11. (c)	**12.** (d)	**13.** (b)	**14.** (c)	**15.** (b)	**16.** (d)	**17.** (b)	**18.** (b)	**19.** (c)	**20.** (d)
21. (a)	**22.** (d)	**23.** (b)	**24.** (c)	**25.** (a)	**26.** (b)	**27.** (d)	**28.** (b)	**29.** (d)	**30.** (c)
31. (d)	**32.** (a)	**33.** (a)	**34.** (d)	**35.** (c)	**36.** (a)	**37.** (c)	**38.** (c)	**39.** (a)	**40.** (a)
41. (c)	**42.** (c)	**43.** (d)	**44.** (d)	**45.** (a)	**46.** (c)	**47.** (c)	**48.** (a)	**49.** (d)	**50.** (b)
51. (c)	**52.** (c)	**53.** (c)	**54.** (b)	**55.** (d)	**56.** (c)	**57.** (d)	**58.** (a)	**59.** (c)	**60.** (b)
61. (d)	**62.** (c)	**63.** (b)	**64.** (b)	**65.** (a)	**66.** (d)	**67.** (d)	**68.** (b)	**69.** (d)	**70.** (b)
71. (c)	**72.** (c)	**73.** (d)	**74.** (c)	**75.** (b)	**76.** (d)	**77.** (b)	**78.** (a)	**79.** (b)	

Module 12

Transmission and Distribution

Multiple Choice Questions

Transmission and Distribution

1. Underground cable is protected by :
   (a) M. S. and G. I. pipe
   (b) granite stone
   (c) a layer of sand and bricks
   (d) a layer of stone slabs.

2. A 11 kV underground cable is layed at a depth of :
   (a) 0.1 m (b) 0.5 m
   (c) 0.9 m (d) 1.5 m.

3. Epoxy compound used with undetground cable does not need :
   (a) heating (b) cooling
   (c) stirring (d) mixing.

4. If the underground cable has a lead tube as metallic sheath, the cable joint made once does not need :
   (a) testing with megger
   (b) separate earthing
   (c) metallic junction box
   (d) any plumbing work.

5. The depth of a pole below the ground level in case of normal soil should be :
   (a) 1/4 th of its full height
   (b) 1/5 th of its full height
   (c) 1/6 th of its full height
   (d) 1/8 th of its full height.

6. Which type of insulator is used in stay wire?
   (a) Pin type
   (b) Egg type
   (c) Shackle type
   (d) Suspension type.

7. If a distribution line and street lighting fixtures are installed on the same pole then their span should not exceed :
   (a) 30 m (b) 45 m
   (c) 60 m (d) 100 m

8. As per I. E. rules, no conductor of overhead line including service lines be erected across a street shall at any part thereof be at a height less than................. metres for low and medium voltage lines.
   (a) 4 (13 feet) (b) 4.5 (14.7 feet)
   (c) 5 (16.4 feet) (d) 5.79 (19 feet)

9. As per IE regulation, when service lines are insulated elsewhere other than along or across any street with insulated conductor, the conductor of the line at any part small have a height not less than a specified value.
   What is this specified height up to 11000 V lines?
   (a) 3.963 m (b) 4.572 m
   (c) 5.486 m (d) 5.791 m
   (I.T.I. July 2014, Electrician)

10. In industries, number of machines are installed in rows. In such places overhead bus system is used for distribution of electrical supply to each machine.

The recommended maximum current density for aluminium bus bar is ....

(a) 118 Amps/cm^2 (b) 156 Amps/cm^2

(c) 165 Amps/cm^2 (d) 181 Amps/cm^2

(I.T.I. January 2013 Electrician)

11. The expansion of XLPE cable is ....

(a) X-Line Power Electrical Cable

(b) Cross Line Polythene Enameled Cable

(c) Cross-Linked Poly Ethylene Cable

(d) X-Layers of poly Ethylene cable

(I.T.I. February 2016, Electrician)

12. The higher short circuit rating of PVC cable is 160°C but for XLPE cable it is ....

(a) 250° C (b) 200° C

(c) 180° C (d) 160° C

(I.T.I. February 2016, Electrician)

13. Main parts of a power system is/are ....

(a) Generating stations

(b) Transmission system

(c) Distributing networks

(d) All of these

(I.T.I. February 2017, Wireman)

14. Conductor used for power transmission & distribution is/are :

(a) PILC cable (b) AVV cable

(c) ACSR conductor (d) All of these

(I.T.I. August 2016, Wireman)

15. Cables generally used beyond 66 kV are .....

(a) Oil filled cable (b) Belted cable

(c) H-type cable (d) S.L. type cable

(I.T.I. August 2016, Wireman)

16. Which of the following faults is most likely to occur in cables?

(a) Cross or short circuit fault

(b) Open circuit fault

(c) Breakdown of cable insulation

(d) All of these

(I.T.I. August 2016, Electrician)

17. In the cables, the location of fault is usually found out by comparing ......

(a) The resistance of the conductor

(b) The inductance of conductor

(c) The capacitances of insulated conductors

(d) All above parameters

(I.T.I. August 2016, Electrician)

18. Why are inner sheaths in the cable used?

(a) Provides proper stress distribution

(b) Minimise the stress

(c) Provides protection against moisture, current and voltage surges

(d) To improve the insulation

(I.T.I. August 2017, Electrician)

19. The bedding on a cable consists of ....

(a) Hessian cloth (b) Jute

(c) Both (a) & (b) (d) None of these

(I.T.I. August 2016, Electrician)

20. Conductors used in H.T. transmission line are stranded because of ....

(a) Increased tensile strength

(b) Flexibility

(c) Cheaper in cost

(d) Reduced resistivity

(I.T.I. August 2016, Wireman)

21. Underground cables are laid at sufficient depth ....

(a) To minimize temperature stresses

(b) To avoid being unearthed easily due to removal of soil

(c) To minimize the effect of shocks and vibrations due to passing vehicles, etc

(d) For all of the above reasons

(I.T.I. August 2016, Electrician)

22. Armouring provides ......

(a) Electrical protection to cables

(b) Protection from mechanical injury

(c) good tensile strength

(d) reduced voltage drop in the cable

(I.T.I. August 2016, Wireman)

23. Metal conductors used for substation grounding systems are made of –

(a) Steel, aluminium, nickel
(b) Steel, copper, brass
(c) Copper, steel, aluminium
(d) Copper, steel, nickel

(I.T.I. February 2017, Electrician)

24. Which materials is used for wiring continuous bus bar?
(a) Aluminium (b) Copper
(c) Both (a) and (b) (d) None of these

(I.T.I. February 2018, Electrician)

25. Advantages of higher transmission voltage is/are ......
(a) Power transfer capability of the transmission line is increased
(b) Transmission of lines losses are reduced
(c) Area of cross section and volume of the conductor is reduced
(d) All of these

(I.T.I. August 2016, Electrician)

26. Following set of voltages is allowed for high voltge power transmission commercially available in India :
(a) 220 kV, 400 kV, 700 kV
(b) 220 kV, 400 kV, 765 kV
(c) 110 kV, 400 kV, 700 kV
(d) 110 kV, 400 kV, 765, kV

(I.T.I. February 2017, Wireman)

27. A radial distribution system is not used to supply in ....
(a) High density load areas
(b) Medium density load areas
(c) Low density load area
(d) None of these

(I.T.I. February 2017, Electrician)

28. It is desired to supply a large load reliably over a large area. Which of the following distribution is recommended?
(a) Direct distribution from the grid
(b) Radial distribution
(c) Loop distribution
(d) Network distribution

(I.T.I. February 2017, Electrician)

29. Following is true about ring main distribution system–
(a) Voltage profile at consumer premises is better compared to radial system
(b) It is similar to two radial systems connected in parallel
(c) More conductor is required per consumer
(d) Both (a) & (b) are correct

(I.T.I. February 2017, Wireman)

30. Low tension cables are operated below the voltage of .....
(a) 500 V (b) 1 kV
(c) 11 kV (d) 33 kV

(I.T.I. February 2017, Wireman)

31. The feeder voltage at which distribution of power is undertaken :
(a) 66 kV (b) 33 kV
(c) 22 kV (d) 11 kV

(I.T.I. February 2017, Electrician)

32. The maximum voltage level upto which pin type insulator can be used is .... kV.
(a) 11 (b) 33
(c) 66 (d) 132

(I.T.I. February 2017, Electrician)

33. The structure of pin type insulator is wavy. It is done to increase ....
(a) Mechanical strength
(b) Puncture strength
(c) Flash over voltage
(d) Thermal strength

(I.T.I. February 2017, Electrician)

34. Transmission line string insulators are mainly made of ......
(a) Glass (b) Porcelain
(c) Iron (d) P.V.C.

(I.T.I. August 2016, Electrician)

35. The number of discs in a string of insulator for 400 kV overhead transmission lines is in the range of ....
(a) 32-33 (b) 22-23
(c) 15-16 (d) 9-10

(I.T.I. February 2017, Electrician)

36. Line insulators are made of –
(a) Porcelain, glass, P.V.C.
(b) Synthetic resin, P.V.C., teak wood
(c) Synthetic resin, glass, porcelain
(d) Porcelain, glass, bone china
(I.T.I. February 2017, Wireman)

37. Suspension type insulators are conventionally used for line voltage –
(a) Upto 50 kV (b) Upto 80 kV
(c) Beyond 50 kV (d) Beyond 80 kV
(I.T.I. February 2017, Wireman)

38. Safety factor of insulator is the ratio of ....
(a) Puncture strength to Flash over voltage
(b) Flash over voltage to puncture strength
(c) Flash over voltage to line voltage
(d) Line voltage to flash over voltage
(I.T.I. August 2016, Electrician)

39. In stay wire, the type of insulator used is .....
(a) Pin insulator
(b) String insulator
(c) Disc insulator
(d) Egg type insulator
(I.T.I. August 2016, Wireman)

40. Isolator can be used to open the circuit ....
(a) On fault
(b) Any time with load
(c) After opening/disconnecting load
(d) None of these
(I.T.I. August 2016, Wireman)

41. Isolator works for .... operation.
(a) On load (b) Semi load
(c) Load ON/OFF (d) All of these
(I.T.I. August 2019 Electrician)

42. What is ACSR stands for?
(a) All Conductors Steel Reinforced
(b) Aluminium Core Steel Reinforced
(c) Aluminium Covered Steel Reinforced
(d) Aluminium Conductor Steel Reinforced
(I.T.I. August 2019 Electrician)

43. Why are sheaths used in cables?
(a) Provide proper insulation
(b) Prove mechanical strength
(c) Prevent ingress of moisture
(d) None of these
(I.T.I. August 2019 Electrician)

44. For what range is the underground service lines used?
(a) Distance more than 100 metres
(b) Distance more than 25 metres
(c) Distance less than 1 kilometre
(d) Distance more than 1 kilometre
(I.T.I. August 2019 Electrician)

45. What is the purpose of bedding insulation in W.G. cable?
(a) Protect the metallic sheath against corrosion
(b) Protect the cable from moisture and gases
(c) Protect armouring from atmospheric condition
(d) Protect the cable from mechanical injury
(I.T.I. August 2019 Electrician, Electroplator, Wireman, Lift & Escalator Mechanic)

46. Which test is conducted to locate the faults in U.G. cables?
(a) Loop test
(b) External growler test
(c) Break down voltage test
(d) Insulation resistance test
(I.T.I. August 2019 Electrician, Electroplator, Wireman, Lift & Escalator Mechanic)

47. What is the size of neutral conductor compared to phase conductor in underground cable?
(a) Same size of phase conductor
(b) $\frac{1}{4}$ size of phase conductor
(c) Half size of phase conductor
(d) $\frac{1}{3}$ size of phase conductor
(I.T.I. August 2019 Electrician, Electroplator, Wireman, Lift & Escalator Mechanic)

48. Name the part marked "X" of belted U.G. cable.

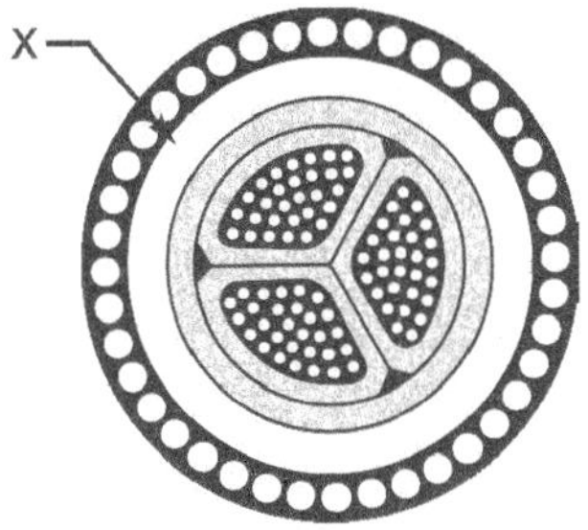

(a) Jute filling (b) Armouring
(c) Paper insulation (d) Lead sheath

(I.T.I. August 2019 Electrician, Electroplator, Wireman, Lift & Escalator Mechanic)

49. What is the purpose of the part marked 'X' of the U.G. cable?

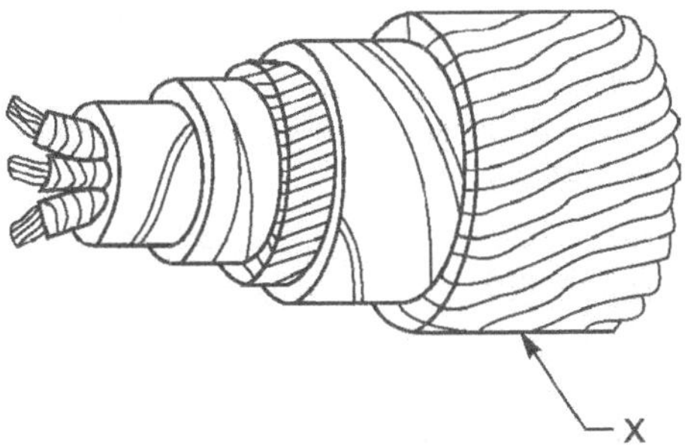

(a) Protect the cable from mechanical injury
(b) Protect the metallic sheath against corrosion
(c) Protect the armouring from atmospheric changes
(d) Protect the cable from moisture and chemical hazards

(I.T.I. August 2019 Electrician)

50. Which test is conducted to locate the short circuit and ground fault in U.G. cables?
(a) Crackle test
(b) Murray loop test
(c) Dielectric strength test
(d) Break down voltage test

(I.T.I. August 2019 Electrician)

51. What is the advantage of A.C. power transmission?
(a) Corona loss negligible
(b) Stress on transmission lines is minimum
(c) Low voltage drop in transmission lines
(d) Voltages can be stepped up and stepped down easily

(I.T.I. August 2019 Electrician)

52. What is the purpose of cross arm in overhead (O.H.) lines?
(a) Provide more support to the O.H. pole
(b) Protect from short between conductors
(c) Reduce the sag of the lines between poles
(d) Holding the insulators on which the conductors are fastened

(I.T.I. August 2019 Electrician)

53. What is the advantage of overhead lines compared to underground cable?
(a) Public safety is more
(b) Faults can be located easily
(c) No interference with the communication lines
(d) Not liable to the hazards from lightning discharges

(I.T.I. August 2019 Electrician)

54. What is the name of line insulator?

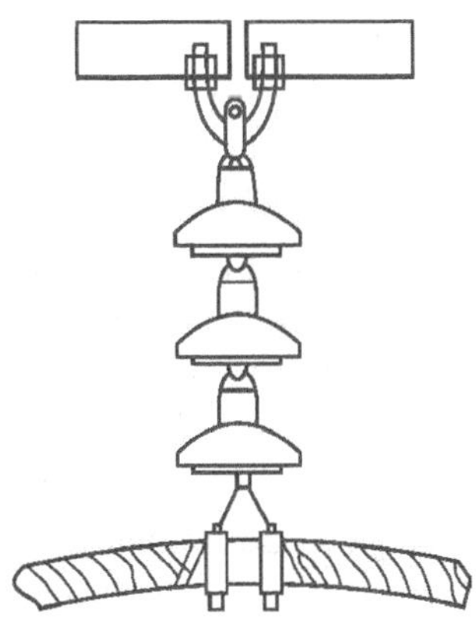

(a) Pin type insulator
(b) Disc type insulator
(c) Shackle type insulator
(d) Suspension type insulator

(I.T.I. August 2019 Electrician)

55. Which type of lightning arrester?

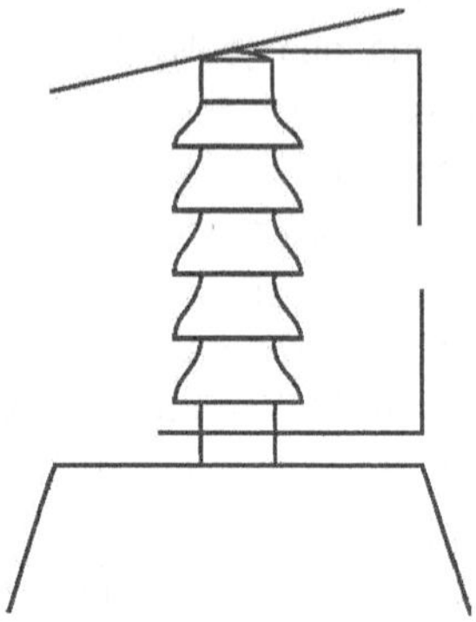

(a) Rod gap arrester
(b) Horn gap arrester
(c) Multi gap arrester
(d) Expulsion type arrester

(I.T.I. August 2019 Electrician)

56. What does section 44 refer to?

(a) Penalty for illegal transmission or use of energy
(b) Penalty for interference with meters
(c) Penalty for maliciously wasting energy
(d) Theft of energy

(I.T.I. August 2019 Electrician)

House Service Connections and Wiring Layout

1. Large industrial consumers are supplied power at

(a) 400 kV (b) 11 kV
(c) 66 kV (d) 132 kV

2. In a transmission system the feeder supplies power to

(a) transformer substations (step up)
(b) service mains
(c) distributors
(d) all of the above

3. The conductor connecting consumer's terminals to the distributor is called the

(a) feeder
(b) distributor
(c) service main
(d) none of the above

4. As per IE rules, the maximum allowable variation between declared and actual voltage at consumer's premises should be

(a) ± 6% (b) ± 8%
(c) ± 5% (d) ± 10%

5. In India, high voltage transmission is mostly by

(a) overhead system
(b) underground system
(c) both A and B
(d) none of the above

6. Overhead system can be designed for operation upto

(a) 400 kV (b) 220 kV
(c) 132 kV (d) 66 kV

7. The underground system cannot be operated above

(a) 220 kV (b) 66 kV
(c) 33 kV (d) 11 kV

8. Steel poles need galvanizing for longer life. Hence galvanizing is the process of applying a layer of

(a) zinc (b) paint
(c) varnish (d) tarcoal

9. Steel poles for transmission lines need protection against

(a) borer (b) termites
(c) corrosion (d) all of above

10. The conductors used in high tension overhead lines are

(a) stranded
(b) solid
(c) both A and B
(d) none of the above

11. ACSR conductors having 70 aluminium conductors and 6 steel conductors will be specified as

(a) 70/6 (b) 6/70
(c) 70/76 (d) 6/76

12. ACSR conductor implies
(a) aluminium conductor steel reinforced
(b) all conductors surface realigned
(c) anodised core steel reinforced
(d) none of the above

13. Galvanised steel wire is usually used as
(a) stay wire
(b) earth wire
(c) structural components
(d) all of the above

14. The minimum clearance between 132 kV transmission line and ground is about
(a) 6.4 metres (b) 3.2 metres
(c) 10.5 metres (d) 7.5 metres

15. If a 132 kV line passes over a residential buiding, the minimum vertical clearance from the roof of the building shall be
(a) 4.57 metres (b) 5 metres
(c) 6 metres (d) 3 metres

16. The colour of the neutral wire of 3-core flexible cable is
(a) blue (b) red
(c) black (d) brown

17. The material used for the manufacture of ground wire is
(a) aluminium (b) galvanised steel
(c) cast iron (d) stainless steel

18. The top most conductor in high voltage transmission line is
(a) R-phase conductor
(b) Y-phase conductor
(c) B-phase conductor
(d) earth conductor

19. Pin type insulators are generally not used for voltages exceeding
(a) 66 kV (b) 33 kV
(c) 25 kV (d) 11 kV

20. Strain type insulators are used
(a) at dead ends
(b) at intermediate anchor towers
(c) on straight runs
(d) any of A or B

21. The insulators may fail due to
(a) flash over
(b) short circuits
(c) deposition of dust
(d) any of the above

22. Copper used as conductor in a cable is
(a) hard drawn
(b) annealed
(c) alloyed with chromium
(d) hardened and tempered

23. Normally expected life of a cable is around
(a) 5 years (b) 15 years
(c) 25 years (d) 50 years

24. The size of conductor used in power cables depends on
(a) operating voltage (b) power factor
(c) current to be carried
(d) type of insulation used.

[I.T.I. August 2017, Electrician, Wireman]

25. Aluminium conductor cable can be joined by
(a) gas welding (b) soldering
(c) compression (d) thermit welding

26. VIR cables are usually used for
(a) internal wiring of buildings and workshops
(b) 6.6 kV
(c) 11 kV
(d) 33 kV

27. The insulation resistance of a cable of length 10 km is 1 MΩ; its resistance for 50 km length will be
(a) 1 MΩ
(b) 5 MΩ
(c) 0.2 MΩ
(d) none of the above

28. With the rise in temperature, the insulation resistivity
(a) remains unchanged
(b) decreases linearly
(c) increases linearly
(d) reduces exponentially

29. The surge impedance of cables is around

(a) 20 W (b) 50 W

(c) 100 W (d) 250 W

30. While laying 11kV underground cables, the minimum bending radius must be

(a) 3d (b) 6d

(c) 12d (d) 18d

Where '*d*' is the diameter of cable.

(I.T.I. August 2016, Wireman)

31. The main switch used in single phase supply is to be provided with ....

(a) fuse in both line and neutral

(b) a fuse in the line and a link in the neutral

(c) a link in the line and a fuse in the neutral

(d) thicker fuse wire in the phase and thinner fuse wire in the neutral

(I.T.I. July 2012, Electrician)

(I.T.I. August 2015, Wireman)

32. What is the main purpose of the Iron clad fuse cut out provided at the incoming side of the power supply with seal?

(a) To protect the incoming supply points

(b) To ensure the line is not loaded beyond the capacity

(c) To protect the distribution board from the short circuit fault

(d) To provide the safety from earth leakage current

(I.T.I. February 2015, Wireman)

33. Faults occur in electrical installations mainly due to failure/deterioration of insulation or bad termination. The type of fault which contributes to majority of electrical shock accidents in electrical installation is due to ....

(a) short circuit fault

(b) open circuit fault

(c) line to neutral fault

(d) line to ground fault

(I.T.I. February 2015, Wireman)

34. A fuse in a circuit provides protection against ....

(a) open circuit

(b) over load

(c) short circuit and over load

(d) open circuit and over load

(I.T.I. February 2015, Electrician)

35. Two pin sockets should not be used in domestic wiring unless the appliance to be connected is ....

(a) double earthed

(b) double insulated

(c) controlled by ELCB

(d) controlled by MCB

(I.T.I. February 2015, Wireman)

(I.T.I August 2015, Electrician)

36. In case of low or medium voltage, voltage at the point of commencement of supply at consumer should not vary from the declared voltage by more than ....

(a) 5% (b) 6%

(c) 8% (d) 10%

(I.T.I. February 2015, Wireman)

37. In residential building wiring, the sub circuit starts from ....

(a) distribution board

(b) supply main board

(c) junction box

(d) switch box

(I.T.I. February 2014, Wireman)

38. How many times the faulty current will be more than the fuse rating current?

(a) At least 2.0 times

(b) At least 3.0 times

(c) At least 3.5 times

(d) At least 4.0 times

(I.T.I. February 2015, Wireman)

39. According to I.E. rules no cutout link or switch other than a linked switch shall be provided on the .... except a link for testing purposes.

(a) phase conductor

(b) live conductor

(c) neutral conductor

(d) earthed conductor

40. The BIS symbol represents ....

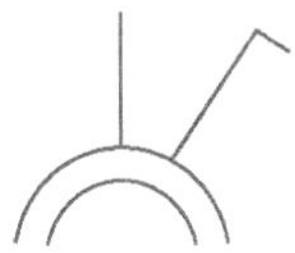

(a) socket outlet 6 A
(b) socket outlet combined with switch 6 A
(c) socket outlet 16 A
(d) socket outlet combined with switch 16 A

(I.T.I. February 2015, Wireman)

41. Which BIS symbol does represent the main fuse board without switches (Lighting circuit fuse boards)?

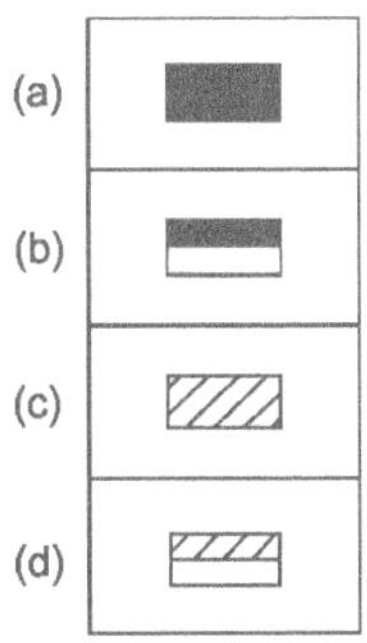

(I.T.I. February 2015, Wireman)

42. Name the device indicated as 'X' in the figure shown?

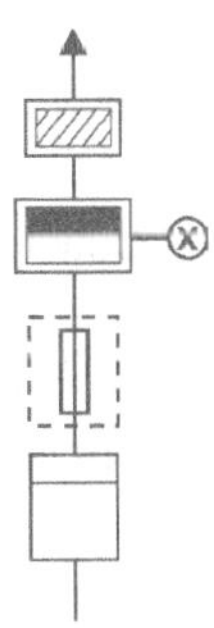

(a) Main switch
(b) IC cut out
(c) Distribution board
(d) Energy meter

(I.T.I. February 2015, Wireman)

43. What is the type of diagram shown in the figure?

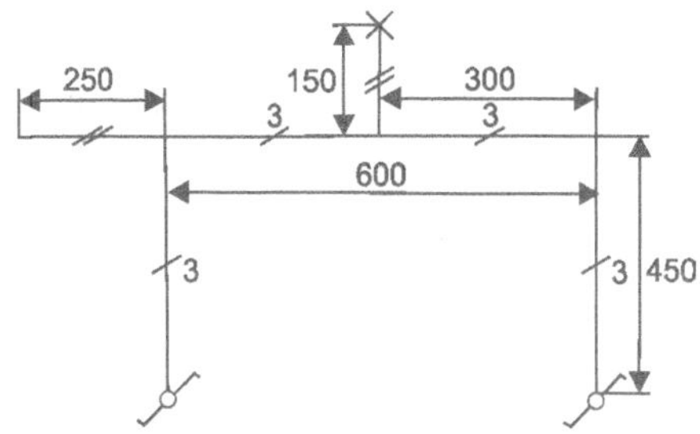

(a) circuit diagram
(b) installation diagram
(c) layout diagram
(d) wiring diagram

(I.T.I. August 2014, Wireman)

44. What will be the permissible leakage current in a wiring installation, having the load current of 10 amps?

(a) 200 mA (b) 20 mA
(c) 2 mA (d) 0.2 mA

(I.T.I. February 2015, Electrician)

45. Which one of the following is the suitable height for fixing a switch board?

(a) 4 metres (b) 3 metres
(c) 2 metres (d) 1.5 metres

46. According to the IS recommendations, the switches should not be fixed at lower than a certain height from floor level. What is the minimum height?

(a) 1.0 metre (b) 1.3 metre
(c) 2.0 metre (d) 2.4 metres

(I.T.I. February 2015, Electrician, Wireman)

47. All the lighting fittings shall be at a height of not less than- from the floor ....

(a) 2.25 m (b) 2.20 m
(c) 2.15 m (d) 2.10 m

(I.T.I. February 2015, Wireman)

48. The space in the back of switch board, width should not be less than ....

(a) 22.88 cm (b) 21.88 cm
(c) 20.88 cm (d) 20.77 cm

(I.T.I. February 2015, Electrician)

49. As per I.E. rule, the space in back of the fixed type metal panel board width from the farthest part of conductor should not be less than ....

(a) 20 cm (b) 21.5 cm
(c) 22.88 cm (d) 24.44 cm

(I.T.I. February 2015, Wireman)

50. According to I.E. rules, a clear space width shall be provided in front of the switch board and the wall not less than ....
(a) 91.44 cm (b) 9.144 cm
(c) 91.44 mm (d) 9.144 cm

(I.T.I. August 2014, Wireman)

51. Above floor level, the light fittings shall be at a height of not less than ....
(a) 2.2 cm (b) 2.25 cm
(c) 2.2 metres (d) 2.25 metres

(I.T.I. August 2014, Wireman)

52. As per BIS 732-1963, the Distribution board shall be fixed at a height from the floor level not more than ....
(a) 3.0 metres (b) 2.5 metres
(c) 2.0 metres (d) 1.5 metres

(I.T.I. February 2014, Wireman)

53. National Electric code recommends a specified minimum distance between the washing unit and the switch board. The recommended minimum distance is ....
(a) 0.6 metres (b) 1.3 metres
(c) 2.5 metres (d) 3.4 metres

(I.T.I. August 2014, Wireman)

54. National Electric code recommends that no accessory should be mounted within a certain minimum distance from the edge of the panel board.

What is the distance?
(a) 1.3 cms (b) 2.5 cms
(c) 3.0 cms (d) 4.0 cms

(I.T.I. February 2015, Electrician)

55. As per IE rules, the permissible voltage drop should not be more than the standard voltage percentage is ....
(a) 2% (b) 3%
(c) 5% (d) 12.5%

(I.T.I. August 2015, Electrician)

56. What is the name of the item indicate f as 'X' in the figure?

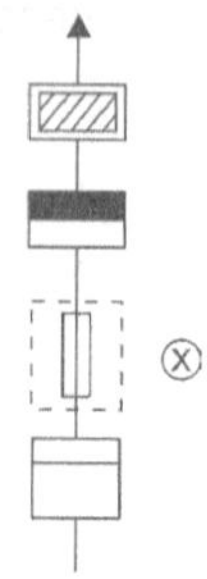

(a) Distribution board
(b) Energy meter
(c) iron clad cut out
(d) Main switch

(I.T.I. August 2015, Wireman)

57. The height of the bottom of switch board installed from the floor shall be not less than ....
(a) 900 mm (3 feet)
(b) 1200 mm (4 feet)
(c) 1350 mm $\left(4\frac{1}{2}\text{ feet}\right)$
(d) 1500 mm (5 feet)

(I.T.I. August 2015, Wireman)

58. Which type of distribution system is suitable for flats and other commercial buildings which is continued through out the height of the building?
(a) Bus bar chamber
(b) Raising mains
(c) Vertical distribution mains
(d) Bare conductors in trunk age

(I.T.I. August 2015, Wireman)

59. Bus bar should be installed at a height from the ground is ....
(a) 2.25 m (b) 2.50 m
(c) 2.75 m (d) 3.00 m

(I.T.I. February 2016, Wireman)

60. The most suitable way of providing electric supply to a multi storey building is by means of ......
(a) Rising main
(b) Over hand line
(c) Paper insulated cable
(d) P.V.C. insulated armoured cable

(I.T.I. February 2016, Wireman)

61. The 1968 guide for electrical layout in residential buildings is given by BIS in .....

(a) IS 4648 (b) IS 9537 (part I)

(c) IS 9537 (part 2) (d) IS 964

(I.T.I. February 2017, Wireman)

62. The insulation of the cable decreases with .....

(a) The increase in length of the insulation

(b) The decrease in length of the insulation

(c) Either (a) or (b)

(d) None of these

(I.T.I. August 2016, Electrician)

63. Service connection is taken from .....

(a) Secondary of pole substation of energy meter

(b) Main substation of pole substation

(c) Energy meter to distribution panel

(d) Single phase supply to neutral of a sub circuit

(I.T.I. February 2017, Wireman)

KEY TO MULTIPLE CHOICE QUESTIONS

Transmission and Distribution

1. (c)	**2.** (c)	**3.** (a)	**4.** (b)	**5.** (c)	**6.** (b)	**7.** (b)	**8.** (d)	**9.** (a)	**10.** (a)
11. (c)	**12.** (a)	**13.** (d)	**14.** (c)	**15.** (a)	**16.** (d)	**17.** (a)	**18.** (a)	**19.** (c)	**20.** (b)
21. (c)	**22.** (b)	**23.** (c)	**24.** (a)	**25.** (d)	**26.** (b)	**27.** (a)	**28.** (d)	**29.** (d)	**30.** (b)
31. (d)	**32.** (b)	**33.** (c)	**34.** (b)	**35.** (b)	**36.** (c)	**37.** (c)	**38.** (a)	**39.** (d)	**40.** (c)
41. (c)	**42.** (d)	**43.** (c)	**44.** (b)	**45.** (a)	**46.** (a)	**47.** (c)	**48.** (d)	**49.** (c)	**50.** (b)
51. (d)	**52.** (d)	**53.** (b)	**54.** (d)	**55.** (a)	**56.** (b)				

House Service Connections and Wiring Layout

1. (b)	**2.** (c)	**3.** (c)	**4.** (c)	**5.** (a)	**6.** (a)	**7.** (b)	**8.** (a)	**9.** (c)	**10.** (a)
11. (a)	**12.** (a)	**13.** (d)	**14.** (a)	**15.** (a)	**16.** (a)	**17.** (b)	**18.** (d)	**19.** (b)	**20.** (d)
21. (d)	**22.** (b)	**23.** (b)	**24.** (c)	**25.** (c)	**26.** (a)	**27.** (c)	**28.** (d)	**29.** (b)	**30.** (c)
31. (b)	**32.** (b)	**33.** (a)	**34.** (c)	**35.** (b)	**36.** (a)	**37.** (a)	**38.** (c)	**39.** (d)	**40.** (d)
41. (a)	**42.** (a)	**43.** (c)	**44.** (d)	**45.** (d)	**46.** (b)	**47.** (a)	**48.** (a)	**49.** (c)	**50.** (a)
51. (d)	**52.** (c)	**53.** (c)	**54.** (b)	**55.** (c)	**56.** (c)	**57.** (b)	**58.** (b)	**59.** (c)	**60.** (a)
61. (a)	**62.** (a)	**63.** (a)							

Module 13

Circuit Breakers

MULTIPLE CHOICE QUESTIONS

1. A circuit breaker normally operates ......
   (a) when the power is to be supplied
   (b) when the line is to be tested
   (c) when the switch is to be put on
   (d) whenever fault occurs in the line.
2. The fluids used in circuit breakers should be of ....
   (a) high dielectric strength and thermal stability
   (b) non inflammability
   (c) arc extinguishing ability
   (d) all of the above
3. Main purpose of oil in OCB is to ......
   (a) provide insulation
   (b) provide cooling of contacts
   (c) quenching arc
   (d) none of the above
4. In a minimum oil circuit breaker, the oil is used ......
   (a) to act as circuit breaking medium only
   (b) for circuit breaking and providing insulation
   (c) for providing insulation only
   (d) for none of the above purposes
5. What is the preferred type of "circuit breaker" to be installed in extra high voltage A.C. System?
   (a) Bulk oil type circuit breaker
   (b) Air blast circuit breaker
   (c) Vacuum circuit breaker
   (d) Sulphur hexafluoride (SF_6) circuit breaker.
6. Which of the following circuit breakers has high reliability and negligible maintenance?
   (a) Air blast circuit breaker
   (b) SF_6 circuit breaker
   (c) Oil circuit breaker
   (d) Vacuum circuit breaker
7. The main purpose of filling oil in a circuit breaker is:
   (a) to produce insulation effect
   (b) to stop the ionisation of the surrounding air
   (c) to ensure the operation of the device at a low temperature
   (d) to suppress the sparking produced across the connector at the time of breaking a circuit.
8. The circuit breaker which does not contain any serviceable part is known as:
   (a) oil circuit breaker (OCB)
   (b) miniature circuit breaker (MCB)
   (c) air circuit breaker (ACB)
   (d) air blast circuit breaker (ABCB).
9. Current rating of MCB selected for an air conditioner of 1 ton is ....
   (a) 10/15 amperes (b) 15/20 amperes
   (c) 10/20 amperes (d) 15/30 amperes

   (I.T.I. February 2015, Wireman)

10. Which one of the following is not applicable to MCB?

(a) Tamper proof as they are sealed
(b) Used for close excess current protection
(c) Used for control as well as to protect the circuit
(d) Used for coarse excess current protection.

(I.T.I. July 2014, Electrician)

11. Which device can be used as switch and as well as protective device?

(a) Triple pole iron clad switch
(b) Intermediate switch
(c) Miniature circuit breaker
(d) Earth leakage circuit breaker

(I.T.I. February 2015, Electrician)

12. Earth Leakage Circuit Breaker (ELCB) works on the principle of ....

(a) over load current
(b) short circuit current
(c) residual current
(d) neutral current

(I.T.I. August 2014, Wireman)
(I.T.I. February 2016, Electronic Mechanic)

13. MCCB's expansion is ....

(a) Moulded Case Circuit Breaker
(b) Main Case Circuit Breaker
(c) Main Control Circuit Breaker
(d) Main Current Circuit Breaker

(I.T.I. February 2015, Wireman)

14. The figure shown is the internal circuit details of a ......

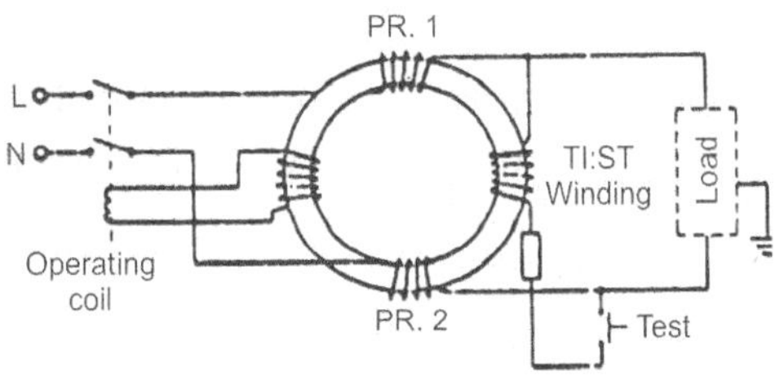

(a) ELCB voltage operated
(b) Over current relay
(c) ELCB current operated
(d) Under voltage trip device.

(I.T.I. January 2014, Electrician)

15. The arrangement of residual current device is shown in the figure. Name the part indicate as "X" is ......

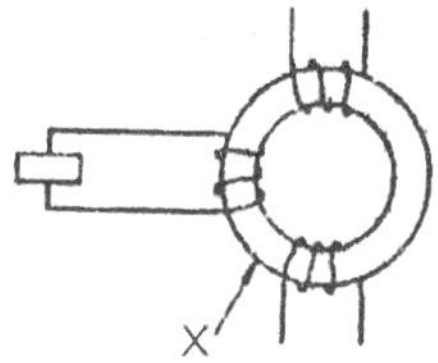

(a) Secondary winding
(b) Primary winding
(c) Torroid magnetic coil
(d) Relay coil

(I.T.I. August 2014, Wireman)

16. Name the part indicated as 'X' in the ELCB circuit shown in the figure?

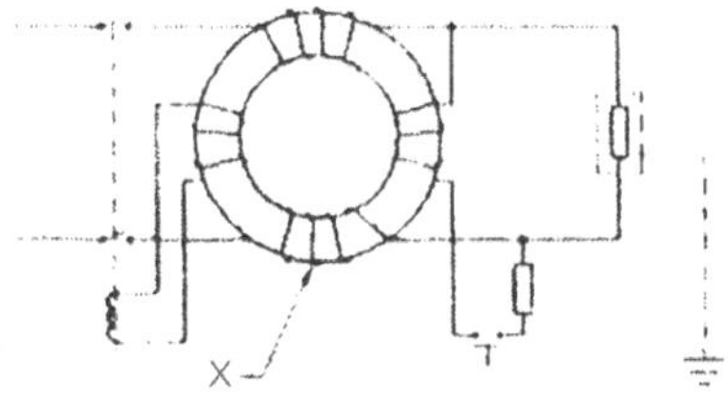

(a) Operating coil (b) Secondary coil
(c) Primary coil (d) Test winding

(I.T.I. August 2014, Electrician)

17. The main disadvantage of ELCB is ....

(a) less reliable in operation
(b) not easy to install
(c) not easy to maintain
(d) more costly

(I.T.I. February 2016, Wireman)

18. In circuit breaker, the maximum power that a circuit breaker can safely interrupt under a fault is known as ....

(a) maximum operating power
(b) minimum operating power
(c) breaking capacity
(d) rupturing capacity

(I.T.I. August 2015, Electrician)

19. The rupturing capacity of circuit breakers is generally expressed in ....

(a) mega volt ampere (MVA)

(b) kilo volt ampere (KVA)
(c) kilo ampere (KA)
(d) volt ampere (VA)

(I.T.I. February 2016, Electrician)

20. Which device is used to protect the circuits from short circuit fault ?
(a) Switch (b) MCB
(c) Meter (d) Isolator

(I.T.I. August 2015, Electrician Wireman)

21. Maximum number of poles available in MCB's is ....
(a) 4 (b) 3
(c) 2 (d) 1

(I.T.I. February 2016, Wireman)

22. For air conditioners of 1.5 ton capacity, sockets with MCB rating is ....
(a) 16 A (b) 18 A
(c) 20 A (d) 22 A

(I.T.I. February 2016, Wireman)

23. The 'L' series MCB's used to protect the equipment is...
(a) geyser (b) air conditioner
(c) locomotives (d) motor

(I.T.I. February 2016, Wireman)

24. MCB stands for...
(a) Miniature Circuit Breaker
(b) Minimum Capacity Breaker
(c) Molded Circuit Breaker
(d) Maximum Capacity Breaker

(I.T.I. August 2016, Electrician)

25. No. of poles in MCB (TPN) is ...
(a) 2 poles (b) 4 poles
(c) 3 poles (d) 5 poles

(I.T.I. August 2016, Electrician)

26. For preventing the passage of full fault current during short circuit, the device used is ...
(a) ELCB (b) MCCB
(c) MCB (d) SSB

(I.T.I. August, Wireman)

27. SF_6 gas is ...
(a) Sulphur fluoride
(b) Sulphur di fluoride
(c) Sulphur hexa di fluoride
(d) Sulphur hexa fluoride

(I.T.I. August 2017, Electrician)

28. Voltage ranges of SF_6 circuit breakers vary from ...
(a) 3 kV to 30 kV
(b) 45 kV to 60 kV
(c) 60 kV to 110 kV
(d) 115 kV to 500 kV

(I.T.I. February 2016, Wireman)

29. The single most property of SF_6 for circuit breaking is ...
(a) No toxic
(b) Non inflammable
(c) High electronegative in character
(d) High insulation

(I.T.I. August 2016, Wireman)

30. RCCB stands for ...
(a) Residual Current Circuit Breaker
(b) Released Current Circuit Breaker
(c) Residual Capacitance Circuit Breaker
(d) Released Capacitance Circuit Breaker

(I.T.I. February 2019, Electrician)

31. Following is/are true about M.C.C.B. :
(a) Trip current is adjustable
(b) Two earth lines are provided
(c) It protects human from electric shock
(d) All of these

(I.T.I. February 2017, Wireman)

32. Air blast circuit breaker is used in substation of ...
(a) Outdoor (b) Indoor
(c) Any where (d) None of these

(I.T.I. February 2016, Wireman)

33. Bulk oil circuit breaker can be used for voltage upto ...
(a) 66 kV (b) 33 kV
(c) 11 kV (d) 1.1 kV

(I.T.I. August 2016, Wireman)

34. A thermal protection switch can protect against ...
(a) short circuit
(b) execssive temperature

(c) overload
(d) over voltage

(I.T.I. February 2017, Electrician)

35. What is the making to breaking current rate for extra high voltage circuit breakers?
(a) More than 1
(b) Less than 1
(c) Equal to 1
(d) A negative value

(I.T.I. February 2017, Electrician)

36. The arcing contact of the circuit breaker is made up of ...
(a) Copper tungsten alloy
(b) Porcelain
(c) Aluminum alloy
(d) Electrolytic copper

(I.T.I. August 2017, Electrician)

37. Which is the application of D.C. series M.C.B.?
(a) A.C. motor (b) Air conditioners
(c) D.C. motor (d) Locomotives

(I.T.I. August 2019 Electrician)

38. E.L.C.B. stands for ......
(a) Electric Leakage Circuit Breaker
(b) Earth Leakage Circuit Breaker
(c) Earth Larger Circuit Breaker
(d) All of these

(I.T.I. August 2019 Electrician)

39. Device used to protect a low voltage circuit in case of a leakage fault is ....
(a) R.C.C.B. (b) E.L.C.B.
(c) M.C.B. (d) M.O.C.B.

(I.T.I. August 2019 Electrician)

KEY TO MULTIPLE CHOICE QUESTIONS

1. (d)	**2.** (d)	**3.** (c)	**4.** (a)	**5.** (d)	**6.** (b)	**7.** (d)	**8.** (b)	**9.** (a)	**10.** (d)
11. (c)	**12.** (c)	**13.** (a)	**14.** (c)	**15.** (c)	**16.** (c)	**17.** (d)	**18.** (d)	**19.** (a)	**20.** (b)
21. (a)	**22.** (c)	**23.** (a)	**24.** (a)	**25.** (b)	**26.** (c)	**27.** (d)	**28.** (d)	**29.** (c)	**30.** (a)
31. (a)	**32.** (b)	**33.** (a)	**34.** (c)	**35.** (a)	**36.** (a)	**37.** (d)	**38.** (b)	**39.** (a)	

Module 14

EV Scenario in India

Multiple Choice Questions

1. Full form of EV is ____________
   (a) Energy voltage (b) Electric vehicles
   (c) Electric voltage (d) Energy vehicles
2. Which Battery is preferred for EV?
   (a) Lead-acid (Pb-acid)
   (b) Lithium-ion (Li-ion)
   (c) Sodium-sulphur (NaS)
   (d) Nickel-cadmium (NiCd)
3. ____________vehicles are powered by battery only.
   (a) Conventional (b) EV
   (c) HEV (d) PHEV
4. Which of the following is not an advantage of hybrid vehicles?
   (a) Environmentally friendly.
   (b) Initial cost is low.
   (c) Better mileage.
   (d) Higher energy conservation.
5. Which of the following is not the type of hybrid vehicle?
   (a) Plug in hybrid
   (b) Parallel hybrid
   (c) Natural gas for vehicle
   (d) Series hybrid
6. Which vehicle uses a high-voltage battery?
   (a) Electric vehicle
   (b) Hybrid vehicle
   (c) Both electric and hybrid vehicles
   (d) None of these

KEY TO MULTIPLE CHOICE QUESTIONS

1.	(b)	**2.**	(b)	**3.**	(b)	**4.**	(b)	**5.**	(c)	**6.**	(c)

2nd Year

Workshop Calculation and Science

1. Friction-Lubrication

1. Graphite is a:
   (a) Liquid lubricant
   (b) Semi-liquid lubricant
   (c) Solid lubricant
   (d) None of the above
2. For effective cutting of steel with hand hacksaw, which one of the following coolant is suitable.
   (a) Cutting oil (b) Grease
   (c) Lard oil (d) Kerosene oil
3. The most important quality of any lubricant is:
   (a) Viscosity (b) Specific gravity
   (c) Specific heat (d) Emulsification
4. Which one of the following is NOT system of lubrication:
   (a) Gravity feed system
   (b) Force feed system
   (c) Splash feed system
   (d) Friction feed system
5. Splash lubrication is also known as:
   (a) Gravity oiling (b) Feed oiling
   (c) Ring oiling (d) None of these
6. As per manufacturer's manual, lubricants are to be applied .... at different points or parts.
   (a) every day (b) once in a month
   (c) daily, weekly, monthly or at regular intervals
   (d) None of the above
7. One of the functions of cutting fluids is to .... the tools as well as the workpiece.
   (a) cut (b) lubricate
   (c) cool (d) None of these
8. Because of the lubricating function, the friction is .... and the heat generated is less.
   (a) generated (b) reduced
   (c) removed (d) None of these
9. A good cutting fluid should be sufficiently ....
   (a) viscous (b) oily
   (c) hot (d) None of these
10. Straight mineral oils are the .... which can be used undiluted.
    (a) coolants (b) lubricants
    (c) Both (a) and (b) (d) None of these
11. Synthetic oils consist of carefully chosen .... in dilute solution with water.
    (a) lubricants (b) cutting oils
    (c) chemicals (d) All of above
12. .... oils are used in automatic lathes.
    (a) Cutting (b) Compounded
    (c) Lubricating (d) Grease

KEY TO MCQs 1. Friction-Lubrication

1. (c) **2.** (a) **3.** (a) **4.** (d) **5.** (c) **6.** (c) **7.** (c) **8.** (b) **9.** (a) **10.** (a)
11. (c) **12.** (b)

2. Algebra

1. What is the value for $(a - b)^2$ if $a = 2$ and $b = 3$?
   (a) 4 (b) 2
   (c) 3 (d) 1
2. What is the value of (12*a*) (5*a*) (7*a*)?
   (a) $390a^3$ (b) $405a^3$
   (c) $420a^3$ (d) $421a^3$
3. If a = 5 and b = 0, then $1 \div (a^b)$ = ...
   (a) 2 (b) 1
   (c) 0.2 (d) 0
4. If a = 2, b = 5, c = 4 and d = 3, then $a^b \div c^d$ = ...
   (a) 5 (b) 2
   (c) 0.5 (d) 0.2
5. 7589 – ? = 3434
   (a) 4242 (b) 4155
   (c) 1123 (d) 11023
6. It we subtract square of 22, from cube of 12, then the value is ....
   (a) 1244 (b) 1344
   (c) 1454 (d) 1354
7. Expand $(a - 2b)^2$
   (a) $a^2 + 4b + 4b^2$ (b) $a^2 - 4ab + 4b^2$
   (c) $a^2 - 2b - 4b^2$ (d) $a^2 - 2b + 4b^2$
8. Simplify: $(x + 8)(x - 8)$ = ...
   (a) $x^2 - 64$ (b) $x^2 + 64$
   (c) $x^2 + 16$ (d) $x^2 - 16$
9. Expand: $(x + 3y)^2$
   (a) $x^2 + 6xy + 9y^2$ (b) $x^2 - 6xy + 9y^2$
   (c) $x^2 + 3xy + 9y^2$ (d) $x^2 - 3xy + 9y^2$
10. $(a + b + c)^2$ = ...
    (a) $a^2 + b^2 + c^2 + 2ac - 2ab - 2bc$
    (b) $a^2 + b^2 + c^2 + 2(ab + bc + ac)$
    (c) $a^2 + b^2 + c^2 - 2ac - 2ab - 2bc$
    (d) $a^2 - b^2 + c^2 + 2(ac + ab - bc)$
11. Choose the correct equation of $(a - b)^2$
    (a) $(a - b)^2 = (a + b)^2 - 4ab$
    (b) $(a - b)^2 = (a + b)^2 + 4ab$
    (c) $(a - b)^2 = (a + b)^2 - 2ab$
    (d) $(a - b)^2 = (a + b)^2 + 2ab$
12. Evaluate : $\dfrac{2.39^2 - 1.61^2}{2.39 - 1.61}$
    (a) 2 (b) 4
    (c) 6 (d) 8
13. What is the value of *x*, if $4x = 3(35 - x)$
    (a) 4 (b) 35
    (c) 10 (d) 15
14. What is the value of *m* if $\frac{8}{5}m + 6 = 8$
    (a) $\frac{4}{5}$ (b) $\frac{5}{4}$
    (c) $\frac{2}{5}$ (d) $\frac{5}{2}$
15. What is the value of *y*, if $(-3y + 2y - 6 - 9y + 6y + 54 - 8) = 0$
    (a) 5 (b) 8
    (c) 10 (d) 12
16. If "– 2" is root of the equation $x^3 + 8x^2 + 21x + 18 = 0$, other roots are ....
    (a) – 1, – 2 (b) – 2, – 3
    (c) – 3, – 3 (d) – 1, – 4
17. By putting value for $x = 3$ in function $y = 10x$, the value of *y* will be ....
    (a) 3.334 (b) 7
    (c) 30 (d) 15
18. Find the value of y if $\frac{y}{91} = \frac{11}{13}$
    (a) 107 (b) 117
    (c) 87 (d) 77
19. Find the value of *y* if $14 + 3y = -5y - 18$.
    (a) 1 (b) 2
    (c) – 4 (d) 4
20. 587 × 999 = ...
    (a) 615173 (b) 614823
    (c) 586413 (d) 587523

21. Find the value of $\left(x + \frac{1}{x}\right)$ if $x = 7 - \frac{4}{\sqrt{3}}$

(a) $3\sqrt{3}$ (b) $8\sqrt{3}$
(c) 14 (d) $14 + 8\sqrt{3}$

22. If (x + 2) is a factor of (x^2 + mx + 14), then the value of "m" is ....

(a) 7 (b) 2
(c) 9 (d) 14

23. The difference between a number and its three fifth is 50. What is the number?

(a) 75 (b) 100
(c) 125 (d) None of these

24. The sum of three consecutive numbers is 126. Numbers are

(a) 39, 40, 41 (b) 42, 43, 44
(c) 40, 41, 42 (d) 41, 42, 43

25. If one third of one fourth of number is 15, then three tenth of that number is ....

(a) 35 (b) 36
(c) 45 (d) 54

26. A number is doubled and 9 is added. If the result is trebled, it becomes 75. What is that number?

(a) 3.5 (b) 6
(c) 8 (d) None of these

27. What least number must be added to 2010 to obtain a number, which is completely divisible by 19?

(a) 1 (b) 2
(c) 3 (d) 4

28. What least number must be subtracted from 1672 to obtain a number, which is completey divisible by 17?

(a) 2 (b) 4
(c) 6 (d) 8

29. What is the name of the concept containing two or more unknown quantities?

(a) Equation
(b) Simple equation
(c) Quadratic equation
(d) Simultaneous equation

30. How many methods are there for solving simultaneous equation?

(a) one (b) two
(c) three (d) four

KEY TO MCQs 2. Algebra

1. (b)	**2.** (c)	**3.** (b)	**4.** (c)	**5.** (b)	**6.** (a)	**7.** (b)	**8.** (a)	**9.** (a)	**10.** (b)
11. (a)	**12.** (b)	**13.** (d)	**14.** (b)	**15.** (c)	**16.** (b)	**17.** (c)	**18.** (d)	**19.** (c)	**20.** (c)
21. (c)	**22.** (c)	**23.** (c)	**24.** (d)	**25.** (d)	**26.** (c)	**27.** (d)	**28.** (c)	**23.** (d)	**30.** (c)

3. Elasticity

1. A steel bar or wire can be extended elastically only about __________of its original length:

(a) 1 percent (b) 2 percent
(c) 5 percent (d) 10 percent

2. Elasticity is the tendency of solid materials to ____________after forces are applied on them.

(a) remain in their original shape
(b) return to their original shape
(c) partly return to their original shape
(d) None of the above

3. When energy goes into changing the shape of some material and it stays changed, that is called as:
   (a) plasticity
   (b) elasticity
   (c) elastic deformation
   (d) plastic deformation
4. Rubber bands are example of :
   (a) plastic material
   (b) elastic material
   (c) Both plastic and elastic
   (d) None of the above
5. Plastic materials are highly formable materials that are artificially made from organic compounds called _______ along with additive components.
   (a) polymers
   (b) polythine
   (c) wax
   (d) None of the above
6. Example of plastic material is :
   (a) Rubber
   (b) Soft wax
   (c) Stretchy materials
   (d) All of above
7. Stress formula is ....
   (a) $\frac{\text{Area of Original Cross-section}}{\text{Force applied}}$
   (b) $\frac{\text{Force applied}}{\text{Area of Original Cross-section}}$
   (c) $\frac{\text{Change in length}}{\text{Original length}}$
   (d) $\frac{\text{Original length}}{\text{Change in length}}$
8. Stress = ...
   (*a*) Area – Load (*b*) Load × area
   (*c*) $\frac{\text{Area}}{\text{Load}}$ (*d*) $\frac{\text{Load}}{\text{Area}}$
9. Modulus of elasticity = ...
   (a) $\frac{\text{Strain}}{\text{Stress}}$ (b) Strain × stress
   (c) $\frac{\text{Stress}}{\text{Strain}}$ (d) Stress – Strain
10. Tensile stress of a tough material is tripled by changing its composition without changing its extensional strain. The Young's modulus ....
    (a) increases to three times the present value
    (b) decreases to three times the present value
    (c) remains unchanged
    (d) is independent of stress and strain
11. What is the strain in the material if a load of 300 kg hanging from a rod of 3 metres length extends it by 4 mm.
    (a) 0.00433 (b) 0.00333
    (c) 0.00233 (d) 0.00133
12. What is the diameter of a M.S. bar to withstand a load of 6720 kg, if the maximum stress of the material is 698.2 kg per sq. cm?
    (a) 2.5 cm (b) 3.5 cm
    (c) 4.5 cm (d) 5.5 cm

KEY TO MCQs 3. Elasticity

1. (a) **2.** (b) **3.** (d) **4.** (b) **5.** (a) **6.** (b) **7.** (b) **8.** (d) **9.** (c) **10.** (a)
11. (d) **12.** (b)

4. Profit and Loss

1. Reduction from original selling price is called ....
   (a) loss (b) list price
   (c) profit (d) mark down
2. A man buys an article for ₹ 27.50 and sells it for ₹ 28.60. Find his gain percent.
   (a) 1% (b) 2%
   (c) 3% (d) 4%
3. A book was sold for ₹ 27.50 with a profit of 10%. If it were sold for ₹ 25.75, what would have been percentage of profit or loss?
   (a) 2% profit (b) 3% profit
   (c) 2% loss (d) 3% loss
4. If the cost price of 12 pens is equal to the selling price of 8 pens, the gain percentage is ....
   (a) 12% (b) 30%
   (c) 50% (d) 60%
5. A man buys an article for ₹ 50 and sells it for ₹ 52. His gain percent is ....
   (a) 1% (b) 2%
   (c) 3% (d) 4%
6. Eleven bags are bought for ₹ 1000 and each bag sold for ₹ 110. Find out gain in percentage.
   (a) 10% (b) 21%
   (c) 35% (d) 20%
7. What is the profit percent if the cost price is 25% of selling price?
   (a) 150 % (b) 200 %
   (c) 300 % (d) 350 %
8. A man buys an item at ₹ 1200 and sells it at the loss of 20 percent. What is the selling price of that item.
   (a) ₹ 660 (b) ₹ 760
   (c) ₹ 860 (d) ₹ 960
9. A person incurs a loss of 5% while selling a watch for ₹ 1140. At what price should the watch be sold to earn profit of 5%.
   (a) ₹ 1200 (b) ₹ 1230
   (c) ₹ 1260 (d) ₹ 1290
10. By selling a bicyle for ₹ 2850 a shopkeeper gains 14%. If the profit is reduced to 8%, then selling price will be ....
    (a) ₹ 2600 (b) ₹ 2700
    (c) ₹ 2800 (d) ₹ 3000
11. If a man was to sell his chair for ₹ 720, he would lose 25%. To gain he should sell it for ....
    (a) ₹ 1200 (b) ₹ 1000
    (c) ₹ 960 (d) ₹ 900
12. If Harsh sold a match ticket for Rs 340 at a profit of 25%, at what price did he purchase the ticket?
    (a) ₹ 280 (b) ₹ 255
    (c) ₹ 300 (d) ₹ 272
13. A tradesman sold an article at a loss of 20%. Had he sold it for ₹ 100 more, he would have gained profit of 5%. Find the cost price of the article.
    (a) ₹ 360 (b) ₹ 400
    (c) ₹ 425 (d) ₹ 450
14. The cost price of 20 articles is the same as the selling price of "x" articles. Find the value of "x", if the profit is 25%.
    (a) 13 (b) 14
    (c) 15 (d) 16
15. A man bought apples at the rate of 8 for ₹ 34 and sold them at the rate of 12 for ₹ 57. How many apples should be sold to earn a net profit of ₹ 45?
    (a) 90 (b) 100
    (c) 135 (d) 150
16. At what percentage above the cost price must an article be marked so as to gain 33% after allowing a customer a discount of 5%?
    (a) 35% (b) 38%
    (c) 40% (d) 42%
17. A shopkeeper takes Rs 2000 loan from Bank with simple interest of 10% per annum. Find the interest paid by him for a period of 6 months.

(a) ₹ 20 (b) ₹ 100
(c) ₹ 200 (d) ₹ 1200

18. Find the simple interest on ₹ 5200 for a period of 2 years at the rate of 6% per annum.
(a) ₹ 450 (b) ₹ 524
(c) ₹ 600 (d) ₹ 624

19. Find the sum of amount recieved due on ₹ 7500 for a period of 1 year 6 moths at the rate of simple interest 8% per annum.
(a) ₹ 8400 (b) ₹ 7200
(c) ₹ 7275 (d) ₹ 6468

20. Vimal invested ₹ 3000 for a period of one year at the rate of 7% interest per annum. Find the sum of amount received.
(a) ₹ 3210 (b) ₹ 3310
(c) ₹ 4140 (d) ₹ 2510

21. ₹ 1200 is lent out at 5% per annum simple interest for a period of 3 years. Find the sum of amount.
(a) ₹ 1290 (b) ₹ 1380
(c) ₹ 1200 (d) ₹ 1470

22. Ramesh invested ₹ 5000 for a period of 2 years at the rate of 11% simple interest per annum. Find the sum of ammount.
(a) ₹ 6100 (b) ₹ 8100
(c) ₹ 5510 (d) ₹ 7246

23. A person paid ₹ 1680 as interest on principal of ₹ 7000 for a period of 16 months. Find the rate of interest per annum.
(a) 18% (b) 15%
(c) 12% (d) 14%

24. Interest obtained on a principal of ₹ 5000 for a period of 3 years is ₹ 1500. Find the rate of interest per annum.
(a) 8% (b) 9%
(c) 10% (d) 11%

25. A person took loan from a Bank at 12% interest rate per annum. After 3 yeas he paid ₹ 5400 as interest fot the period. Find the principal amount borrowed by him.
(a) ₹ 2000 (b) ₹ 15000
(c) ₹ 1000 (d) ₹ 20000

26. A sum amounts to ₹ 3584 in 219 days at interest rate of 4% per annum. Find the principal amount.
(a) ₹ 3400 (b) ₹ 3260
(c) ₹ 3150 (d) ₹ 3500

27. How much time will it take for an mount of ₹ 450 to yield ₹ 81 at 4.5% simple interest rate per annum.
(a) 3 years (b) 4 years
(c) 2 years (d) 5 years

28. At what rate of compound interest per annum will a sum of ₹ 1200 become ₹ 1348.32 in 2 years?
(a) 6% (b) 6.5%
(c) 7% (d) 7.5%

29. At a certain rate of compound interest ₹ 15320 becomes ₹ 30239 in 6 years. What is the rate of interest per annum?
(a) 11% (b) 12%
(c) 13% (d) 14%

30. In what time will ₹ 1000 becomes ₹ 1331 at interest rate of 10% per annum compounded annually?
(a) 2 years (b) 3 years
(c) 4 years (d) 7 years

31. A bank offers 5% compound interest calculated on half yearly basis. A customer deposited ₹ 1600 each on 1st January and 1st July of a year. How much amount he would have gained by way of interest at the end of the year?
(a) ₹ 120 (b) ₹ 121
(c) ₹ 122 (d) ₹ 123

32. Find the difference between the compound interest and simple interest at 5% per annum for 2 years on principal of ₹ 2000?
(a) ₹ 5 (b) ₹ 10.50
(c) ₹ 4.50 (d) ₹ 5.50

33. What will be the difference between the simple interest and compound interest at 10% per annum on principal of ₹ 1000 for a period of 4 years?
(a) ₹ 31 (b) ₹ 32.10
(c) ₹ 40.40 (d) ₹ 64.10

KEY TO MCQs 4. Profit and Loss

1. (a)	**2.** (d)	**3.** (b)	**4.** (c)	**5.** (d)	**6.** (b)	**7.** (c)	**8.** (d)	**9.** (c)	**10.** (b)
11. (a)	**12.** (d)	**13.** (b)	**14.** (d)	**15.** (a)	**16.** (c)	**17.** (b)	**18.** (d)	**19.** (a)	**20.** (a)
21. (b)	**22.** (a)	**23.** (a)	**24.** (c)	**25.** (b)	**26.** (d)	**27.** (b)	**28.** (a)	**29.** (b)	**30.** (b)
31. (b)	**32.** (a)	**33.** (d)							

5. Estimation and Costing

1. What is the total estimation cost of M.S. rod dia 50 mm and length 500 mm, if cost of matarial ₹ 80/kg and the density of the M.S. rod is 7.8 gm/cm^3, Lathe time taken to complete the component is 1 hr 30 minutes and lathe charge is ₹ 80/hr.

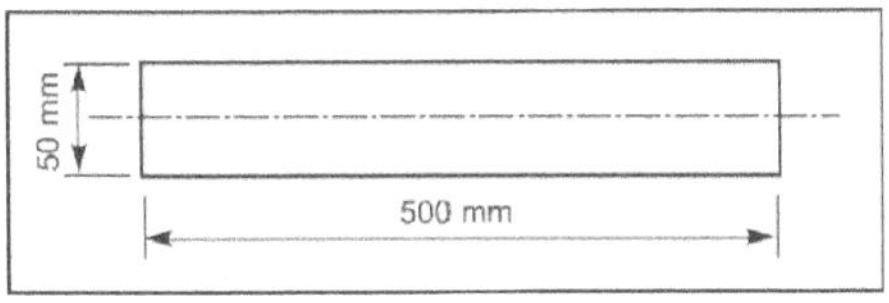

(a) ₹ 733 (b) ₹ 763

(c) ₹ 850 (d) ₹ 500

2. Find the estimation of manufacturing of the component shown in the figure. Density of material is 9.8 grams per c.c. and machining cost is ₹ 380 per hour. Machining time is 4 hours 15 minutes, Cost of material is ₹ 90 per kg.

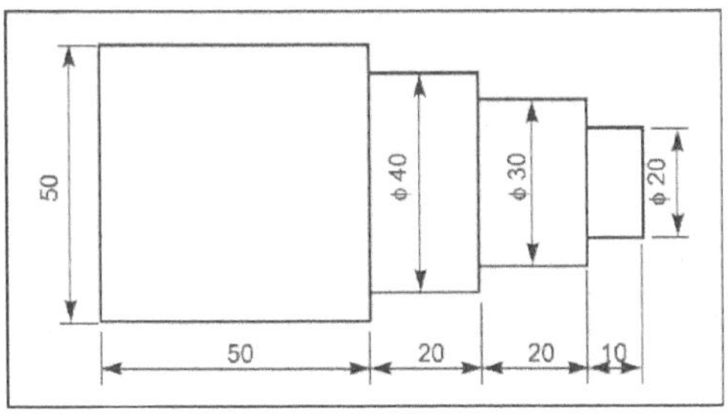

(a) ₹ 1750 (b) ₹ 1825

(c) ₹ 1714 (d) ₹ 1545

3. What is the material cost of component shown in the figure. Density of material is 7.8 grams per c.c. and cost of material is ₹ 180 per kg.

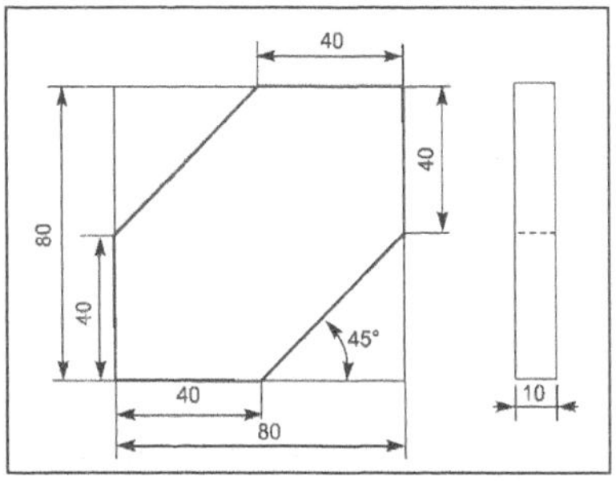

(a) ₹ 75.40 (b) ₹ 67.39

(c) ₹ 55.45 (d) ₹ 40.75

4. What is the total estimation cost of job as shown in the figure. Density of material is 7.8 grams per c.c. Time taken by milling machine to complete the job is 2 hours. Material cost is ₹ 60 per kg. Charge for the milling is ₹ 200 per hour.

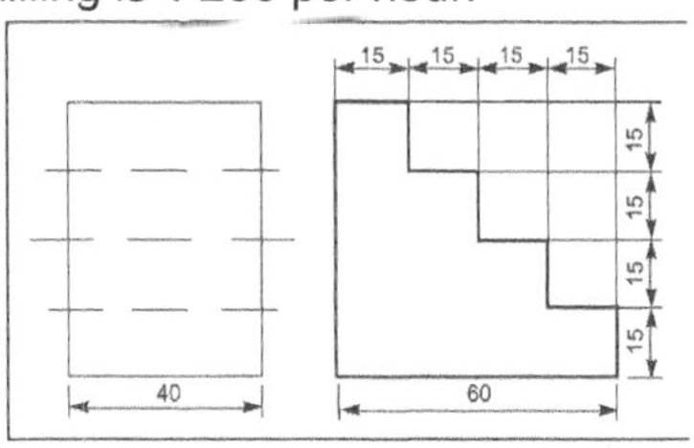

(a) ₹ 542 (b) ₹ 642

(c) ₹ 342 (d) ₹ 442

KEY TO MCQs 5. Estimation and Costing

1. (a)	**2.** (c)	**3.** (b)	**4.** (d)

2nd Year

Engineering Drawing

1. Electrical Signs and Symbols

1. Identify the symbol of 3-pole contractor:

(a) 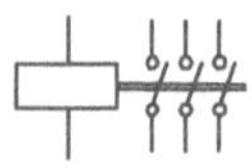

(b) 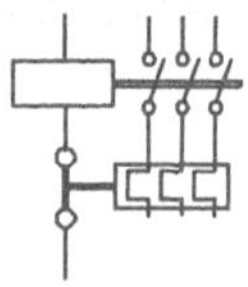

(c) 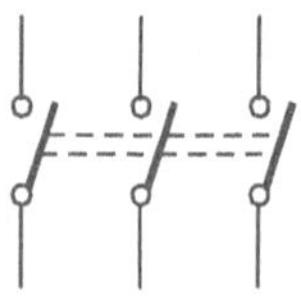

(d) 

2. Which one is pulsating wave form out of the following?

(a) A.C 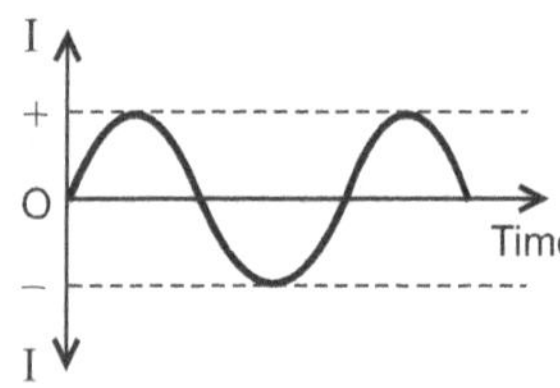

(b) 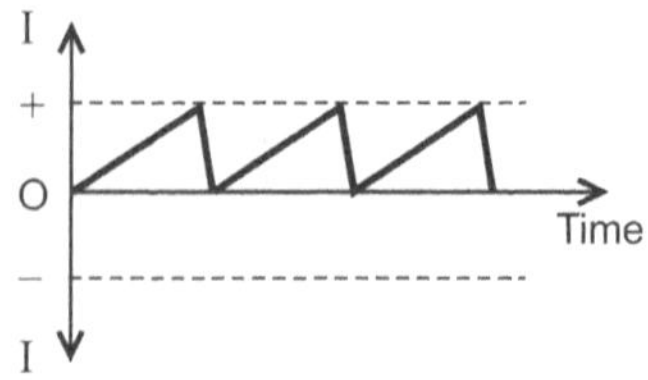

(c) 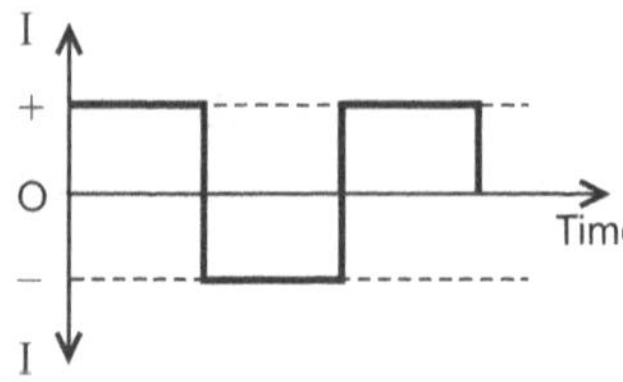

(d) D.C 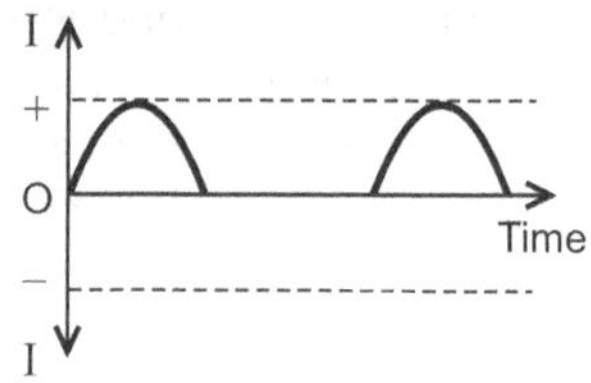

3. Identify the type of shackle shown in the figure given:

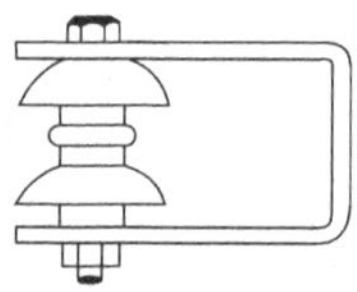

(a) Double shackle
(b) Single shackle
(c) Pin type single shade
(d) Pin type double shade

4. What symbol is used to show a zener diode out of the following symbols?

(a) 

(b) 

(c) 

(d) 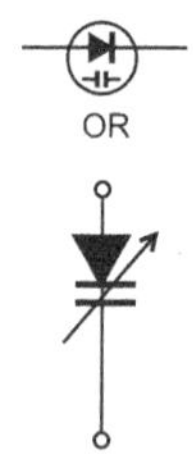

5. Identify the symbol of a variable inductor:

(a) 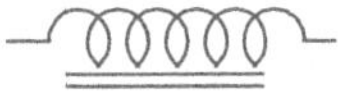 (a) 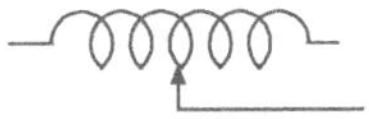

(c) 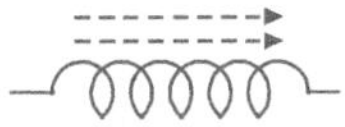 (d) 

6. Identify the symbol of 1 NO + 1 NC push button:

(a) 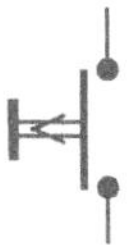 (b) 

(c) 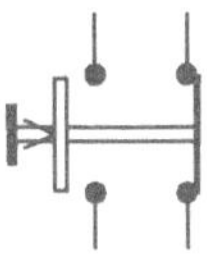 (d) 

7. Identify symbol of UJT out of the given symbols:

(a) 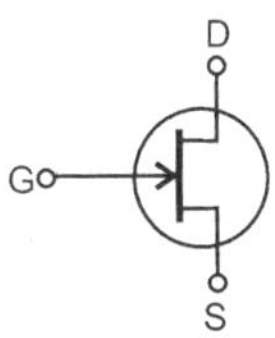

(b) 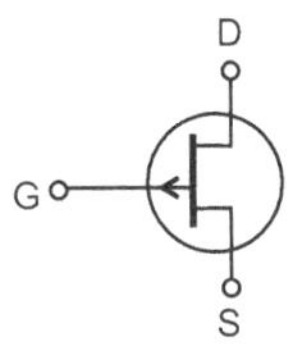

(c) 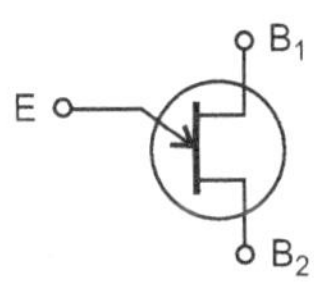

(d) 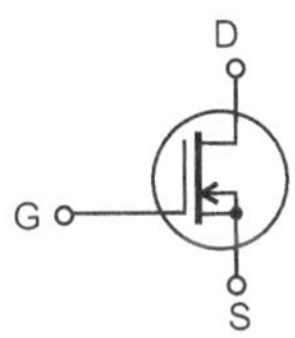

8. Identify the symbol of IGBT out of the given symbols:

(a) 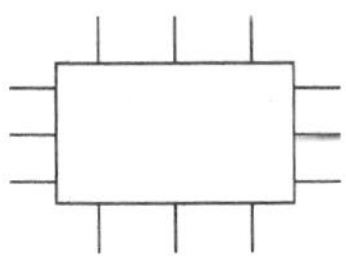 (b) 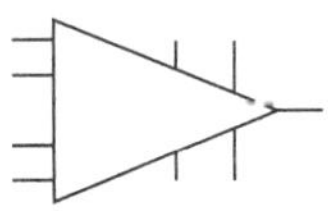

(c) 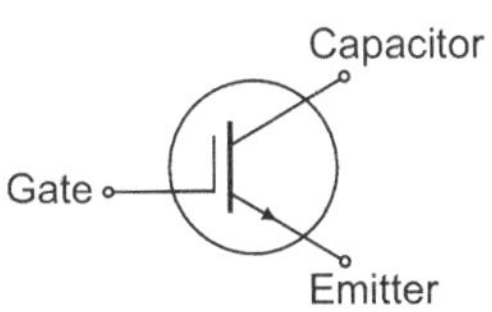

(d) 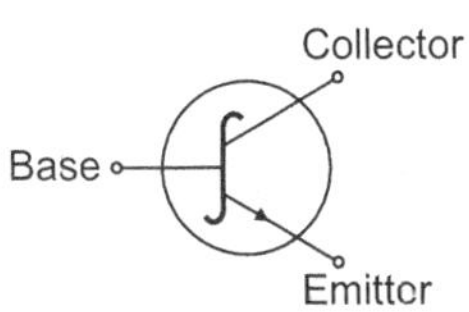

9. The symbol for N-channel MOSFET is:

(a) 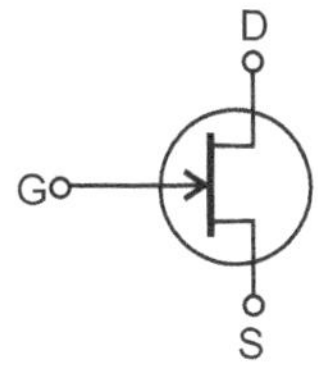

(b) 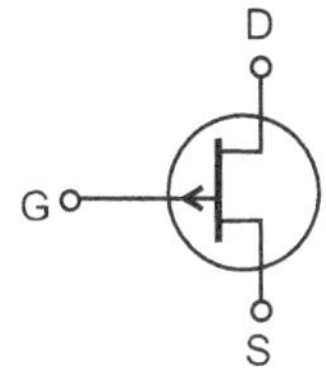

(c) 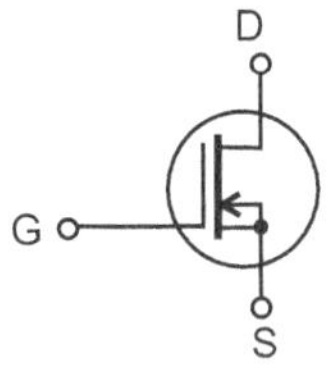

(d) 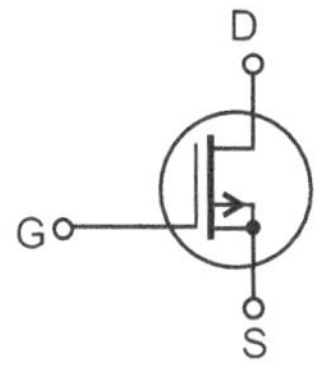

10. Identify LED symbol out of the following:

(a)  (b) 

(c) 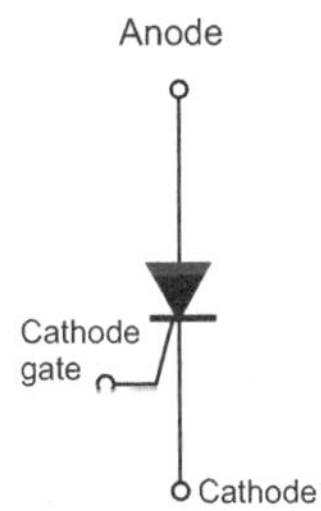

(d) 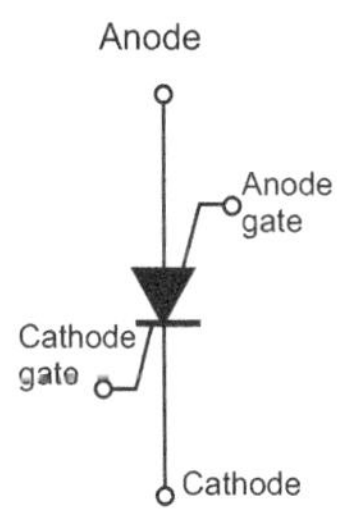

11. Identify OP-Amp. symbol out of the following:

(a) 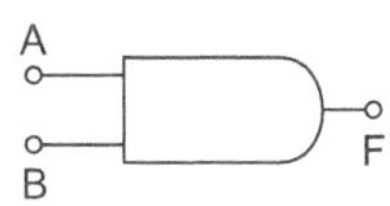

(b) 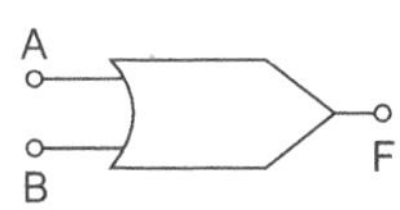

(c) 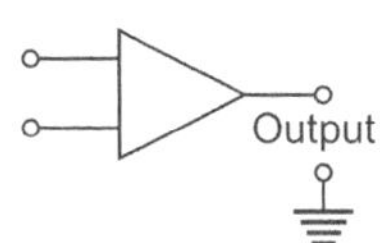

(d) 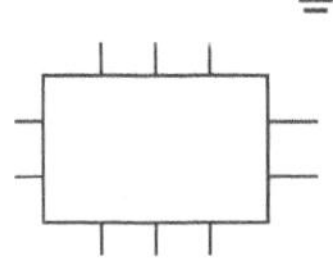

12. What symbol is used to show a thermistor?

(a) 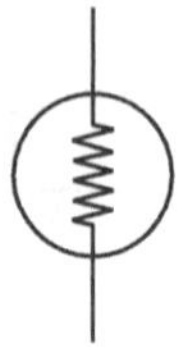

(b) 

(c) 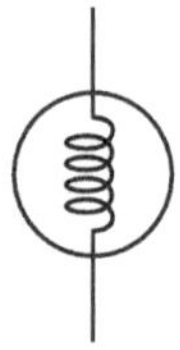

(d) 

13. What symbol is used to show a crystal microphone?

(a) 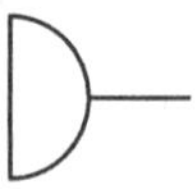

(b) 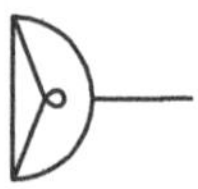

(c) 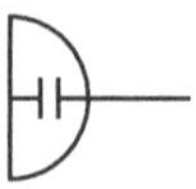

(d) 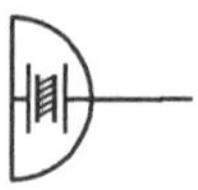

14. Identify the symbol shown in the figure given:

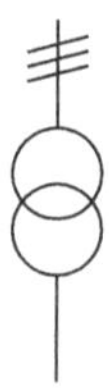

(a) Single phase power transformer
(b) Single phase transformer
(c) 3-phase power transformer
(d) 2-phase power transformer

15. Identify the symbol of NPN transistor out of the following symbols:

(a) 

(b) 

(c) 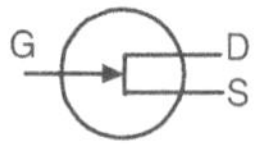

(d) 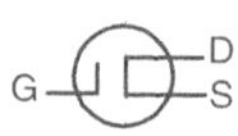

16. Identify the symbol of a 3-phase slip-ring motor:

(a) 

(b) 

(c) 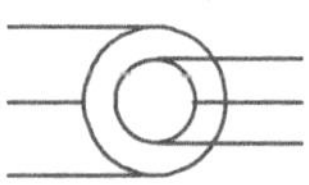

(d) None of these

17. Identify the symbol of 3-phase squirrel cage motor:

(a) 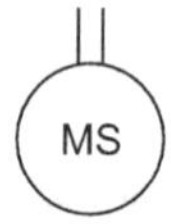

(b) 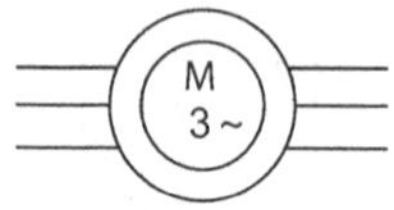

(c) 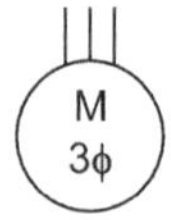

(d) 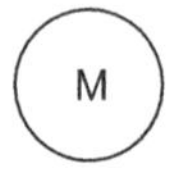

18. Identify pure DC waveform out of the following waveforms:

(a) 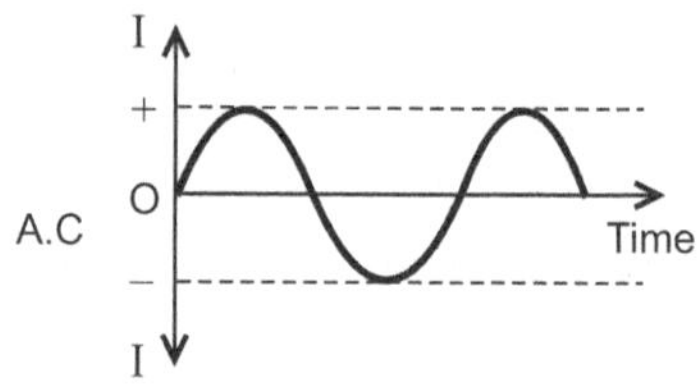

(b) 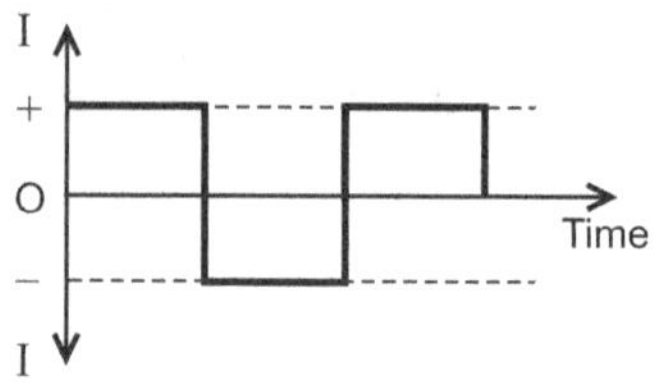

(c) 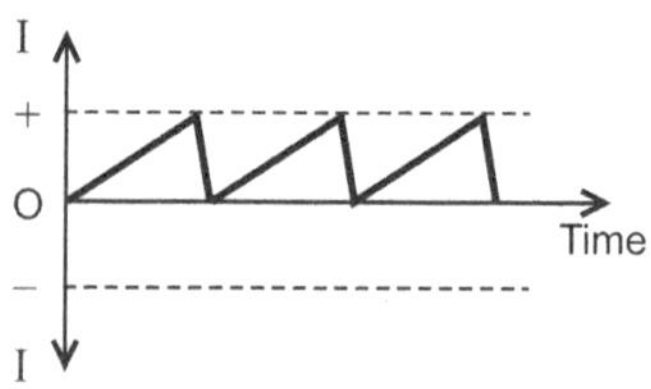

(d) 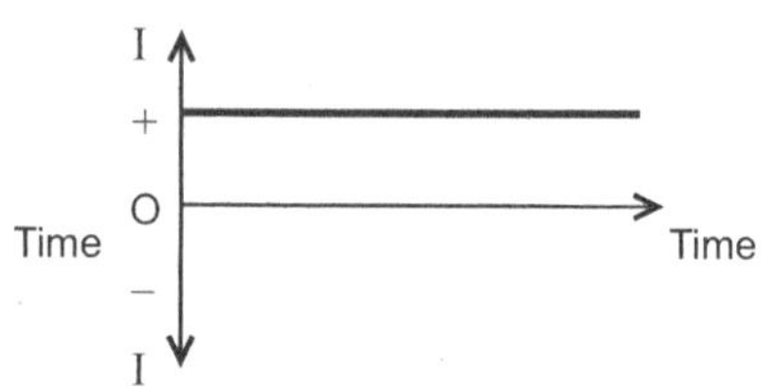

19. Identify the symbol of P-channel MOSFET:

(a) 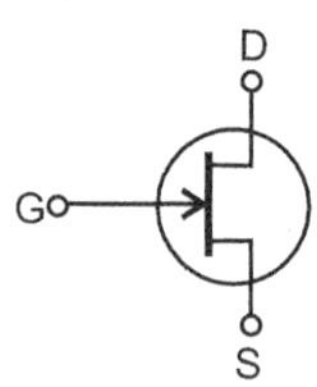

(b) 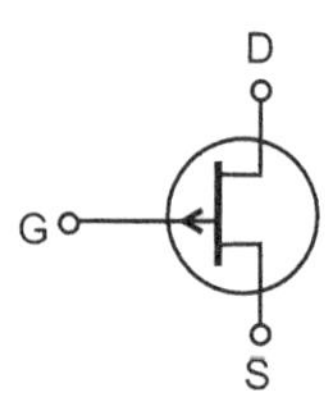

(c) 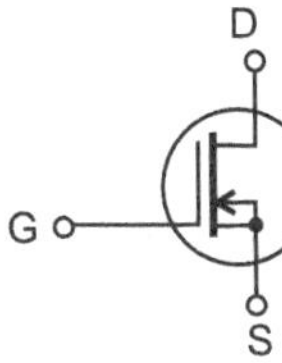

(d) 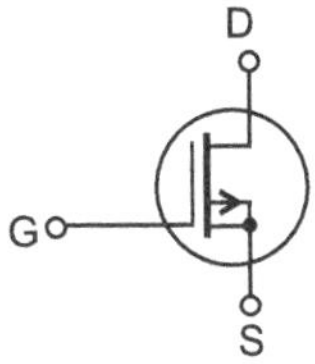

20. Identify the symbol of IF transformer out of the following.

(a) 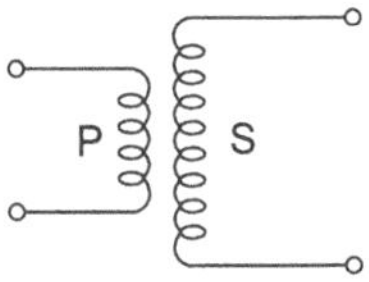

(b) 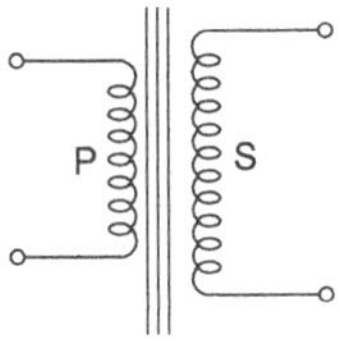

(c) 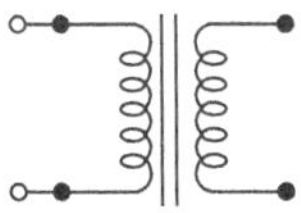
(d) 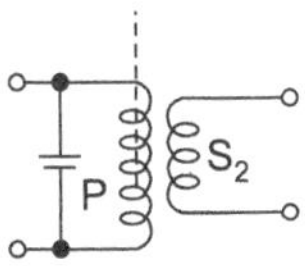

21. Identify the symbol of main fuse board with switches:

(a) 
(b) 

(c) 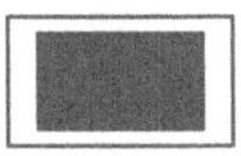
(d) 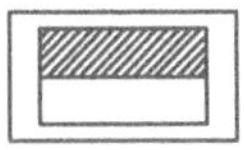

22. Identify the symbol of a power factor meter:

(a) 

(b) 

(c) 

(d) 

23. The symbol for series dc motor is:

(a) 

(b) 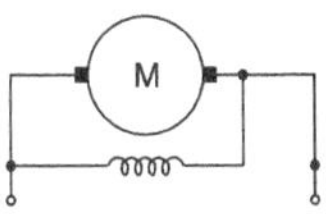

(c) 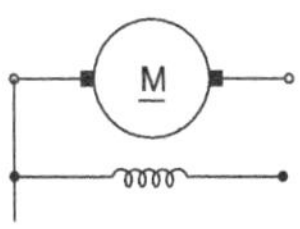

(d) 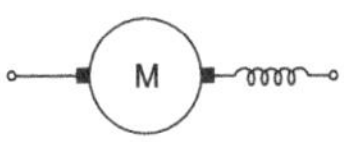

24. Identify the symbol of 3-pin, 5A, wall socket:

(a) 
(b) 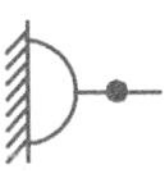

(c) 
(d) 

25. Identify the symbol given:

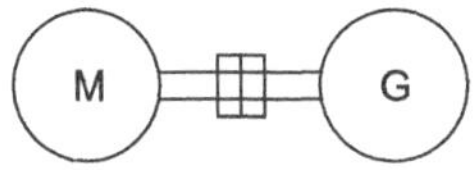

(a) Motor generator set
(b) DC motor
(c) DC generator
(d) None of these

26. Identify the symbol of a DIAC out of the following symbols.

(a) 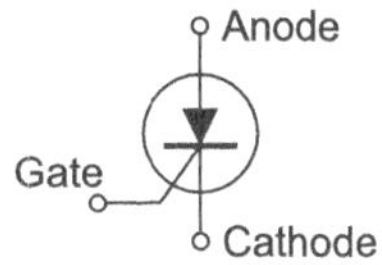

(b) 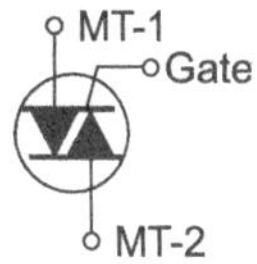

(c) 

(d) 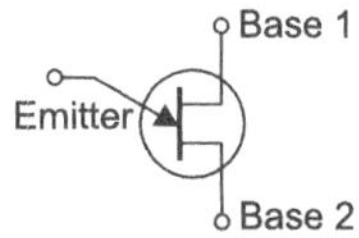

27. What is the symbol for VDR (voltage dependent resistor)?

(a) 
(b) 

(c) 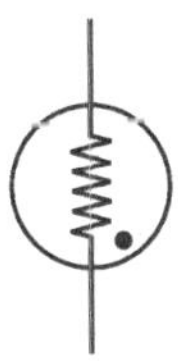
(d) 

28. Identify the symbol for Double pole double throw switch.

(a) 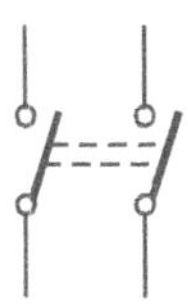
(b) 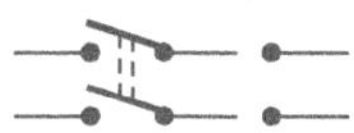

(c) 
(d) 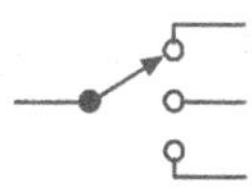

29. Identify the symbol of light sensitive diode.

(a) 

(b) 

(c) 

(d) 

30. Identify the symbol of FET N-channel:

(a) 

(b) 

(c) 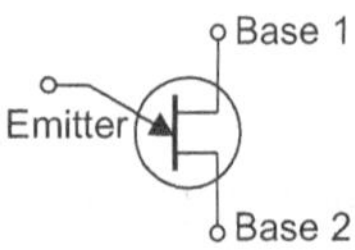

(d) 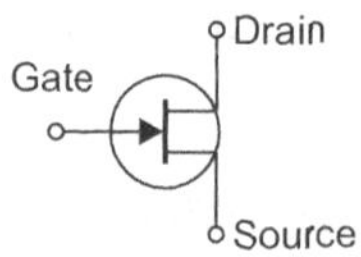

31. Identify the symbol of 3-phase transformer:

(a) 

(b) 

(c) 

(d) None of these

32. Identify the symbol of Isolator with fuse:

(a) 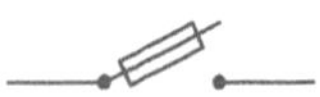

(b) 

(c) 

(d) 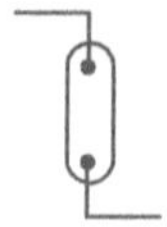

33. Identify the symbol of a SCR:

(a) 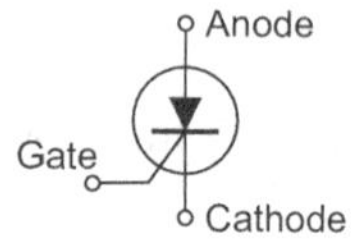

(b) 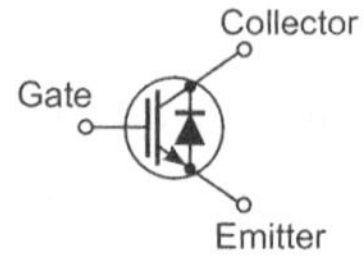

(c) 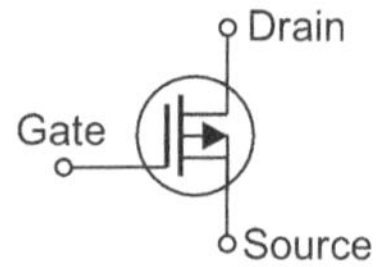

(d) 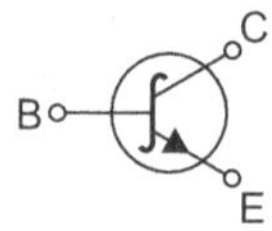

34. Identify the symbol of varactor diode:

(a) 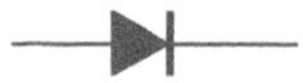

(b) 

(c) 

(d) 

35. The symbol for separately excited generator is:

(a) 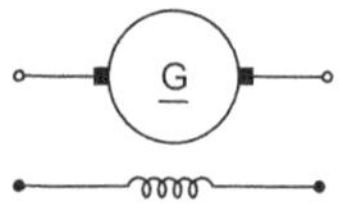

(b) 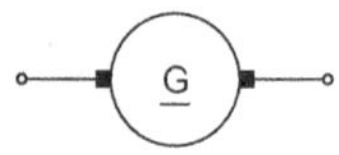

(c) 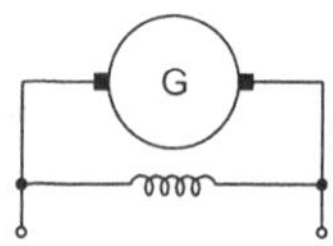

(d) 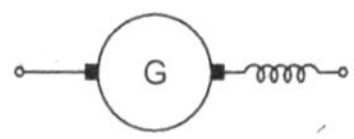

36. Which of the following symbol is used to denote a fixed value resistor?

(a) 

(b) 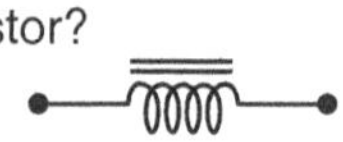

(c) 

(d) 

KEY TO MCQs (Electrical Signs and Symbols)

1. (a)	2. (d)	3. (a)	4. (c)	5. (b)	6. (c)	7. (c)	8. (c)	9. (c)	10. (a)
11. (c)	12. (b)	13. (d)	14. (c)	15. (b)	16. (c)	17. (c)	18. (d)	19. (d)	20. (d)
21. (b)	22. (b)	23. (d)	24. (c)	25. (a)	26. (c)	27. (d)	28. (b)	29. (d)	30. (d)
31. (b)	32. (a)	33. (a)	34. (c)	35. (a)	36. (d)				

2. Electrical Components

1. Read the ohmic value of the body type carbon resistor shown in the figure given:

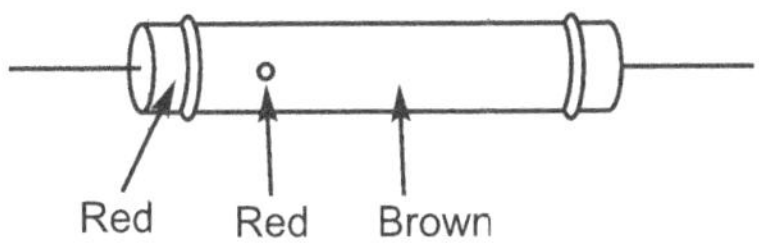

(a) 1.22 ohms (b) 12.2 ohms
(c) 122 ohms (d) 1200 ohms

2. Identify the type of shackle shown in the figure?

(a) Double shade
(b) Single shade pin type shackle
(c) Single shackle
(d) None of these

3. Which type of rotor pole is shown in the given figure. Identify it:

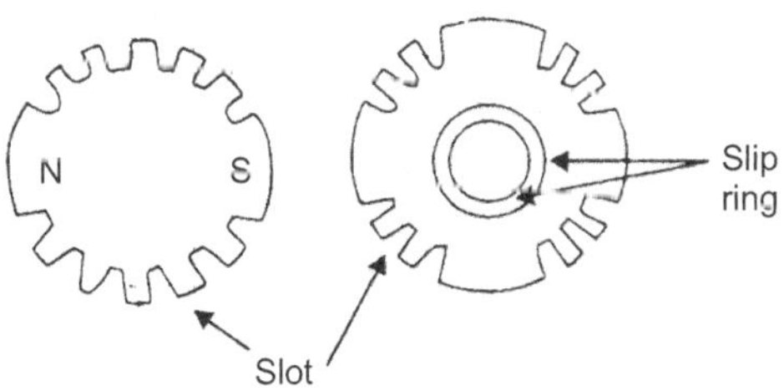

(a) Salient pole (b) Cylindrical pole
(c) Dust core pole (d) None of these

4. What type of insulator is shown in the figure given?

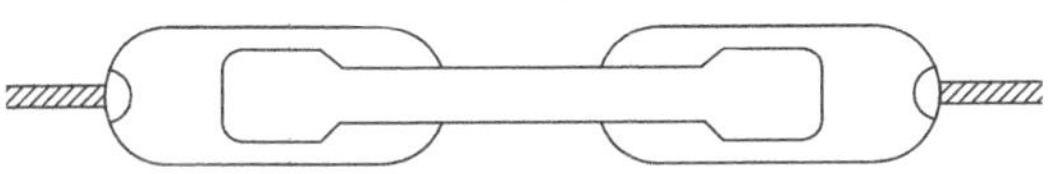

(a) Stay rod insulator
(b) Egg type insulator
(c) Suspension insulator
(d) Disc type insulator

5. What is the ohmic value of the band type carbon resistor shown in the figure given?

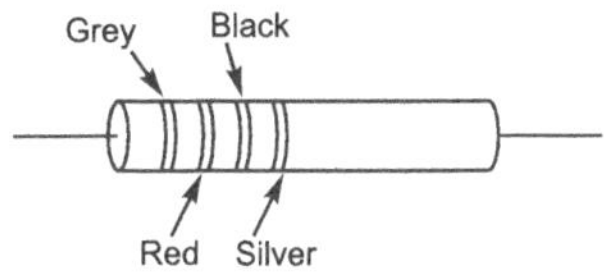

(a) 82 ohms (b) 82 ohms 5%
(c) 82 ohms 10% (d) 8.2 ohms 5%

6. The figure given shows the construction of:

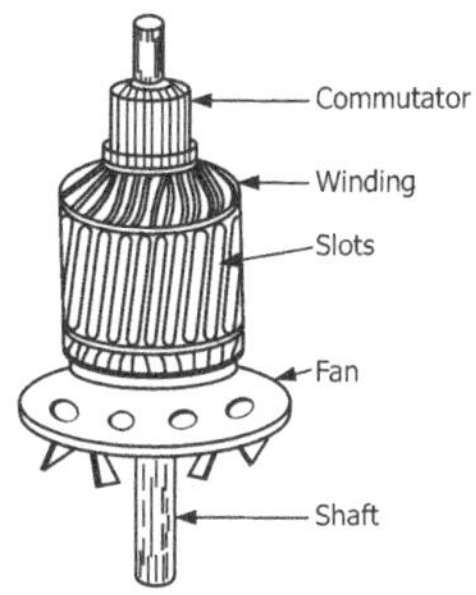

(a) Armature (b) Rocker arm
(c) Brush holder (d) None of these

7. The cross-sectional view of an electric lamp is shown in the figure given. Which type of bulb is this?

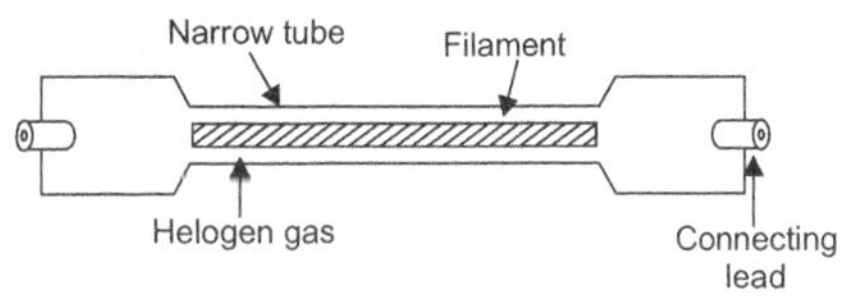

(a) Carbon filament lamp
(b) Vacuum type metal filament
(c) LED lamp
(d) Helogen lamp

8. The figure given shows:

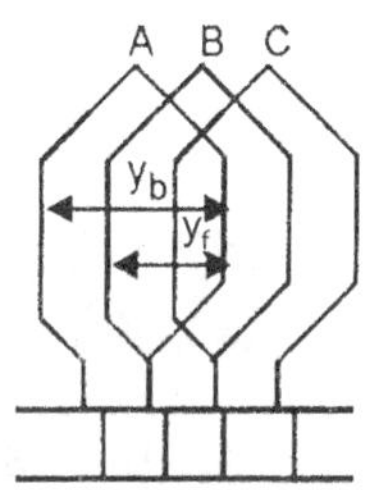

$Y_b = Y_f \pm 2$

(a) Gramme winding (b) Lap winding
(c) Wave winding (d) None of these

9. A sort of suspension insulator is shown in the figure given. Identify it:

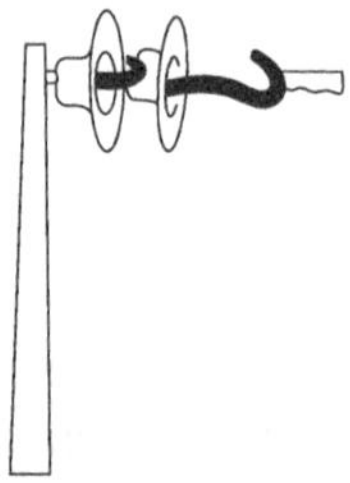

(a) 5-disc type suspension insulator
(b) Cross-sectional view of suspension insulator
(c) 2-disc side suspension insulator
(d) Simple suspension insulator

10. The figure given shows:

(a) Brush on commutator
(b) Brush on slip-ring
(c) Brush holder
(d) None of these

11. The figure given shows ....

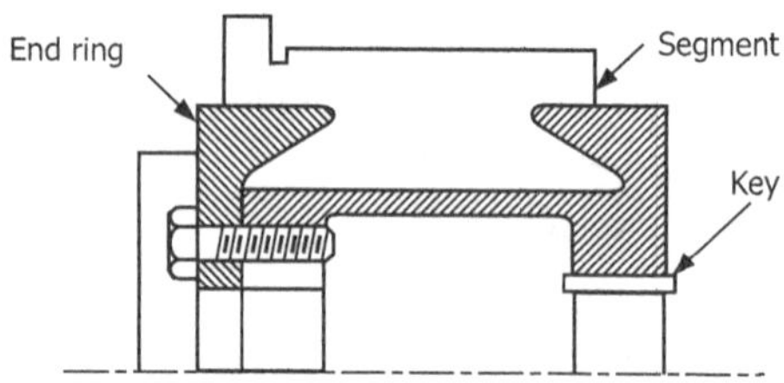

(a) Brush holder (b) Rocker arm
(c) Armature
(d) Commutator segment

12. A type of rotor is shown is the figure given. Identify it.

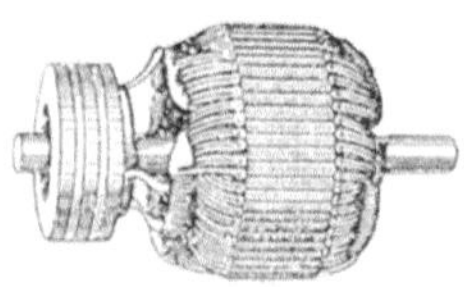

(a) Single cage squirrel cage motor's rotor
(b) Slip-ring motor's rotor
(c) Double cage squirrel-cage motor's rotor
(d) Salient-pole rotor

13. Read the ohmic value of a carbon resistor shown in the figure given:

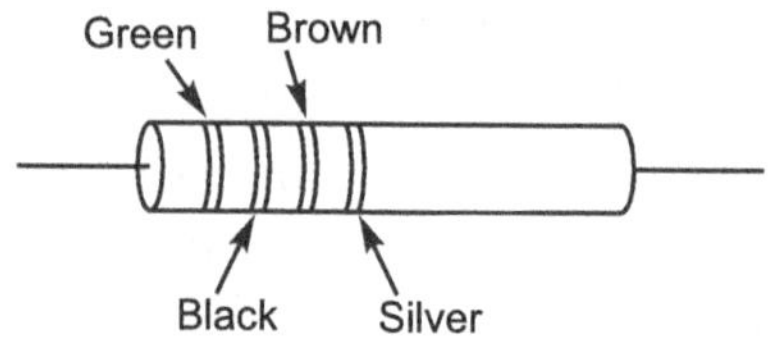

(a) 50 ohms 5% (b) 500 ohms 10%
(c) 5 k ohms 10% (d) 5 ohms 10%

14. Read the value of carbon resistor shown in the figure given.

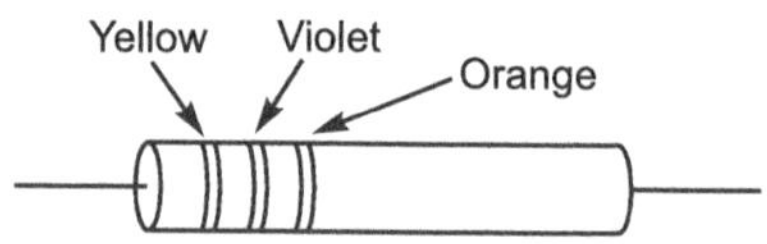

(a) 47 ohms (b) 470 ohms
(c) 4700 ohms (d) 47000 ohms

15. The construction of a rotor is shown in the given figure. Which type of rotor is this?

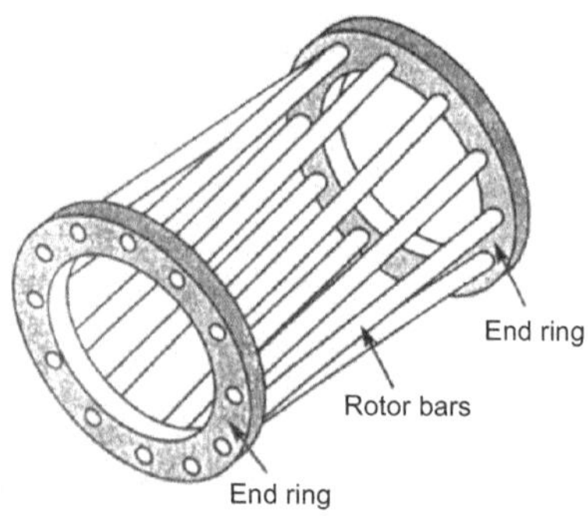

(a) Slip-ring motor rotor
(b) Squirrel-cage motor rotor
(c) Double squirrel-cage motor rotor
(d) None of these

KEY TO MCQs (Electrical Components)

1. (d)	**2.** (b)	**3.** (b)	**4.** (b)	**5.** (b)	**6.** (a)	**7.** (d)	**8.** (b)	**9.** (c)	**10.** (a)
11. (d)	**12.** (b)	**13.** (b)	**14.** (d)	**15.** (b)					

3. Electrical Wiring and Earthing

1. The figure given shows:

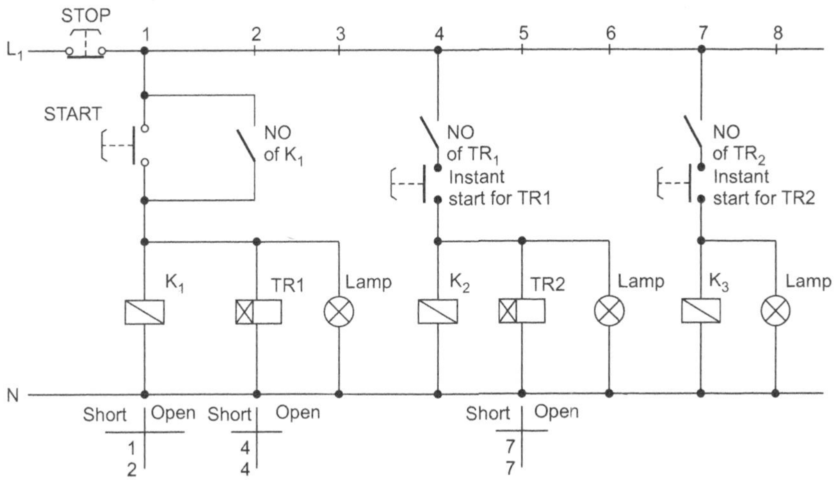

(a) Semi-automatic

(b) Automatic star delta starter

(c) Plugging circuit

(d) Sequential control circuit for 3 motors

2. A battery charging circuit is shown in the figure given. What type of circuit is it?

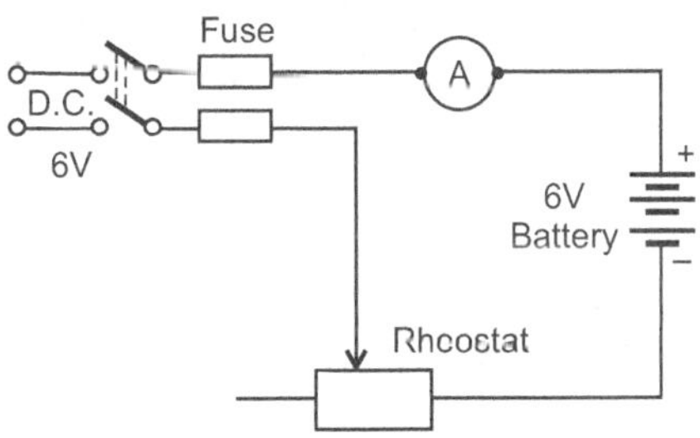

(a) Constant current battery charging circuit

(b) Constant voltage battery charging circuit

(c) Battery charging by DC shunt generator

(d) Battery charging by a battery charger

3. The figure given shows a ....

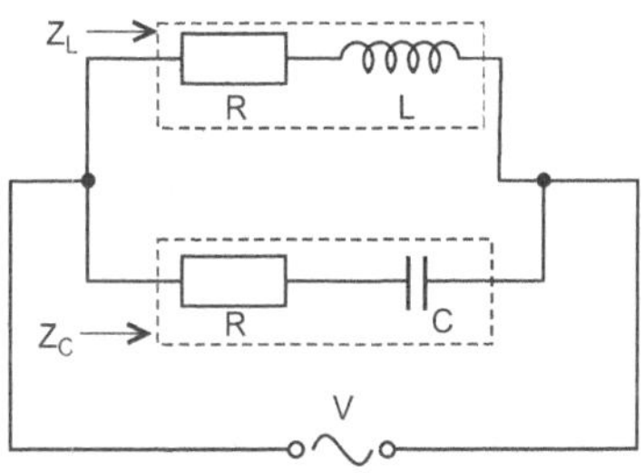

(a) Series RL circuit

(b) Series RC circuit

(c) Series RLC circuit

(d) Parallel RLC circuit

4. Identify the method of alternators synchronization?

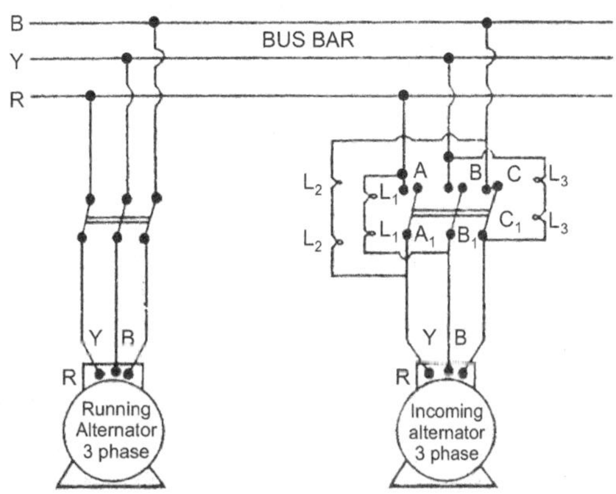

(a) Dark lamp method
(b) Bright lamp method
(c) Dark and bright lamp method
(d) None of these

5. The figure given shows a type of power distribution system. Which type of system is this?

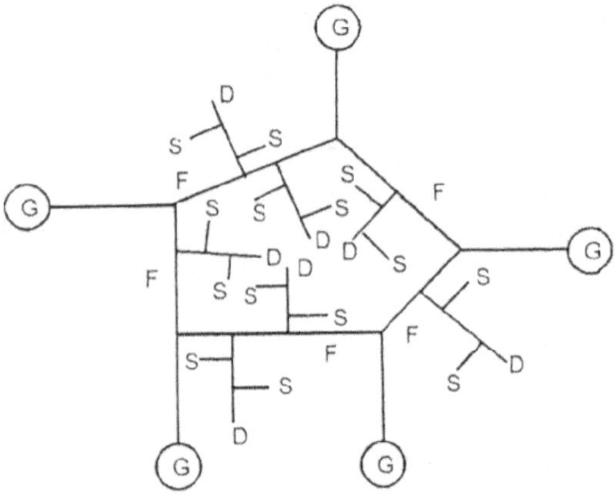

(a) Radial distribution system
(b) Ring distribution system
(c) Interconnected distribution system or Grid System
(d) None of these

6. The figure given shows electrical equivalent of a/an .... gate logic.

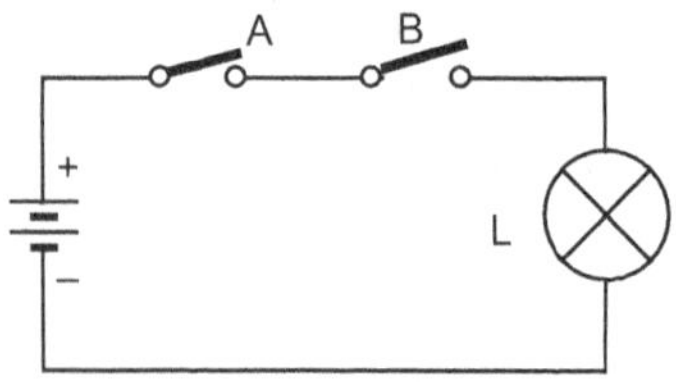

(a) NOT (b) OR
(c) NAND (d) AND

7. An outline of a sub-station is shown in the figure given. Identify it:
(a) Pole-mount sub-station
(b) Indoor sub-station
(c) Out-door sub-station
(d) None of these

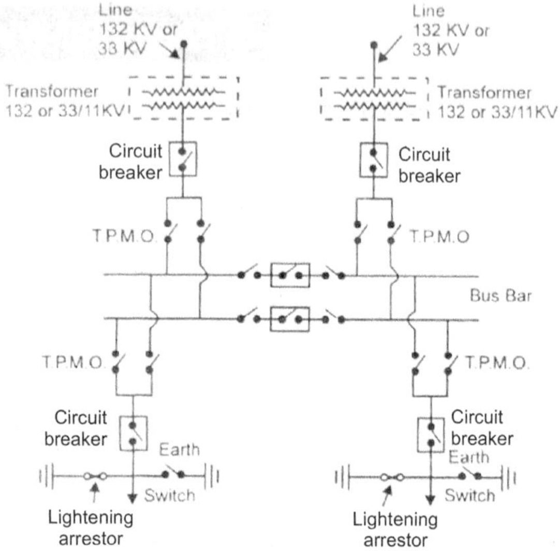

8. The constructional diagram of a DC motor starter is shown in the given figure. What type of starter is this?

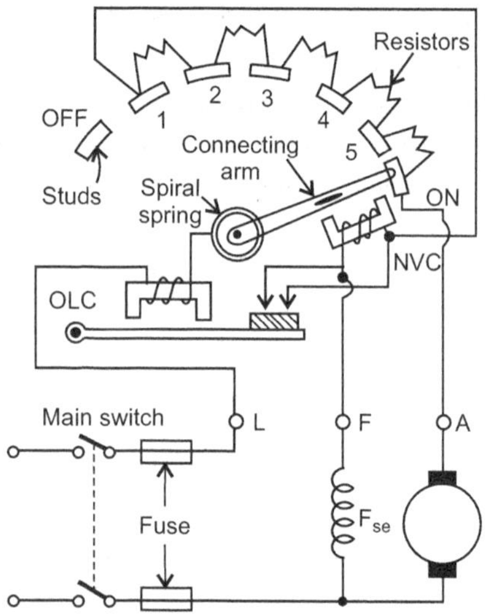

(a) Series DC motor starter
(b) 4-point DC starter
(c) 3-point DC starter
(d) None of these

9. Identify the DC motor starter shown in the figure given.
(a) 3-point DC starter
(b) 4-point DC starter
(c) Series DC motor starter
(d) Series motor starter with NVC only

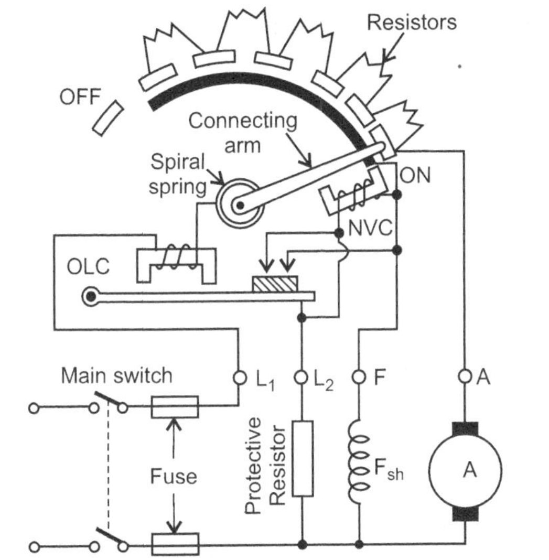

10. The figure given show the connection of ....

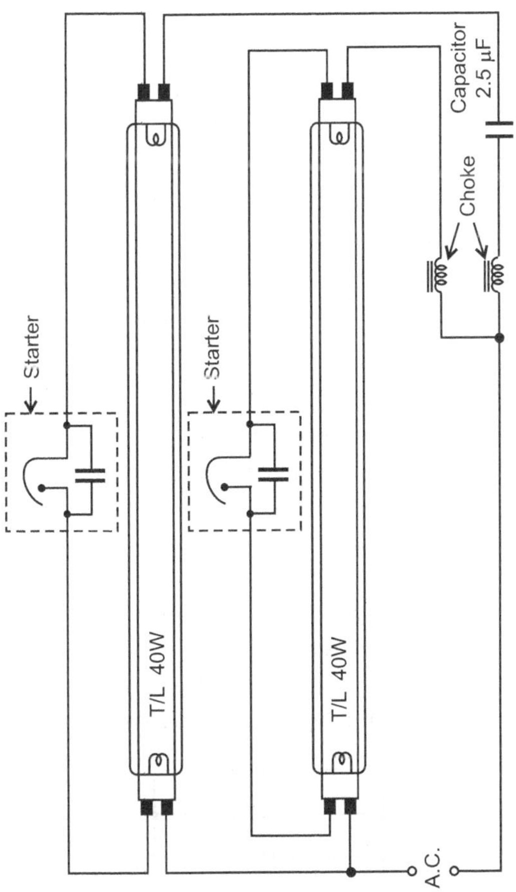

(a) Fluorescent tube

(b) Twin fluorescent tube

(c) Construction of a fluorescent tube

(d) Fluorescent tube with electronic choke

11. A DC series motor starter is shown in the given figure. Identify its speciality?

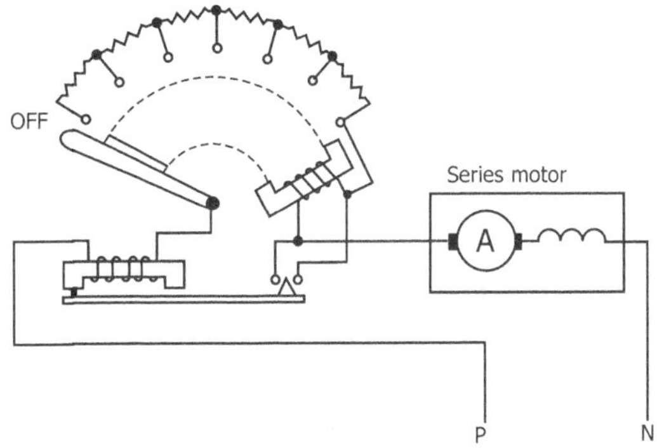

(a) It is provided with NVC

(b) It is provided with NVC and OLC

(c) It is not provided with NVC

(d) It is not provided with OLC

12. The connection diagram of a 3-phase induction motor is shown in the figure given. Identify it:

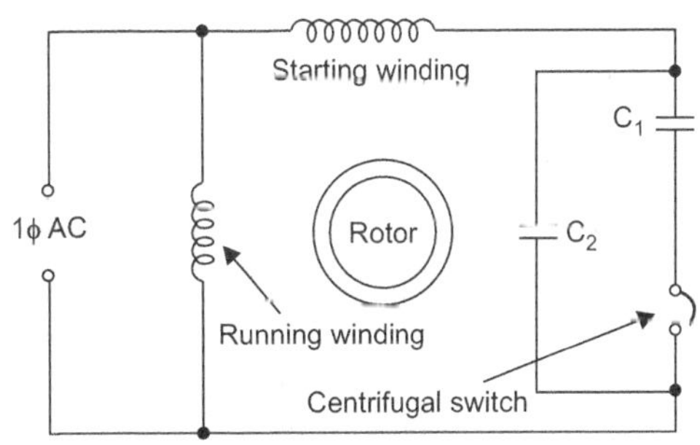

(a) Split phase inductor

(b) Permanent capacitor motor

(c) Capacitor-start motor

(d) Capacitor-start capacitor-run motor

13. Three capacitors are connected in parallel as shown in the figure given. What will be the effective capacitance?

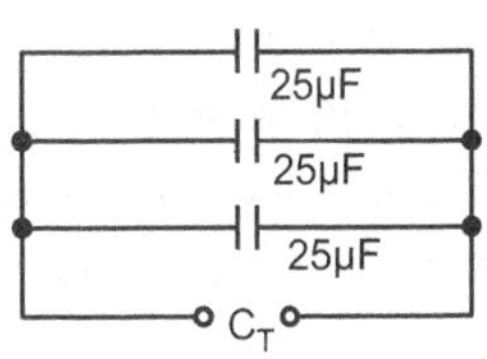

(a) 25 μF (b) 75 μF
(c) 25/3 μF (d) None of these

14. Identify the circuit of staircase wiring:

(a)
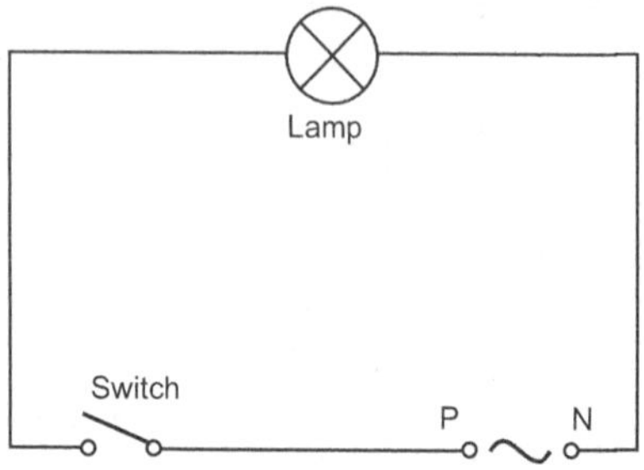

(b)
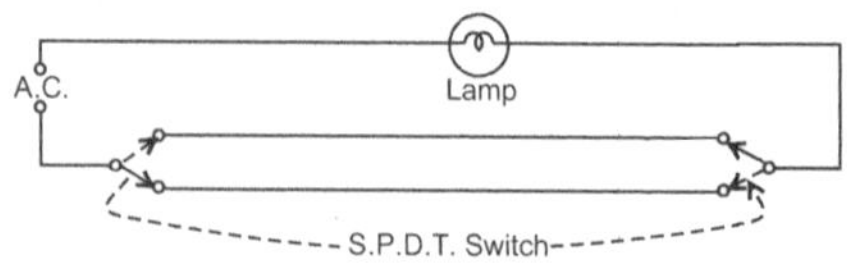

(c)
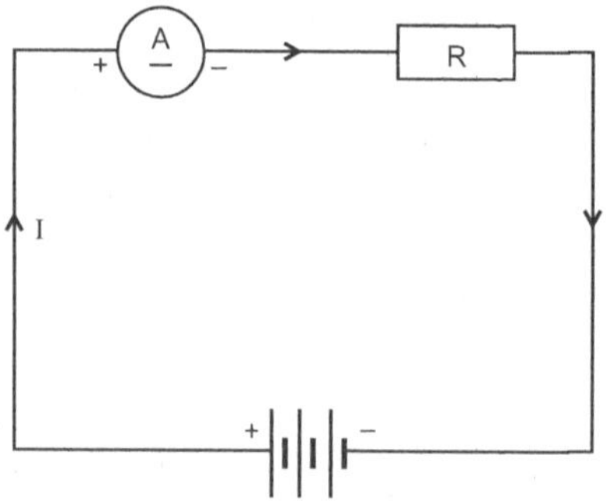

(d) None of these

15. The connection diagram of a 3-phase induction motor is shown in the given figure. Which type of motor is this?

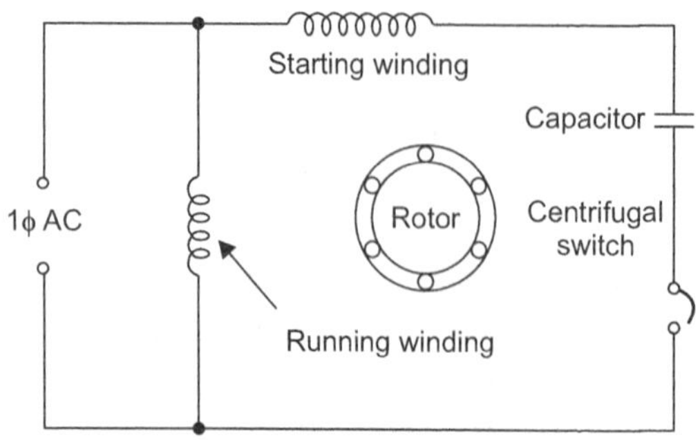

(a) Capacitor-start motor
(b) Permanent capacitor motor
(c) Split-phase induction motor
(d) Capacitor start capacitor-run motor

16. The figure given shows a tube light connections. Give its name.

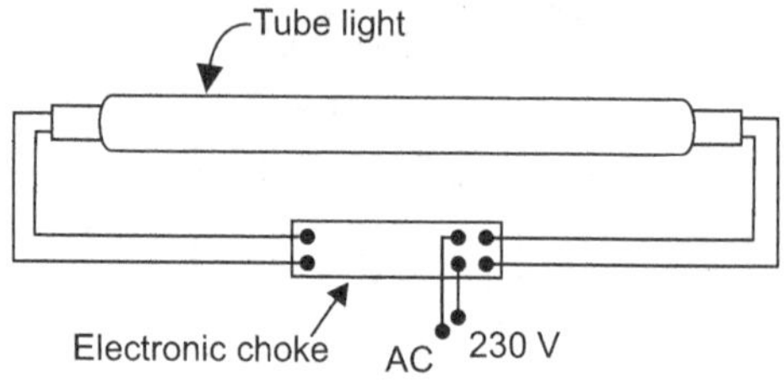

(a) Twin tube light
(b) Ordinary tube light
(c) Electronic choke type tube light
(d) None of these

17. A method of alternators synchronization is shown in the figure. Identify it.

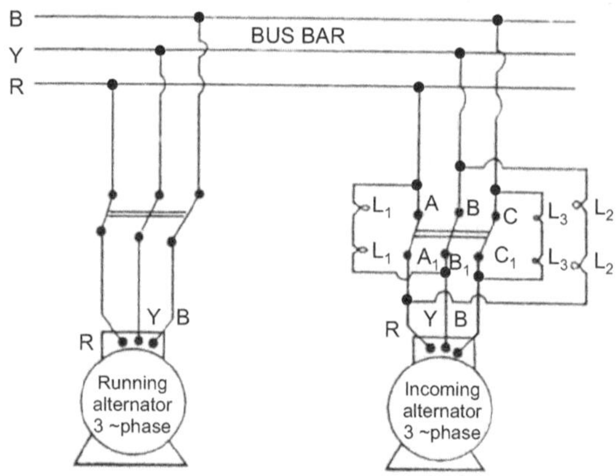

(a) Dark lamp method
(b) Bright lamp method
(c) Dark and bright lamp method
(d) None of these

18. At what depth the copper plate is buried in plate earthing?

(a) 100-200 cm (b) 200-300 cm
(c) 300-400 cm (d) 400-500 cm

19. The figure given shows a connection of cells. Which type of connection is this?

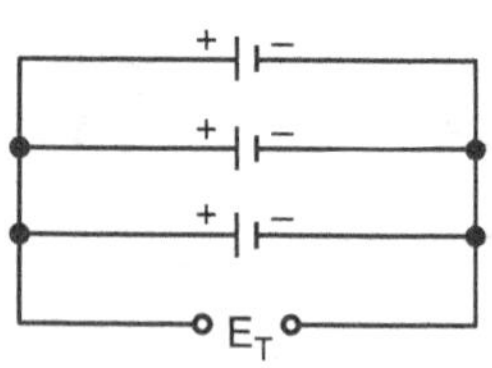

(a) Cells in series
(b) Cells in parallel
(c) Cells in mixed group
(d) None of these

20. A sort of pole-connection is shown in the figure given. What sort of pole connection is it?

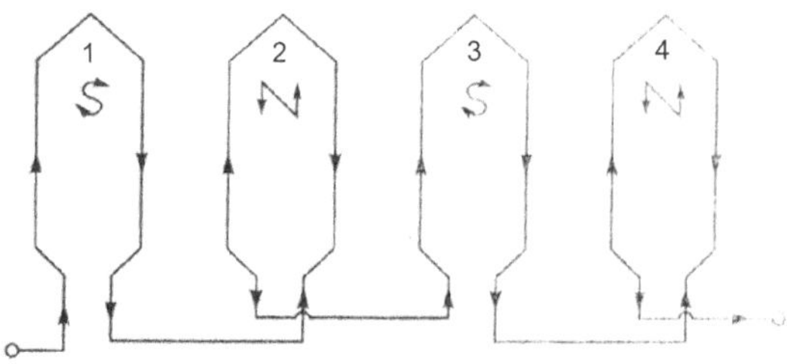

(a) Coil and pole-connection
(b) Adjacent pole-connection
(c) Alternate pole-connection
(d) None of these

21. What type of speed control method is shown is the given figure for a DC shut motor?

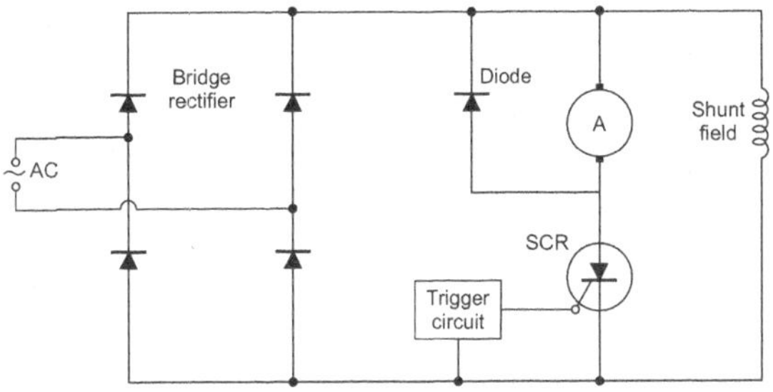

(a) SCR control method
(b) Armature control method
(c) Field control method
(d) Supply voltage variation method

KEY TO MCQs (Electrical Wiring and Earthing)

1. (d)	**2.** (b)	**3.** (d)	**4.** (b)	**5.** (c)	**6.** (d)	**7.** (c)	**8.** (c)	**9.** (b)	**10.** (b)
11. (b)	**12.** (d)	**13.** (b)	**14.** (b)	**15.** (a)	**16.** (c)	**17.** (c)	**18.** (c)	**19.** (b)	**20.** (b)
21. (a)									

4. Electrical Circuit Diagrams

1. Identify the circuit shown in the given figure:

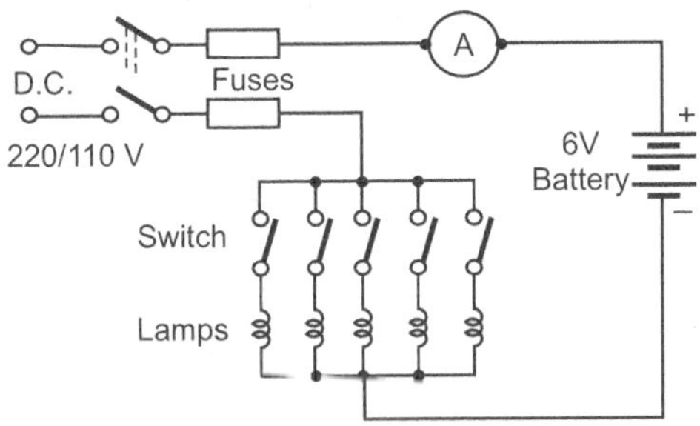

(a) Constant voltage battery charging
(b) Constant current battery charging
(c) Battery charger
(d) Battery charging by DC shunt generator

2. A sort starter circuit is shown is the figure given. What type of starter is this?

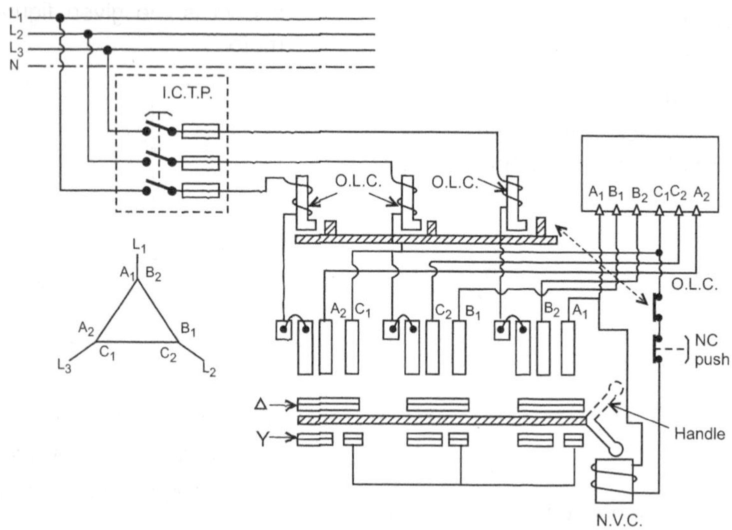

(a) Auto transformer starter
(b) Star delta starter (manual)
(c) DOL starter
(d) None of these

3. The control circuit for .... starter is shown is the given figure. Identify it.

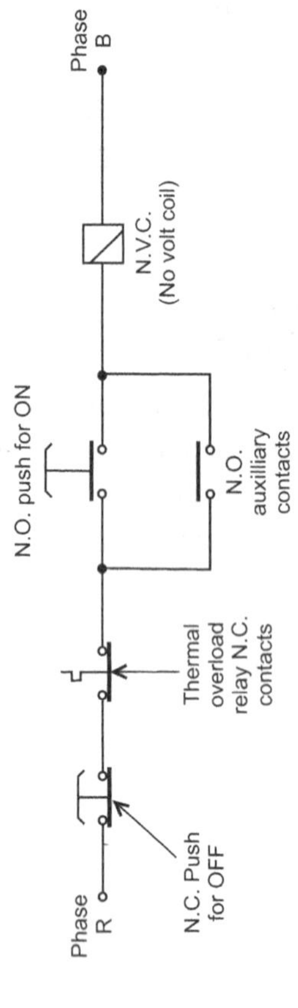

(a) DOL
(b) Auto transformer
(c) Star-delta
(d) Slip-ring induction motor

4. A simple control circuit is shown is the given figure. For what purpose is it used?

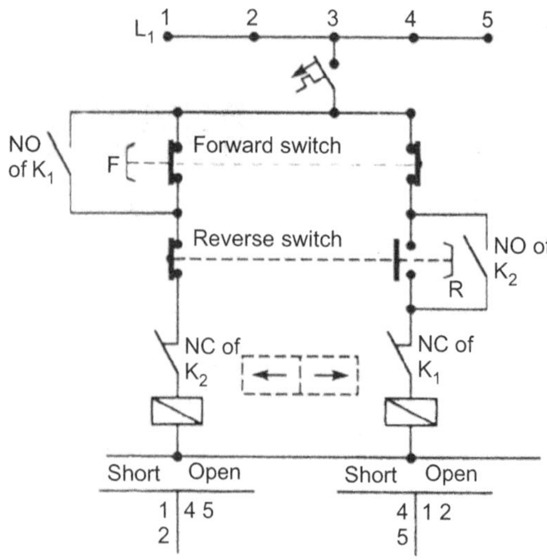

(a) Semi-automatic
(b) Automatic Star-delta starter
(c) Plugging circuit
(d) Forward ↔ Reverse operation control

5. The circuit shown in the figure given can be used as ....

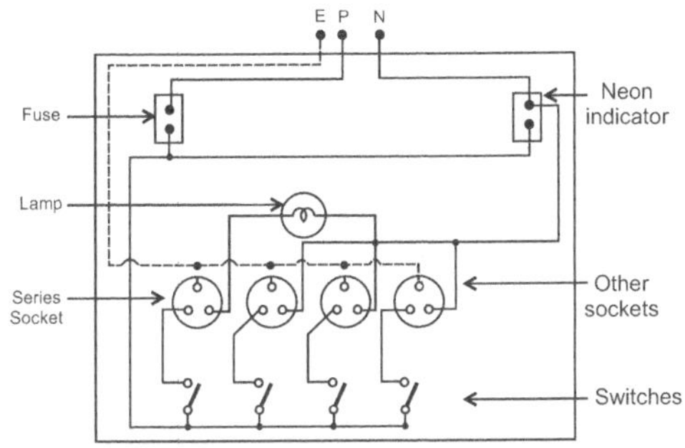

(a) A testing board

(b) A battery charger board

(c) A panel board for a shunt generator

(d) None of these

6. The figure given shows .... winding.

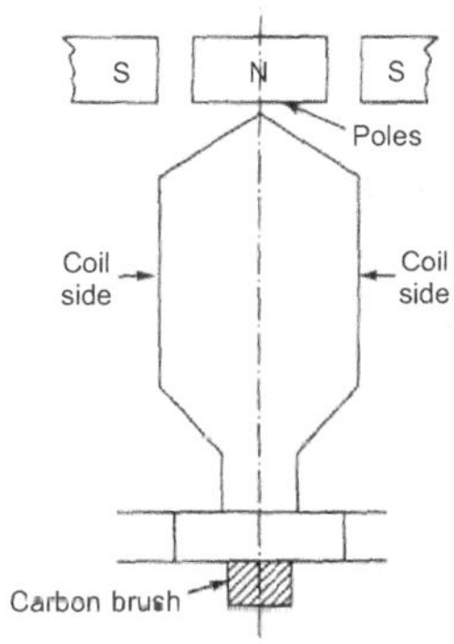

(a) progressive (b) retrogressive

(c) symmetrical (d) unsymmetrical

7. The circuit diagram of a starter is shown in the given figure. Which type of starter is this?

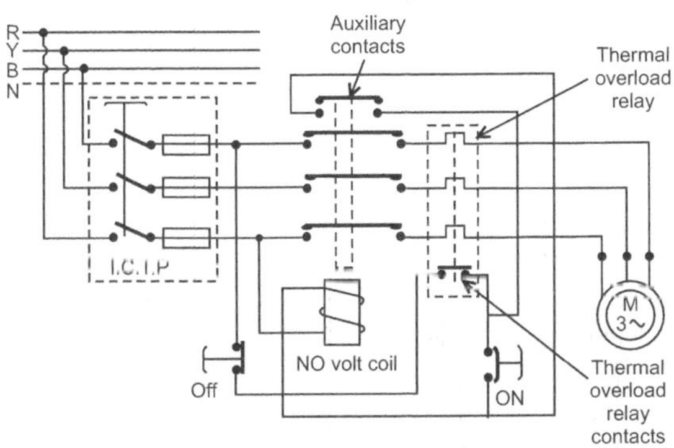

(a) Transformer type

(b) Slip-ring induction motor type

(c) DOL type

(d) Star-delta type

8. A rectifier circuit is shown in the given figure. Which type of rectifier is this?

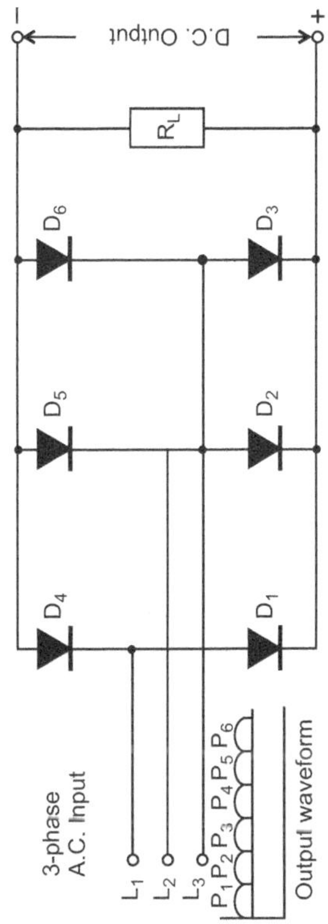

(a) 3-phase full wave

(b) Bridge

(c) Full-wave

(d) Half-wave

9. The figure given shows a ....

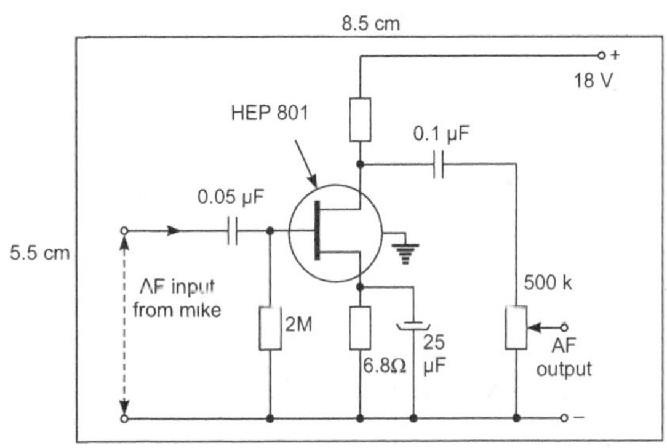

(a) 2-stage RC coupled amplifier circuit
(b) UJT triggered SCR circuit
(c) FET pre-amplifier circuit
(d) 2-stage transformer coupled amplifier circuit

10. In the given figure a rectifier circuit is shown. Which type of rectifier is this?

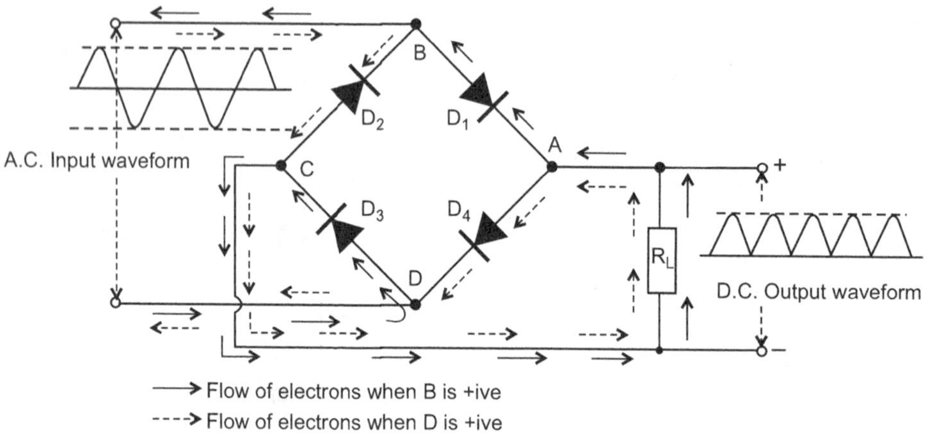

(a) Half-wave (b) Full-wave
(c) Bridge (d) 3-phase

11. An amplifier circuit is shown in the figure given. Which type of amplifier is this?

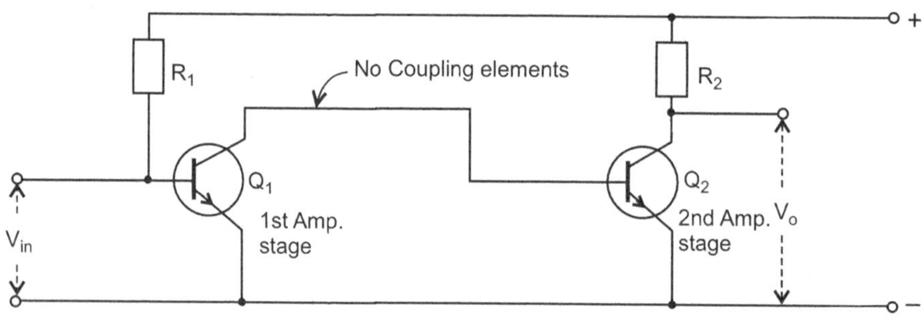

(a) RC coupled (b) Transformer coupled
(c) Single stage (d) Direct couple

12. A logic circuit diagram is shown in the figure given. Which one is this out of the following?

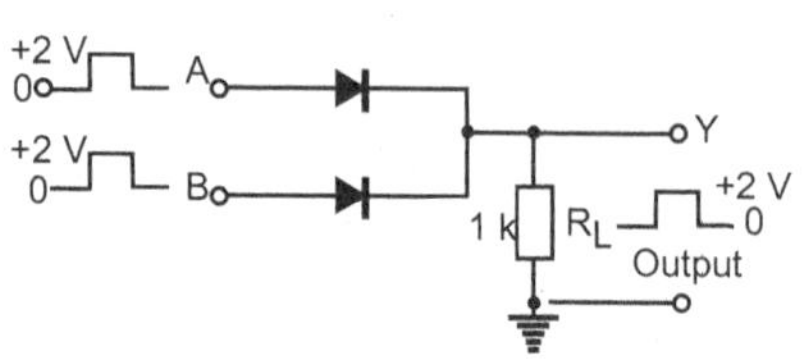

(a) OR logic (b) NOT logic
(c) AND logic (d) NOR logic

13. The figure given shows a ....

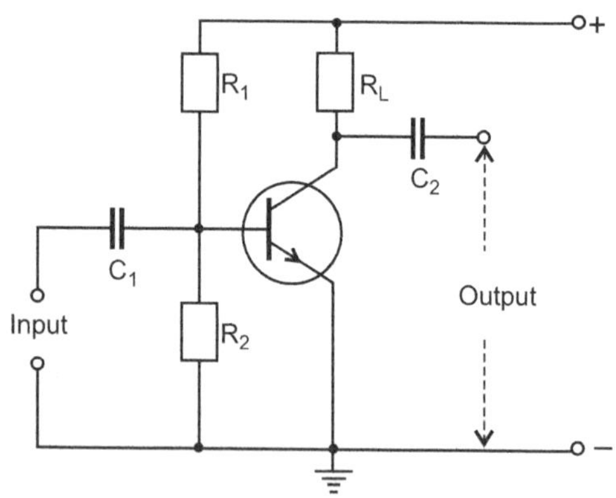

(a) Single stage amplifier circuit
(b) 2-stage RC coupled amplifier circuit
(c) 2-stage transformer coupled amplifier circuit
(d) Direct coupled amplifier circuit.

14. The circuit diagram of an amplifier is shown in the figure given. Which type of amplifier is this?

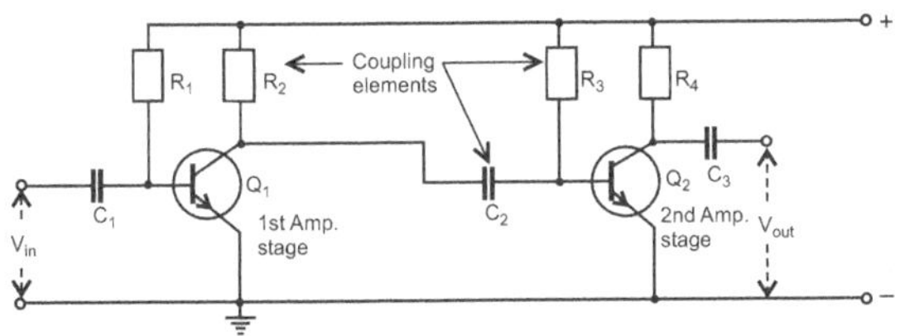

(a) Single stage
(b) RC coupled
(c) Transformer coupled
(d) Direct coupled

15. The figure given shows a ....

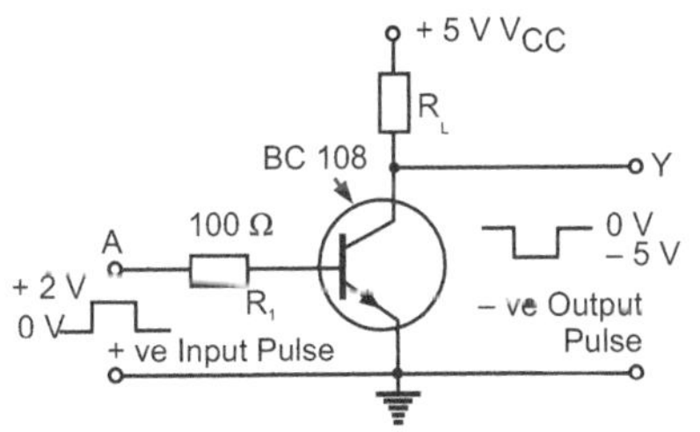

(a) NOT logic circuit
(b) OR logic circuit
(c) AND logic circuit
(d) NOR logic circuit

16. Which one is a series resonance circuit out of the following circuits?

(a)

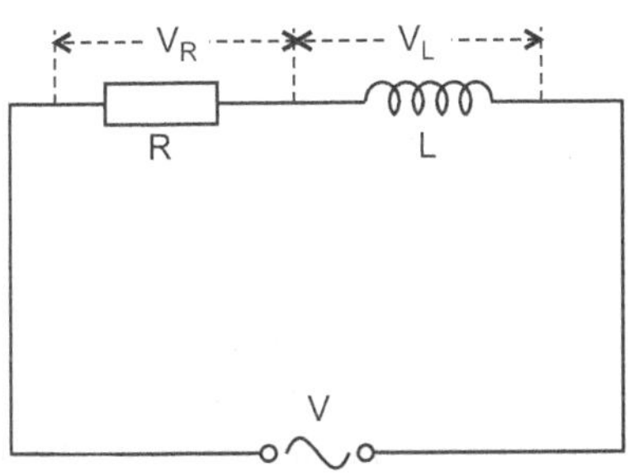

(b)

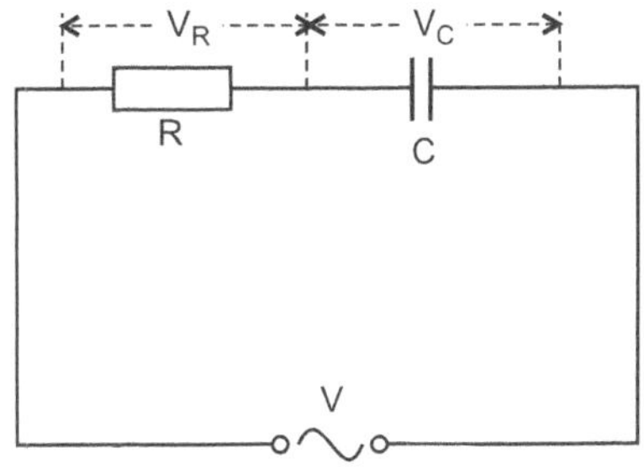

(c)

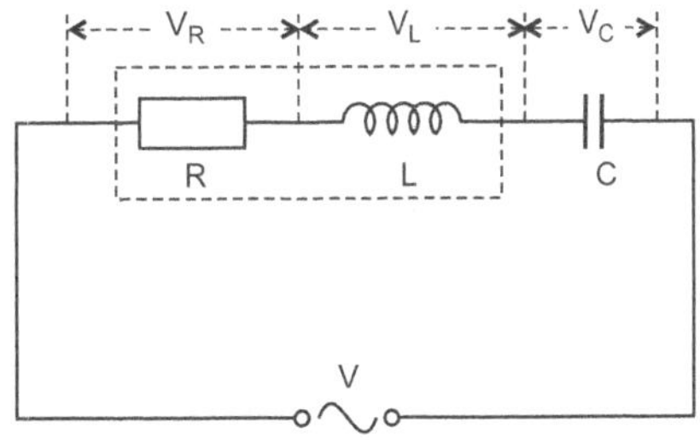

(d)

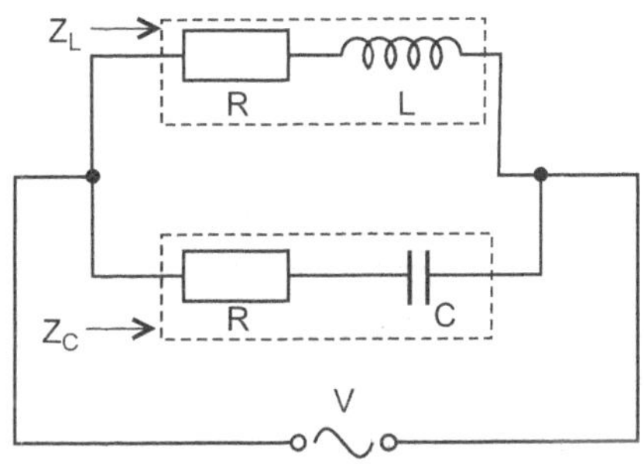

17. The figure given shows the circuit of a:

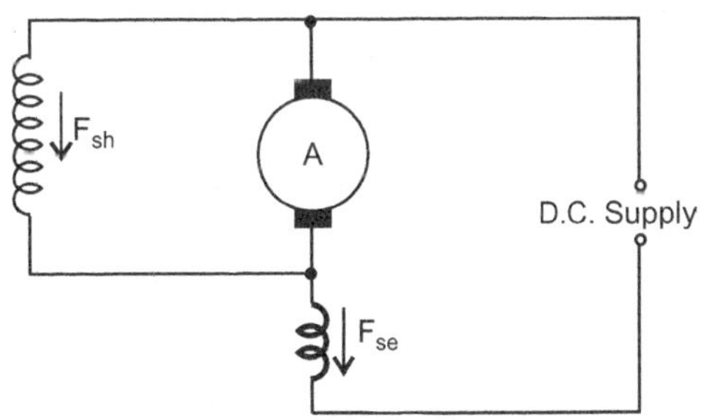

(a) DC short-shunt compound motor
(b) DC series motor
(c) DC shunt motor
(d) DC long-shunt compound motor

18. The figure given shows a DC .... motor diagram.

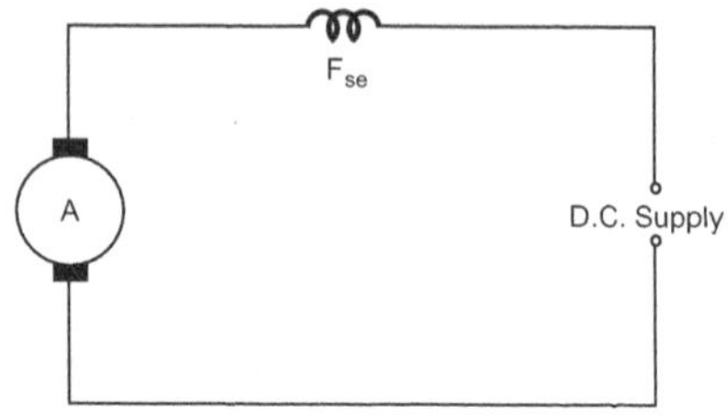

(a) series (b) shunt
(c) compound (d) None of these

19. A type of AC induction motor is shown in the figure given. What type of motor is this?

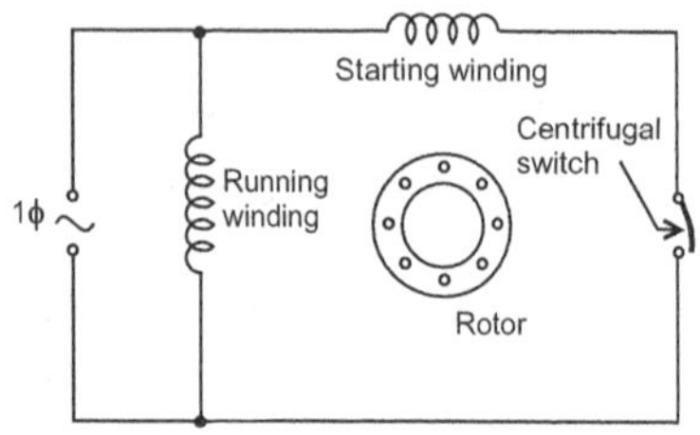

(a) Permanent capacitor motor
(b) Capacitor start motor
(c) Split-phase motor
(d) None of these

20. The figure given shows a/an ....

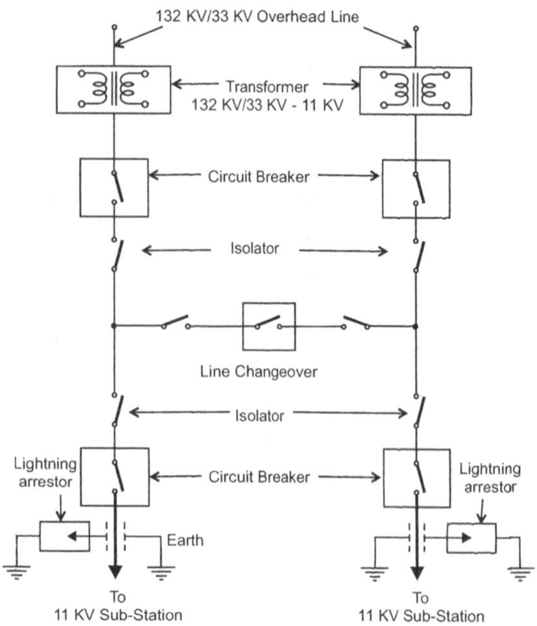

(a) Power control circuit
(b) Outdoor sub-station
(c) indoor sub-station
(d) None of these

21. The figure given shows:

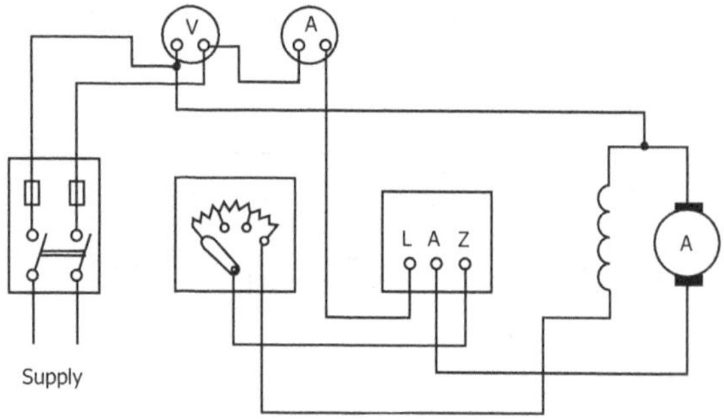

(a) Panel board for AC motor
(b) DC shunt generator panel board
(c) Panel board of a 5HP DC motor
(d) Panel board for a 5HP AC motor

22. The total capacitance of the circuit given is:

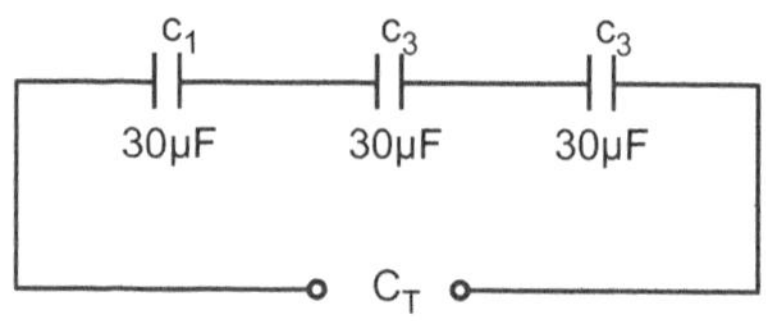

(a) 90 μF (b) 30 μF
(c) 60 μF (d) 10 μF

23. The figure given shows:

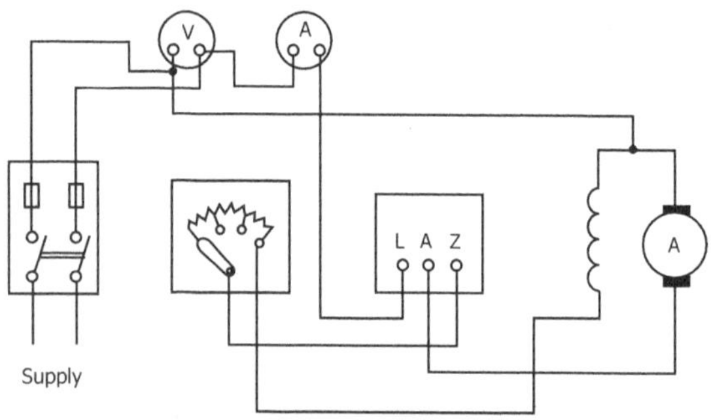

(a) Layout of DC shunt generator for battery charging
(b) Panel board for DC compound motor
(c) Panel board for a 5 HP DC shunt motor
(d) None of these

24. A rectifier circuit is shown in the figure given. What is its name?

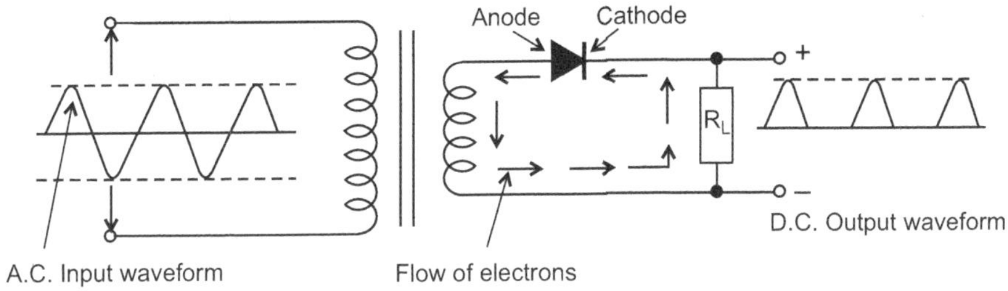

(a) Full-wave
(b) Half-wave
(c) Bridge
(d) 3-phase

25. A lead-acid battery of 12 volts can be charged easily by .... as shown in the figure given.

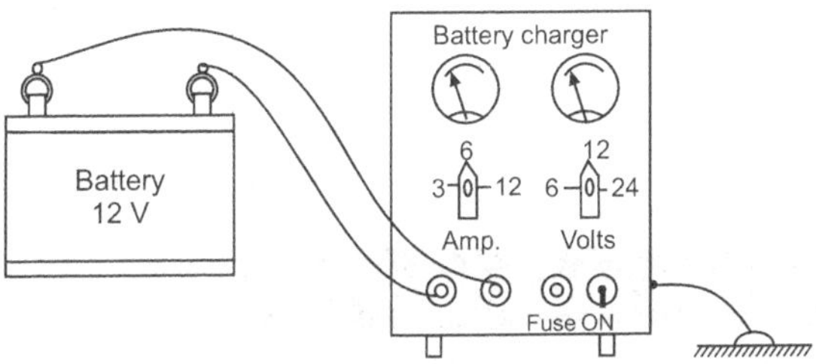

(a) a battery charger
(b) a UPS
(c) DC shunt motor
(d) a SMPS

26. What type of speed control method is shown in the figure given?

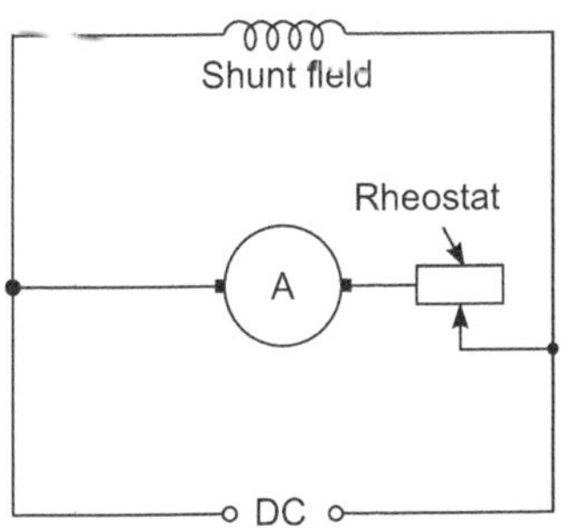

(a) Field control method
(b) Armature control method
(c) Supply voltage control method
(d) Tapped field control method

27. A starter circuit is shown in the figure given. Identify it:

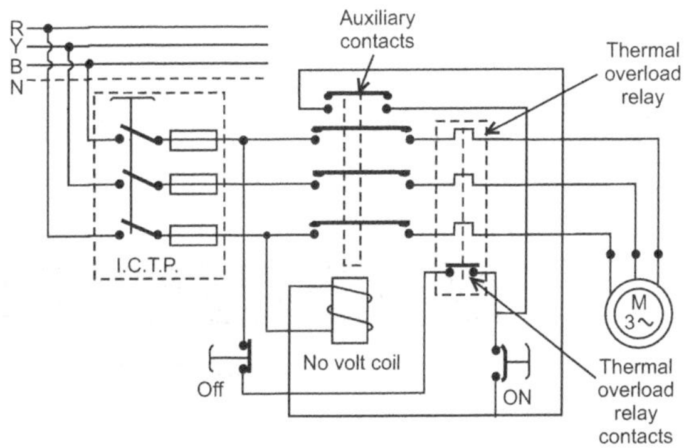

(a) Star-delta starter
(b) Auto transformer starter
(c) DOL starter
(d) Slip-ring motor starter

28. The figure given shows a control circuit. What type of control circuit is shown?

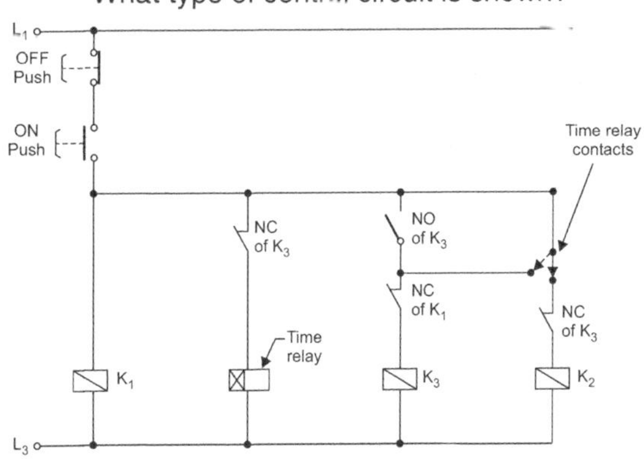

(a) Automatic star-delta starter
(b) Semi automatic star delta starter
(c) Plugging circuit
(d) Forward ↔ Reverse operation control

29. A circuit is shown in the figure given. What for it is meant?

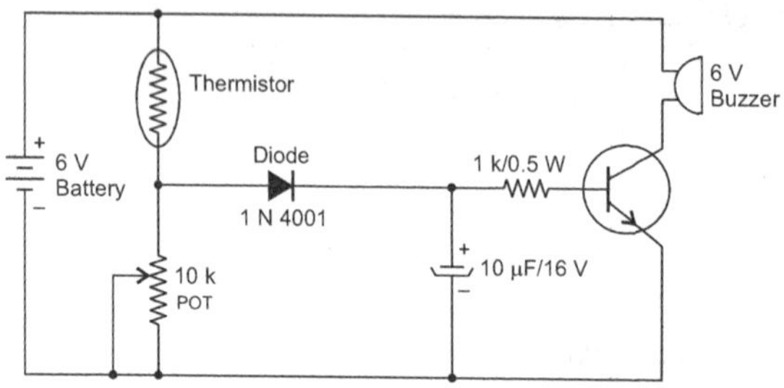

(a) Fire alarm circuit
(b) Stair case wiring
(c) Battery charging circuit
(d) Pre amplifier circuit

KEY TO MCQs (Electrical Circuit Diagrams)

1. (b)	**2.** (b)	**3.** (a)	**4.** (d)	**5.** (a)	**6.** (c)	**7.** (c)	**8.** (a)	**9.** (c)	**10.** (c)
11. (d)	**12.** (a)	**13.** (a)	**14.** (b)	**15.** (a)	**16.** (c)	**17.** (a)	**18.** (a)	**19.** (c)	**20.** (a)
21. (c)	**22.** (d)	**23.** (c)	**24.** (b)	**25.** (a)	**26.** (b)	**27.** (c)	**28.** (a)	**29.** (a)	

5. Block Diagrams of Instruments and Equipment

1. A block diagram is shown in the given figure. Idenfity it:

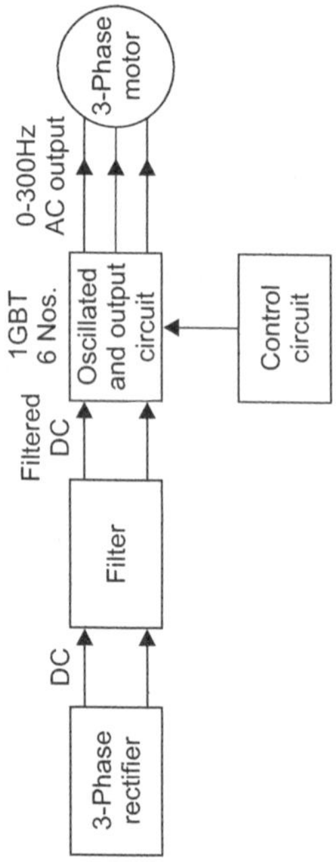

(a) AC drive (b) DC drive
(c) AC/DC drive (d) None of these

2. A domestic appliance is shown is the figure given. What is it?

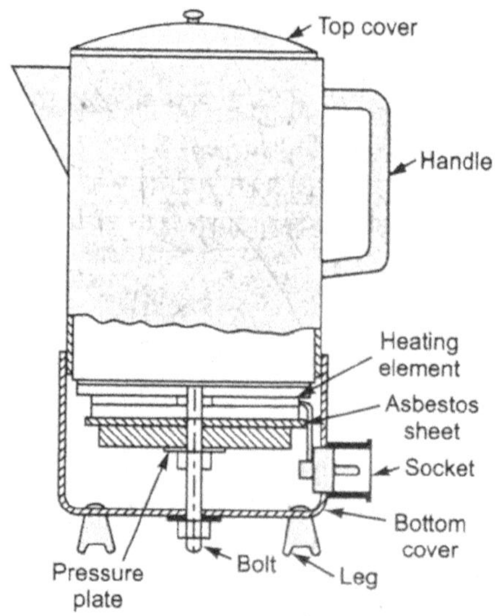

(a) Immersion heater
(b) Electric iron
(c) Electric kettle
(d) Hot plate

3. The figure given shows:

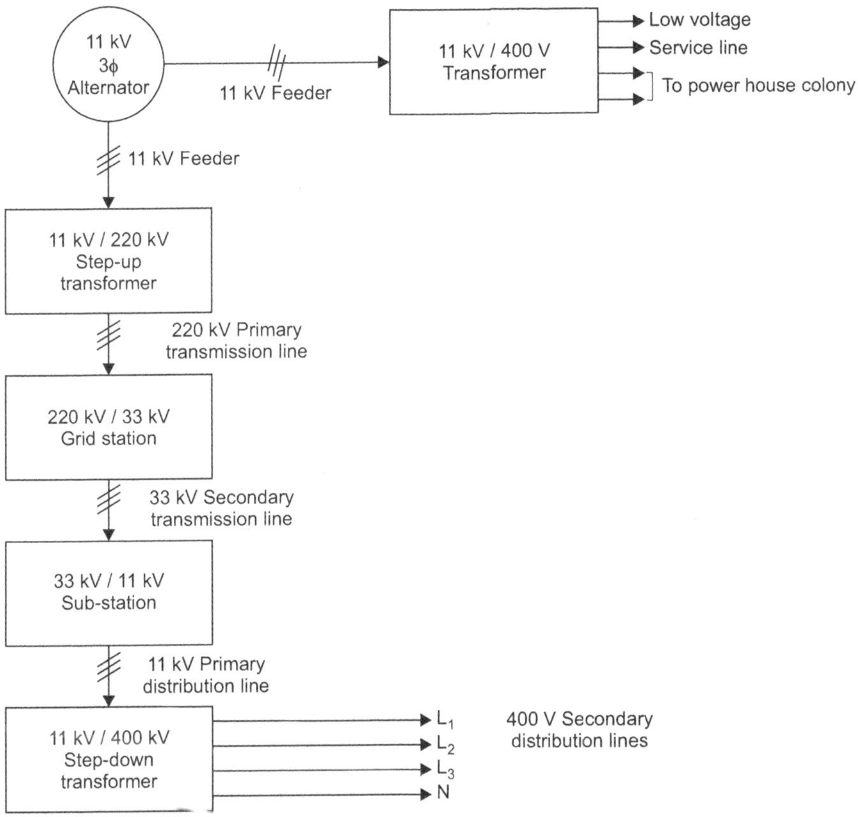

(a) outline of an outdoor sub-station
(b) Outline of an indoor sub-station
(c) Outline of power transmission and distribution
(d) Outline of pole mount sub-station

4. The figure given shows the method of connecting a/an ....

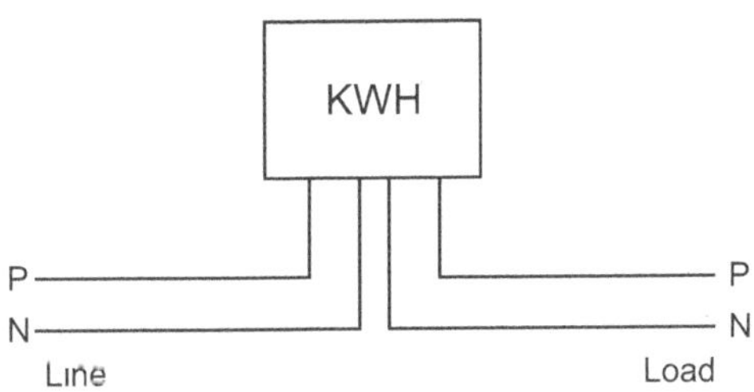

(a) voltmeter
(b) ammeter
(c) energy meter
(d) power factor meter

5. An electrical device is shown in the given figure. Identify this:

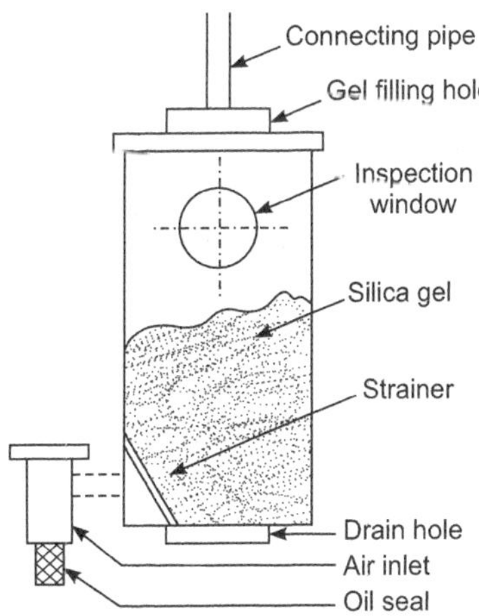

(a) Bucchloz relay
(b) Breather
(c) Temperature gauge
(d) Pressure release valve

6. A power distribution system is shown in the figure given. Identify it:

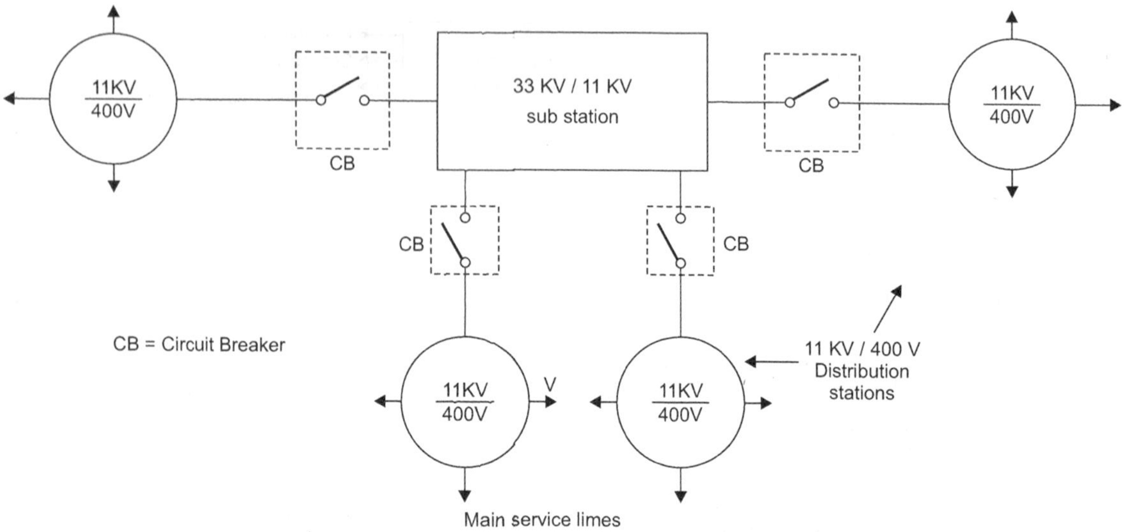

(a) Ring system (b) Radial system
(c) Grid system (d) None of these

7. The construction of a lamp is shown is the figure given. Identify it.

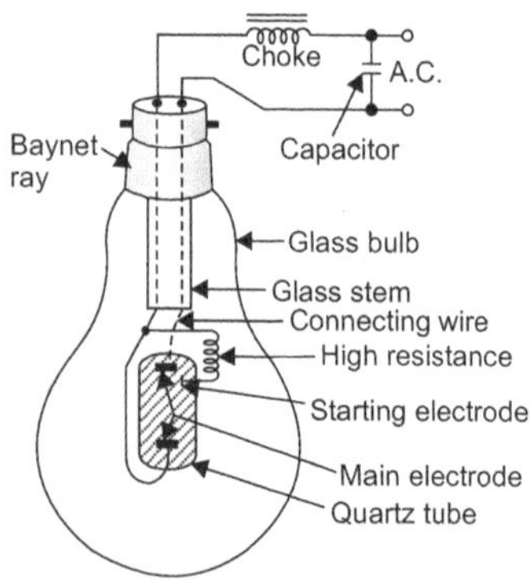

(a) Mercury vapour lamp
(b) Neon sign lamp
(c) Neon lamp
(d) Sodium vapour lamp

8. A type of lamp is shown is the figure given. Identify it.

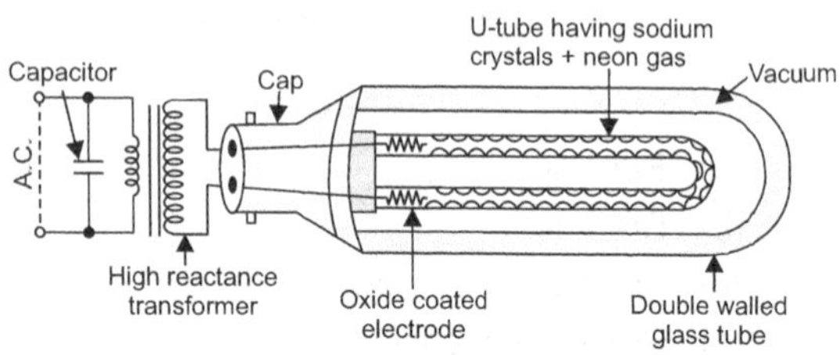

(a) Sodium vapour lamp
(b) Neon lamp
(c) Helogen lamp
(d) Mercury vapour lamp

9. The construction of a lamp is shown in the figure given. What type of lamp is this?

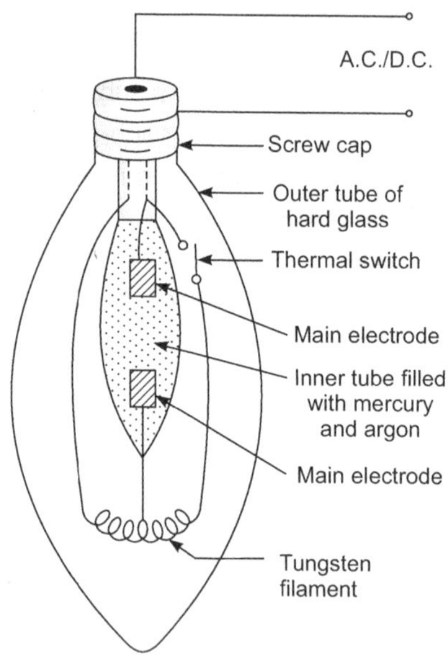

(a) Metal filament lamp
(b) Carbon filament lamp
(c) M.A.T. lamp
(d) None of these

10. The figure given shows an electrical circuit for a domestic appliance. Identify it.

(a) Electric cooking range (b) Electric heater

(c) Automatic electric iron (d) None of these

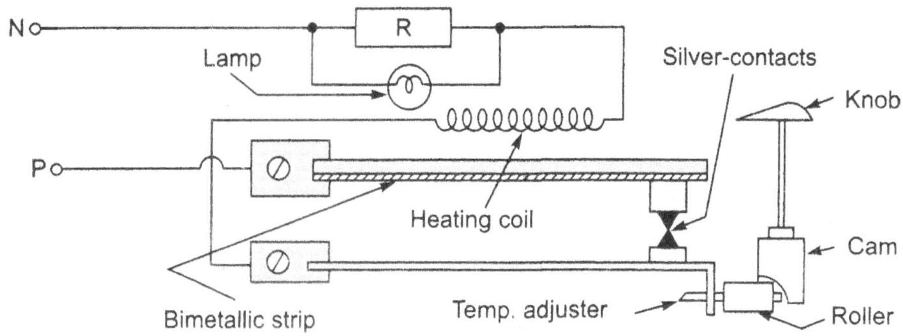

11. The figure given shows a ....

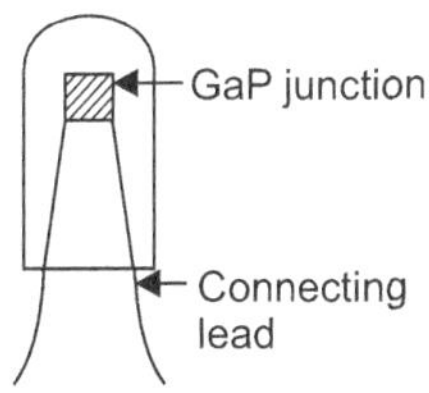

(a) Neon lamp (b) LED lamp

(d) Helogen lamp (d) Carbon filament lamp

12. The figure given shows a blocks diagram. Identify it:

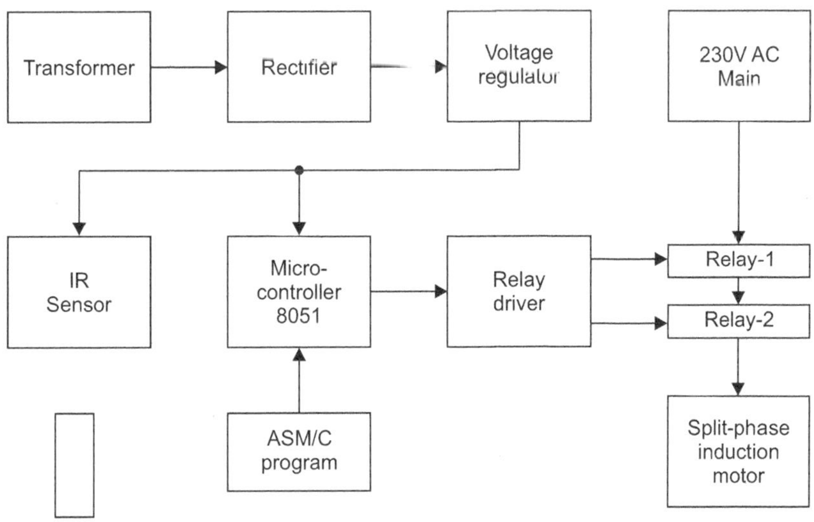

(a) Remote control for an induction motor (b) CRO

(c) AC drive (d) Remote control for DC generator

13. The figure given shows a cooking range. How many positions of working it contains?

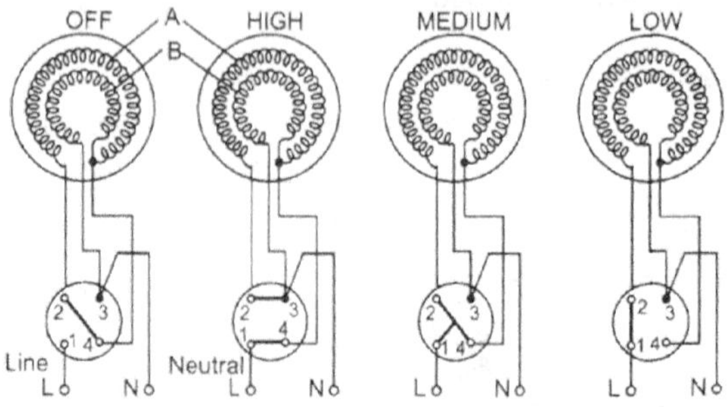

(a) 4 (b) 3
(c) 2 (d) 1

14. The figure given shows a type of lamp. What is the name of the lamp?

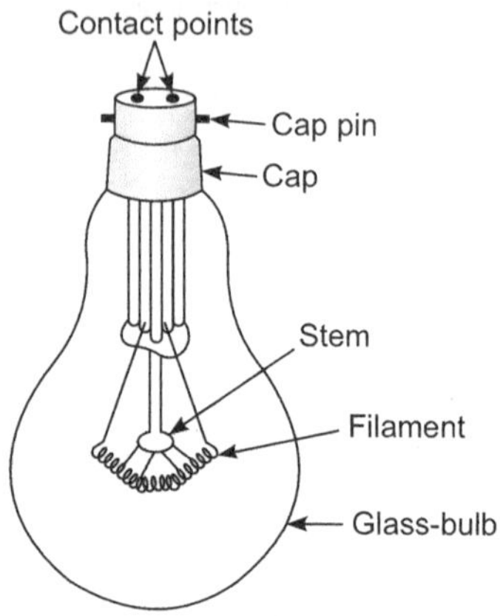

(a) Carbon filament (b) Metal filament
(c) MA lamp (d) MAT lamp

KEY TO MCQs (Electrical Circuit Diagrams)

1. (a) **2.** (c) **3.** (c) **4.** (c) **5.** (b) **6.** (b) **7.** (a) **8.** (a) **9.** (c) **10.** (c)
11. (b) **12.** (a) **13.** (b) **14.** (a)

ALL INDIA TRADE TEST

Electrician Trade Theory

Latest Question Bank Set–1

First Year

Trade Theory

1. What is the name of the safety sign as shown in the figure?

(a) Warning sign
(b) Mandatory sign
(c) Prohibition sign
(d) Information sign

2. What is smothering in extinguishing of fire?

(a) Adding the fuel element to the fire
(b) Removing the fuel element from the fire
(c) Using of water to lower the temperature
(d) Isolating the fire from the supply of oxygen

3. Which step of the 5S-concept refers "Standardization"?

(a) Step – 1 (b) Step – 2
(c) Step – 3 (d) Step – 4

4. Which pliers is used for making the wire hooks and loops?

(a) Flat nose pliers
(b) Long nose pliers
(c) Round nose pliers
(d) Diagonal cutting pliers

5. What is the use of the pincer?

(a) Twisting the flexible wires
(b) Cutting larger diameter of wires
(c) Extracting the pin nails from the wood
(d) Holding small objects, where finger cannot reach

6. What is the back ground colour of the warning signs in the safety sign category?

(a) Blue
(b) White
(c) Yellow
(d) Green

7. Which is the waste disposal method that produces the heat?

(a) Landfill
(b) Composting
(c) Incineration
(d) Waste compaction

8. Which disposal method of waste saves lot of energy?

(a) Land fill
(b) Recycling
(c) Incineration
(d) Composting

9. What is the name of the tool as shown in the figure?

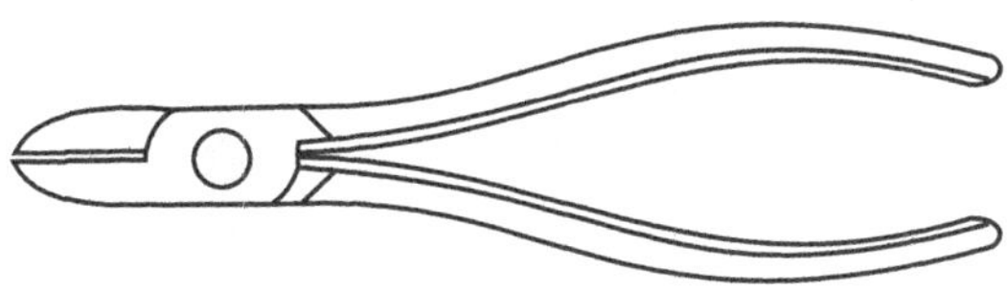

(a) Nose pliers
(b) Crimping tool
(c) Combination pliers
(d) Diagonal cutting pliers

10. What is the name of the file as shown in the figure?

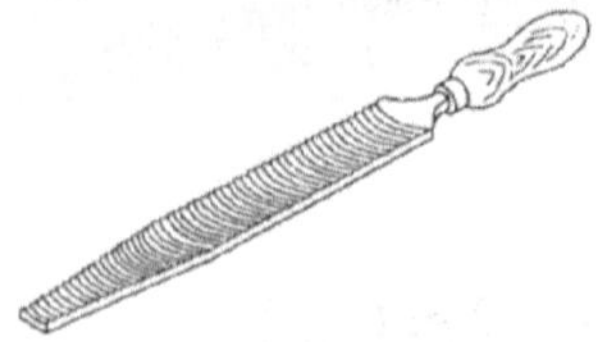

(a) Rasp cut file (b) Single cut file
(c) Double cut file (d) Curved cut file

11. What is the name of the drill bit as shown in the figure?

(a) Flat drill bit
(b) Taper shank bit
(c) Countersink bit
(d) Straight shank bit

12. Which file is used for sharpening the blunt teeth of a tenon saw?

(a) Square file (b) Round file
(c) Triangular file (d) Half round file

13. What is the name of the part marked 'X' as shown in the figure?

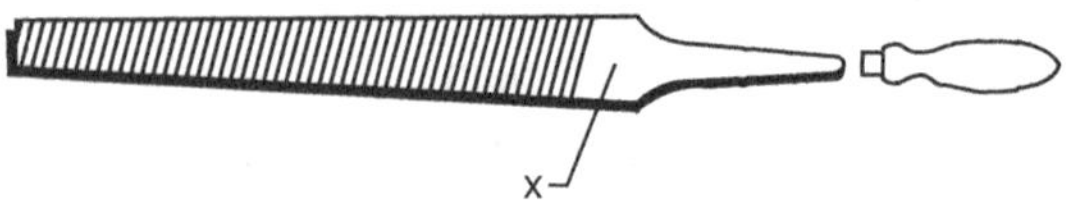

(a) Edge (b) Heel
(c) Tang (d) Point

14. What is the name of the hammer as shown in the figure?

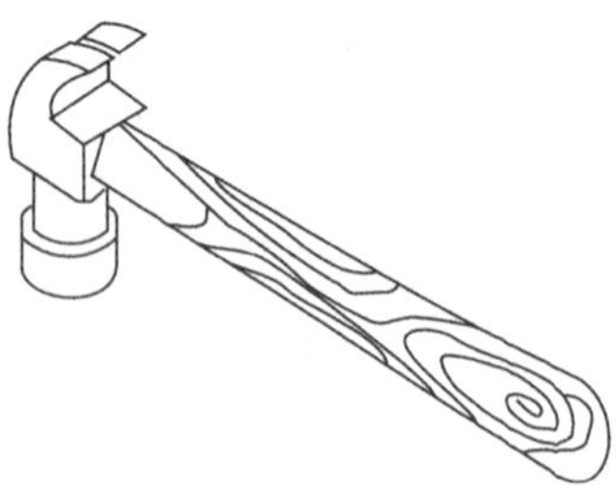

(a) Claw hammer
(b) Ball pein hammer
(c) Cross pein hammer
(d) Straight pein hammer

15. What is the name of the tool as shown in the figure?

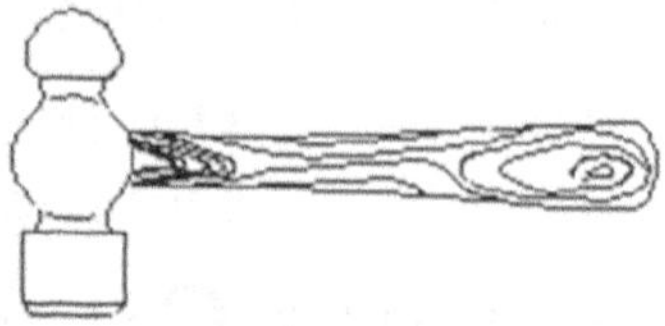

(a) Claw hammer
(b) Tack hammer
(c) Cross pein hammer
(d) Ball pein hammer

16. Which is the semiconductor material?

(a) Eureka
(b) Ebonite
(c) Manganin
(d) Germanium

17. What is the reading of the micrometer as shown in the figure?

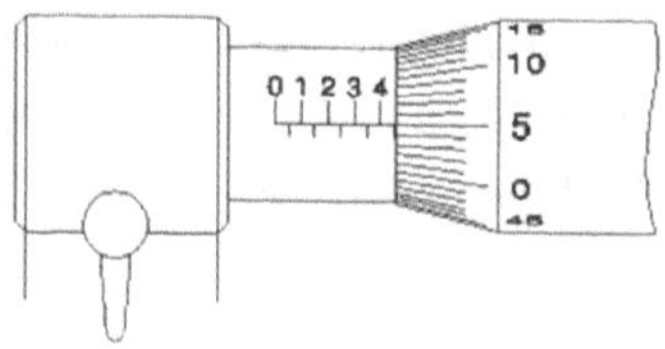

(a) 5.05 mm (b) 5.00 mm
(c) 4.55 mm (d) 4.05 mm

18. Which is used to measure the size of the conductor?

(a) Bevel gauge
(b) Vernier caliper
(c) Standard wire gauge
(d) Depth gauge

19. What is the purpose of 'serving' layer in underground cable?

(a) Protect the cable from mechanical vibration
(b) Protect the cable from mechanical injury
(c) Protect metallic sheath against corrosion
(d) Protect armouring from atmospheric condition

20. Which type of joint is used for extending the length of conductor in overhead lines?
(a) Scarfed joint
(b) Aerial tap joint
(c) Britannia "T" joint
(d) Western Union joint

21. What is the effect on molten solder due to repeated melting?
(a) Tin content reduced
(b) Lead content reduced
(c) Prevent slug formation
(d) Even solder flowing in joints

22. Which type of soldering method is used for servicing and repairing of electronics items?
(a) Dip soldering
(b) Soldering with a flame
(c) Soldering with soldering gun
(d) Machine soldering

23. Which method of soldering is used for repairing the vehicle body?
(a) Dip soldering
(b) Soldering with flame
(c) Soldering with soldering iron
(d) Machine soldering

24. What is the disadvantage of solid conductor compared to stranded conductor?
(a) Less rigidity
(b) Less flexibility
(c) Low melting point
(d) Low mechanical strength

25. Why the soldering iron must be kept into a stand that is not in use while soldering?
(a) It prevents burns and fire
(b) To control the excessive heat
(c) To save the time of soldering process
(d) To avoid dry soldering

26. What is the cause for cold solder defect in soldering?
(a) Excessive heating
(b) Insufficient heating
(c) Incorrect usage of flux
(d) High wattage soldering iron

27. What is the value of current (I) flowing through the circuit?

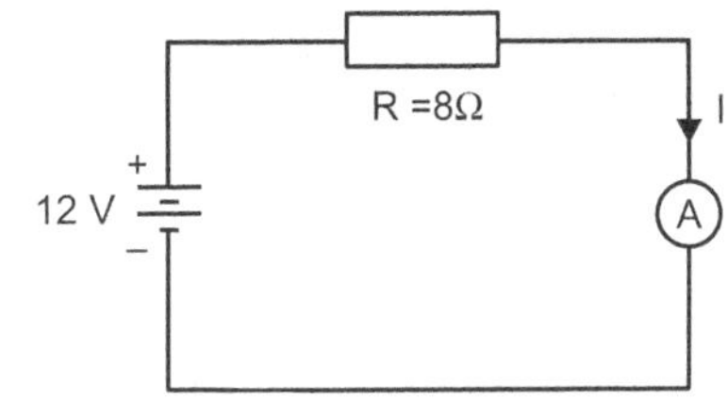

(a) 6 A (b) 3 A
(c) 2 A (d) 1 A

28. What is the purpose of additional covering over the insulation of insulated conductor?
(a) To increase dielectric strength
(b) To reduce the conductor resistance
(c) To reduce the power loss
(d) To increase the voltage drop

29. Which type of cables are used for underground services from 33 kV to 66 kV?
(a) High tension cables
(b) Extra super voltage cables
(c) Super tension cables
(d) Extra- high tension cables

30. Which is the property of a good conductor?
(a) Low specific resistance
(b) High dielectric strength
(c) High specific resistance
(d) Low mechanical strength

31. What is the name of the resistor if its resistance value increases with increase in temperature?
(a) Varistors
(b) Sensistors
(c) Thermistors
(d) Light Dependent Resistor (LDR)

32. What is the voltage drop in resistor 'R_2' in the series circuit?

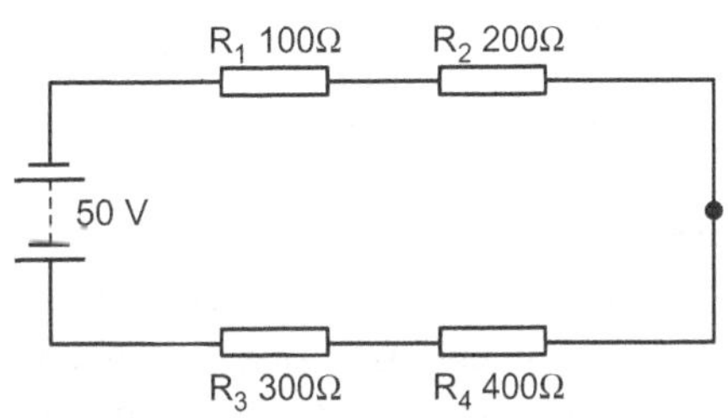

(a) 5 volt (b) 10 volt
(c) 15 volt (d) 20 volt

33. Which method is used for measuring 1 ohm to 1 megohm range resistance?
(a) Substitution method
(b) Kelvin bridge method
(c) Wheat stone bridge method
(d) Voltmeter and ammeter method

34. What is the purpose of the shunt resistor 'R_2' used in series type Ohm meter circuit?

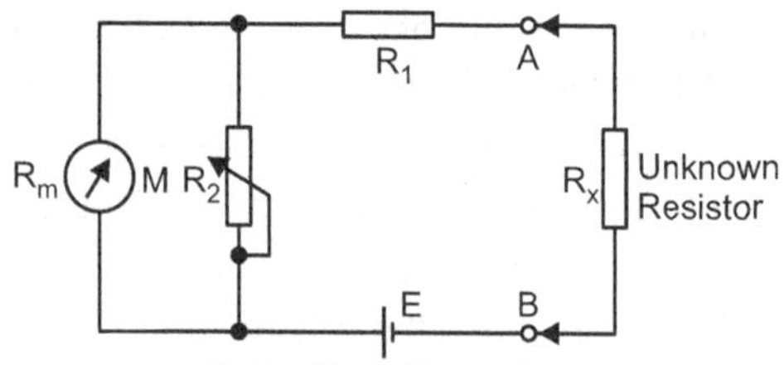

Series Type Ohmmeter

(a) To limit the current in the circuit
(b) To increase the value of meter resistance
(c) To adjust the zero position of the pointer
(d) To prevent the excess current in the circuit

35. Which electrical quantity causes for the heat generation in a conductor?
(a) Voltage
(b) Square of the current
(c) Square of the resistance
(d) Square of the voltage

36. What is the change in value of resistance of the conductor, if its cross section area is doubled?
(a) No change
(b) Decreases 2 times
(c) Increases 2 times
(d) Decreases 4 times

37. What is the resistance of Light Dependent Resistor (LDR), if the intensity of light is increased?
(a) Increases
(b) Decreases
(c) Remains same
(d) Becomes infinity

38. Calculate the total capacitance value in the circuit.

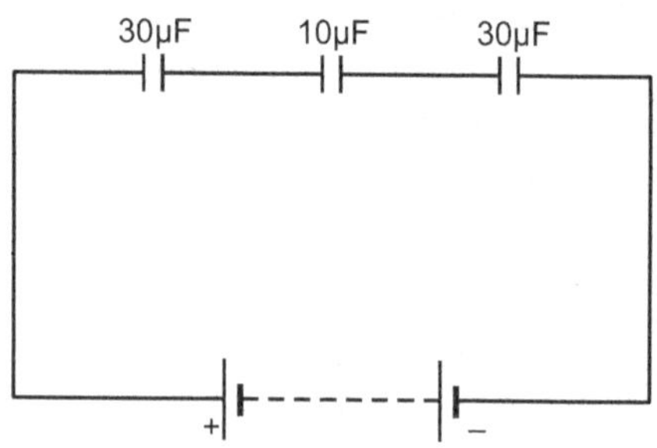

(a) 0.16 μF (b) 6 μF
(c) 30 μF (d) 70 μf

Workshop Calculation and Science

39. What is the value for specific heat of water?
(a) 4 (b) 3
(c) 2 (d) 1

40. What is the boiling point of water?
(a) 0°C (b) 32°C
(c) 100°C (d) 212°C

41. What is the side of a square whose area is 625 mm^2?
(a) 15 mm (b) 20 mm
(c) 25 mm (d) 30 mm

42. What is the perimeter of a rectangle whose length and breadth are 20 cm and 18 cm?
(a) 56 cm (b) 66 cm
(c) 76 cm (d) 86 cm

43. Which is example for first order lever?
(a) A wheel barrow
(b) A pair of scissors
(c) Fire tongs
(d) Lime squeezer

44. Which is example for second order lever?
(a) Common balance
(b) A pair of scissors
(c) Bottle opener
(d) Human forearm

Engineering Drawing

45. Identify the name of tool?

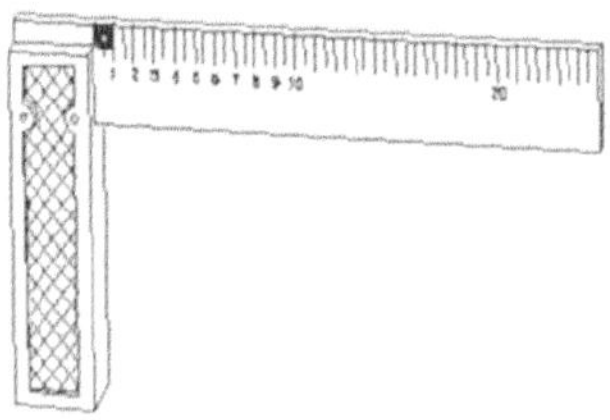

(a) Try square (b) Surface gauge
(c) Steel rule (d) Feeler gauge

46. Identify the conventional symbol of material?

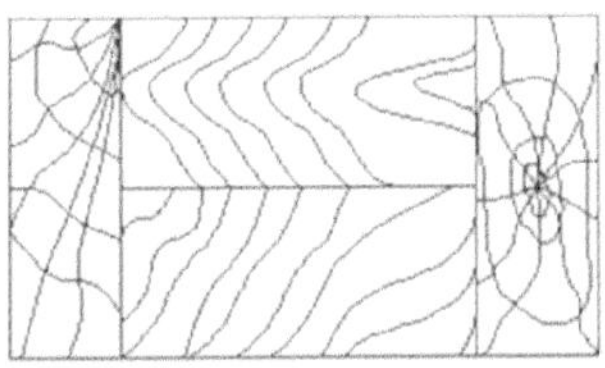

(a) Lead (b) Glass
(c) Wood (d) Paper

47. Identify the name of instrument?

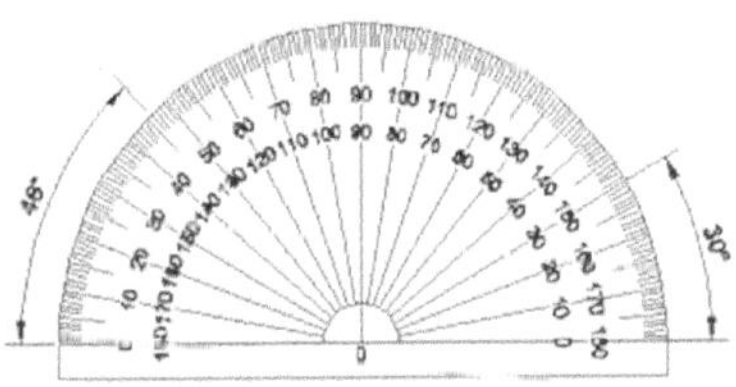

(a) Set square (b) Protractor
(c) French curve (d) ' T ' square

48. Identify the name of set square?

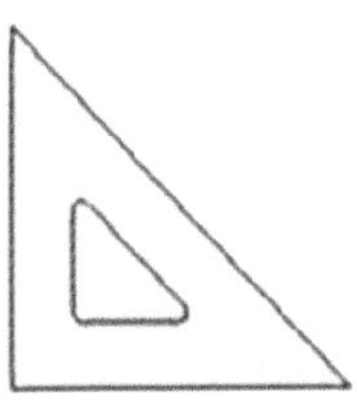

(a) 15° set square (b) 60° set square
(c) 30° set square (d) 45° set square

49. Compass is used to draw___.

(a) rectangle (b) square
(c) circle (d) triangle

50. Which title block position is correct as standard practice?

(a)

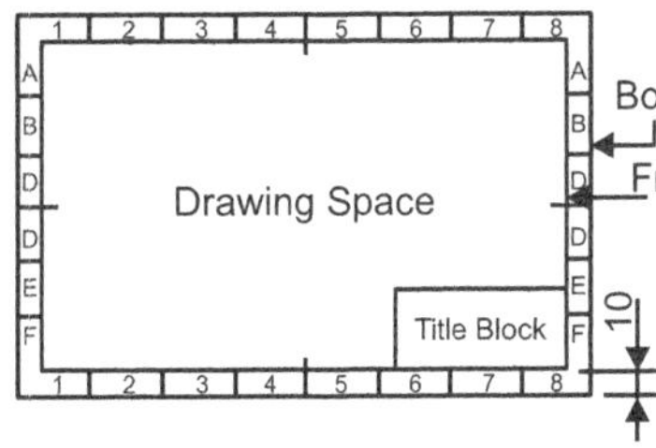

(b)

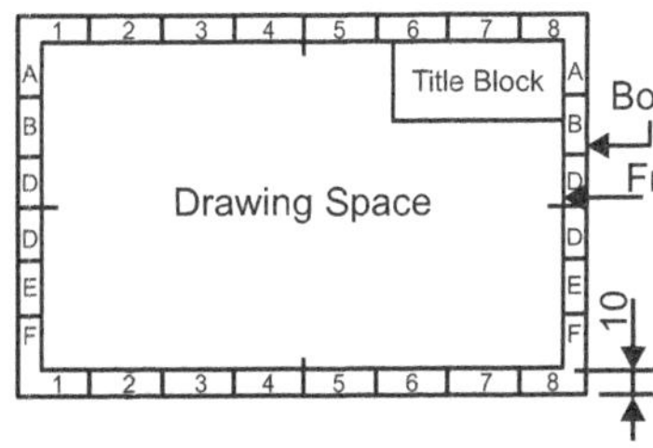

(c)

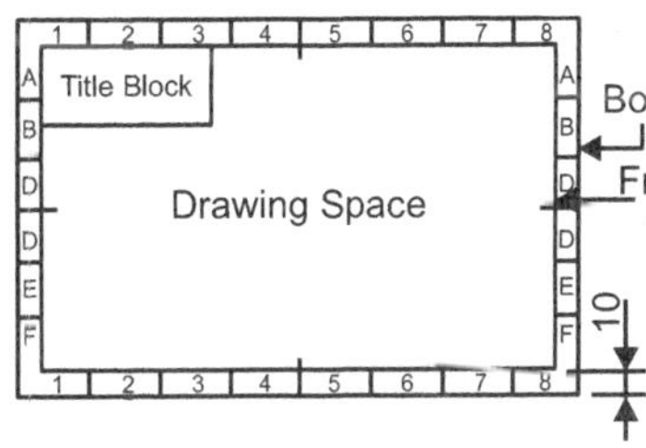

(d)

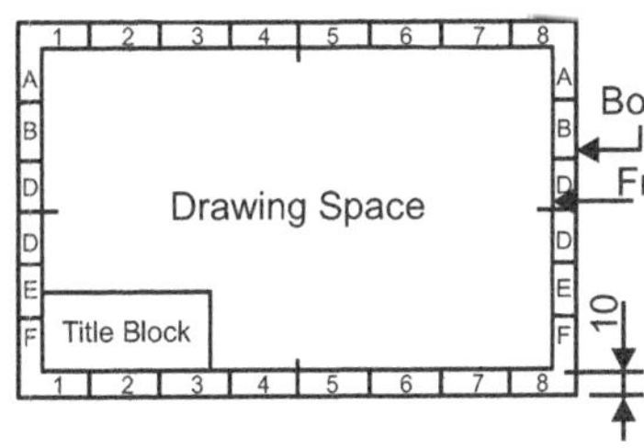

KEY Electrician 1st Year, Set–1

1. (d)	**2.** (d)	**3.** (d)	**4.** (c)	**5.** (c)	**6.** (c)	**7.** (c)	**8.** (b)	**9.** (d)	**10.** (a)
11. (b)	**12.** (c)	**13.** (b)	**14.** (a)	**15.** (d)	**16.** (d)	**17.** (c)	**18.** (c)	**19.** (d)	**20.** (d)
21. (a)	**22.** (c)	**23.** (b)	**24.** (b)	**25.** (a)	**26.** (b)	**27.** (c)	**28.** (a)	**29.** (d)	**30.** (a)
31. (b)	**32.** (b)	**33.** (c)	**34.** (c)	**35.** (b)	**36.** (b)	**37.** (b)	**38.** (b)	**39.** (d)	**40.** (c)
41. (c)	**42.** (c)	**43.** (b)	**44.** (c)	**45.** (a)	**46.** (c)	**47.** (b)	**48.** (d)	**49.** (c)	**50.** (a)

ALL INDIA TRADE TEST

Electrician Trade Theory

Latest Question Bank Set–2

First Year

Trade Theory

1. Which test is conducted to locate the faults in U.G. cables?
   (a) Loop test
   (b) External growler test
   (c) Break down voltage test
   (d) Insulation resistance test
2. What does the number 1.40 represent if a stranded conductor is designated as 7/1.40?
   (a) Area of cross section
   (b) Radius of one conductor
   (c) Diameter of all conductors
   (d) Diameter of each conductor
3. What is the size of the neutral core of a 90 sq mm 31/2 core U.G cable?
   (a) 45 sq mm (b) 50 sq mm
   (c) 70 sq mm (d) 90 sq mm
4. What is the name of the tool as shown in the figure?

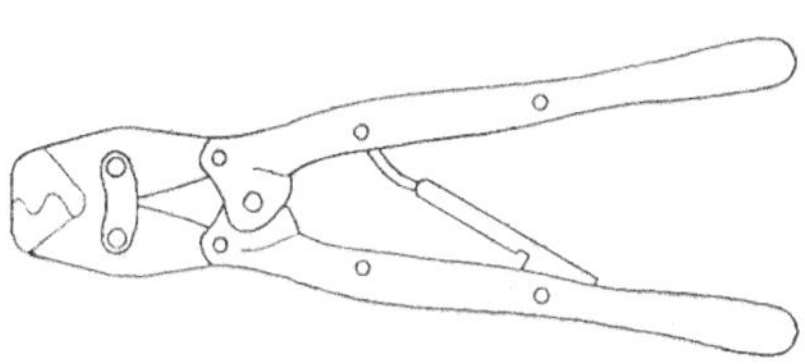

   (a) Cutting plier
   (b) Wire stripper
   (c) Crimping tool
   (d) Side cutting plier
5. Which type of joint is used in overhead lines for high tensile strength?
   (a) Scarfed joint
   (b) Married joint
   (c) Western union joint
   (d) Rat-tail joint
6. What is the unit for Quantity of electricity?
   (a) mho
   (b) coulomb
   (c) volt / second
   (d) ampere / second
7. What is the name of the soldering method as shown in the figure?

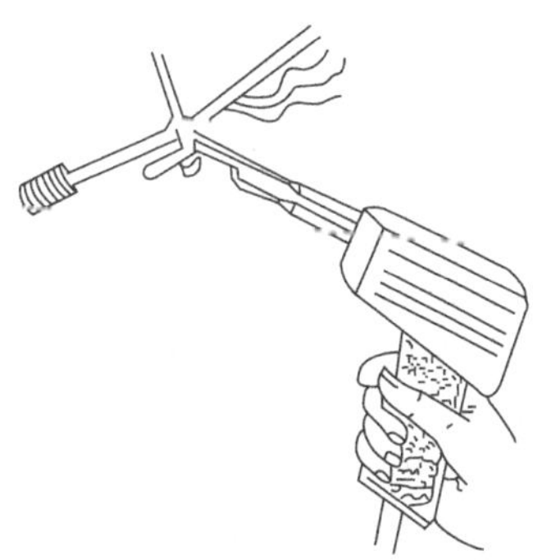

   (a) Dip soldering
   (b) Soldering iron
   (c) Soldering gun
   (d) Soldering with flame
8. What is the name of the soldering method as shown in the figure?

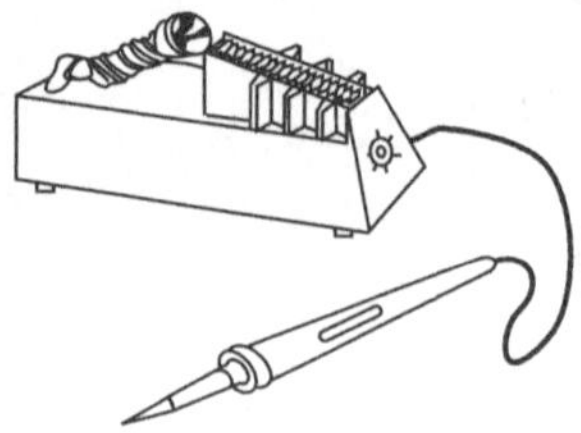

(a) Dip soldering
(b) Soldering with blow lamp
(c) Soldering with soldering gun
(d) Temperature controlled soldering

9. How many electrons are there in the third shell of the copper atom?
(a) 8 (b) 13
(c) 18 (d) 29

10. What is the formula for the Quantity of electricity (Q)?
(a) Current × Time
(b) Voltage × Current
(c) Current × Resistance
(d) Voltage × Resistance

11. Which formula is used to calculate the power of a DC circuit?
(a) Voltage × time
(b) Current × voltage
(c) Current × resistance
(d) Voltage × resistance

12. What is the advantage of the stranded conductor over the solid conductor?
(a) Cost is less
(b) More flexible
(c) Less voltage drop
(d) More insulation resistance

13. What is the advantage of the crimping terminations?
(a) Increase the voltage
(b) Reduce the load current
(c) Avoid loose connections
(d) Easy to replace

14. Which part of the underground cable is protecting the metallic sheath against corrosion?
(a) Serving (b) Bedding
(c) Armouring (d) Lead sheath

15. What is the use of dip soldering method?
(a) Hard soldering
(b) Piping and cable soldering work
(c) Soldering miniature components on PCB
(d) Soldering of tin sheets

16. Which is the physical hazard?
(a) Smoking (b) Vibration
(c) Corrosive (d) Bacteria

17. Which type of fire extinguisher is used for fire on electrical equipment?
(a) Halon type
(b) Foam type
(c) Gas cartridge water type
(d) Stored pressure water type

18. What is the name of the tool as shown in the figure?

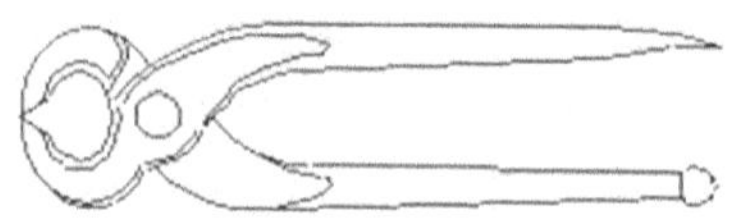

(a) Wire stripper (b) Side cutter
(c) Pincer (d) Gimlet

19. Which is the correct sequence of operation to be performed while using the fire extinguisher?
(a) Pull, Aim, Squeeze, Sweep
(b) Pull, Aim, Sweep, Squeeze
(c) Push, Arrange, Squeeze, Sweep
(d) Push, Arrange, Sweep, Sequence

20. Which Personal Protective Equipment (PPE) is used for the protection from fumes?
(a) Apron (b) Goggles
(c) Ear muff (d) Respirator

21. What is starving in extinguishing of fire?
(a) Adding fuel to the fire
(b) Using water to cool the fire
(c) Removing fuel element from the fire
(d) Preventing oxygen supply to the fire

22. Which artificial respiration method is to be performed to the victim with injuries on the chest and belly?

(a) Schafer's method
(b) Mouth to mouth method
(c) Holger-Nelson's back pressure method
(d) Nelson's arm-lift back pressure method

23. Which type of occupational health hazard is cause for infection due to bacteria?
(a) Electrical hazard
(b) Biological hazard
(c) Physiological hazard
(d) Psychological hazard

24. Which is the golden hour for the victim injured on head with risk of dying?
(a) First 15 minutes (b) First 30 minutes
(c) First 45 minutes (d) First 60 minutes

25. Which condition of the victim is referred as COMA stage?
(a) Unconscious but can respond to calls
(b) Conscious but cannot respond to calls
(c) Breathing and respond to calls
(d) Lie totally senseless and do not respond to calls

26. Which criteria must be satisfied for the quality of PPE's?
(a) It should withstand the hazards
(b) It cannot be used for long period
(c) Made by non standard material
(d) Low cost

27. What is the possible range to measure the size of the wire in a Standard Wire Gauge (SWG)?
(a) 0-44 (b) 0-42
(c) 0-38 (d) 0-36

28. What is the name of the wire joint as shown in the figure?

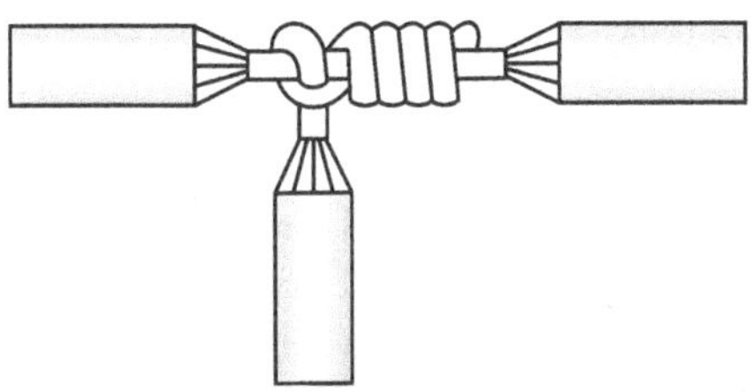

(a) Aerial tap joint
(b) Knotted tap joint
(c) Duplex cross tap joint
(d) Double cross tap joint

29. What is the full form of "XLPE' Cable?
(a) Cross Line Poly Ethylene
(b) X'ess Line Phase Earthing
(c) Cross Linked Poly Ethylene
(d) Excess Length Paper and Ebonite

30. How many electrons are there in the valence shell of a copper atom?
(a) 1 (b) 2
(c) 8 (d) 18

31. Which material is used as conductors for wiring?
(a) Mild steel (b) Copper
(c) Silver (d) Galvanized iron

32. Which layer is provided above the metallic sheath in UG cable?
(a) Armouring (b) Bedding
(c) Serving (d) Paper insulation

33. What is the formula to find the Electro Motive Force (EMF)?
(a) EMF = Potential difference – voltage drop
(b) EMF = Potential difference + voltage drop
(c) EMF = Potential difference + voltage drop/2
(d) EMF = Potential difference + 2 × voltage drop

34. Which law states that the applied voltage in a closed circuit, is equal to the sum of the voltage drops?
(a) Ohm's law
(b) Laws of resistance
(c) Kirchhoff's first law
(d) Kirchhoff's second law

35. What is the formula for the equivalent resistance (R_T) of the three resistors R_1, R_2 & R_3 are connected in parallel?
(a) $R_T = R_1 + R_2 + R_3$
(b) $R_T = \frac{1}{R_1} + \frac{1}{R_2} + \frac{1}{R_3}$

(c) $R_T = \dfrac{1}{R_1 + R_2 + R_3}$

(d) $R_T = \dfrac{1}{\dfrac{1}{R_1} + \dfrac{1}{R_2} + \dfrac{1}{R_3}}$

36. What is the reading of the Ohmmeter connected across the opened 'R_2' resistor as shown in the circuit?

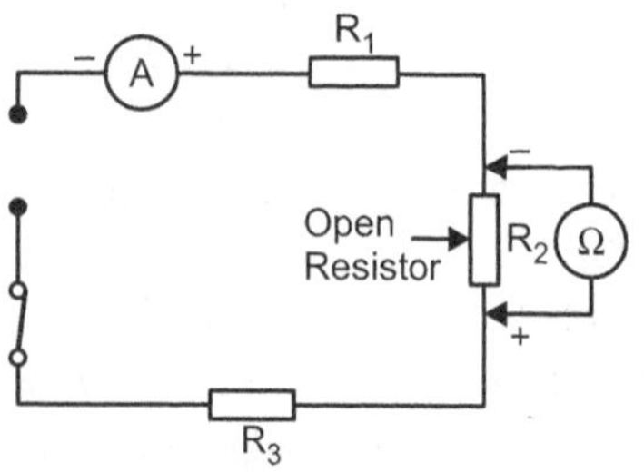

(a) Indicate zero reading

(b) Indicate infinite resistance

(c) Difference in the value of the resistance of R_1 and R_3 only

(d) Sum of the resistance of R_1 and R_3 only

37. What is the name of the resistor as shown in the figure?

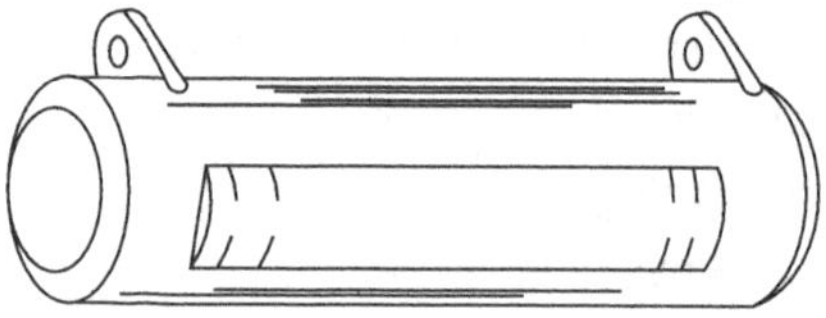

(a) Metal film resistor

(b) Wire wound resistor

(c) Carbon – film resistor

(d) Carbon composition resistor

38. Which material is having the negative temperature co-efficient property?

(a) Mica (b) Eureka

(c) Copper (d) Manganin

Workshop Calculation and Science

39. Convert 0.456 decimal fraction into percentage?

(a) 45.6% (b) 4.56%

(c) 0.456% (d) 0.0456%

40. What is the square of 0.01?

(a) 0.001 (b) 0.0001

(c) 0.00001 (d) 0.000001

41. Which metal contains iron as a major content?

(a) Brass metal (b) Bronze metal

(c) Zinc (d) Ferrous metal

42. What is termed as the quantity of matter contained in a body?

(a) Density (b) Volume

(c) Mass (d) Specific gravity

43. What is called mass per unit volume of a substances?

(a) Mass (b) Weight

(c) Density (d) Volume

44. What is the S.I unit of heat?

(a) calorie (b) joule

(c) centigrade heat unit

(d) british thermal unit

Engineering Drawing

45. Identify the conventional symbol of material?

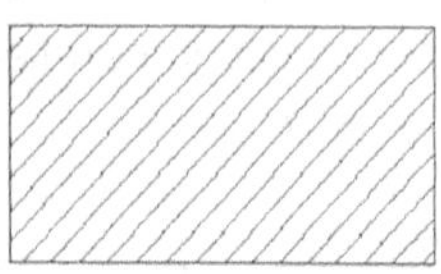

(a) Lead (b) Steel

(c) Glass (d) Porcelain

46. Identify the name of tool?

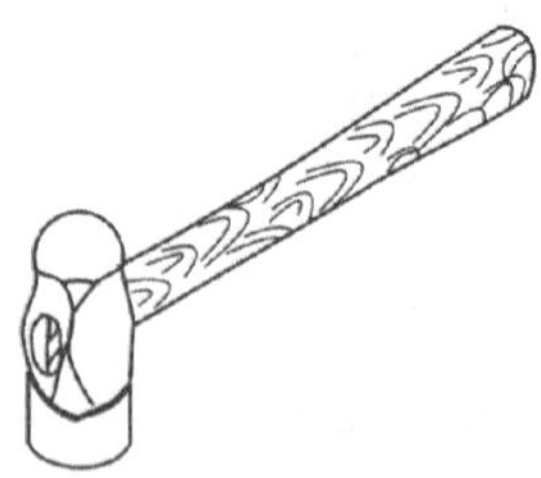

(a) Ball pein hammer
(b) Cross pein hammer
(c) Straight pein hammer
(d) Sledge hammer

47. Identify the name of tool?

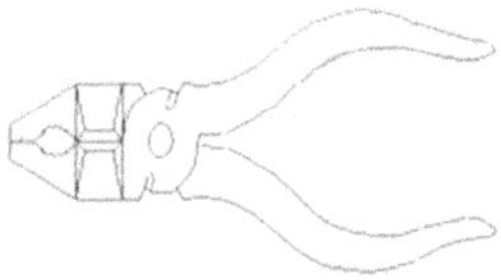

(a) Hawk-billed shear
(b) Aviation shear
(c) Cutting plier
(d) Double cutting shear

48. Identify the name of tool?

(a) Ball pein hammer
(b) Mallet
(c) Cross pein hammer
(d) Staright pein hammer

49. Identify the name of tool?

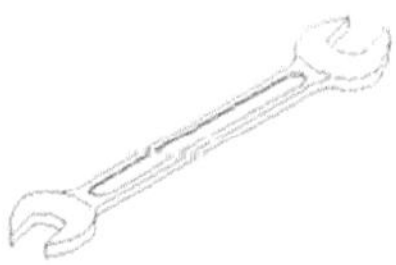

(a) Open end spanner
(b) Ring spanner
(c) Combination spanner
(d) Adjustable spanner

50. Which dimension correctly marked in the circle as per standard?

(a)

(b)

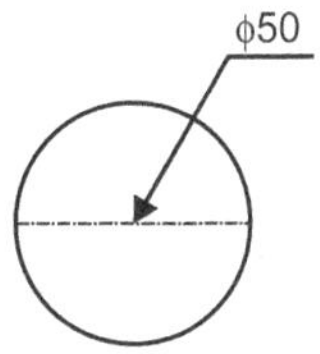

(c)

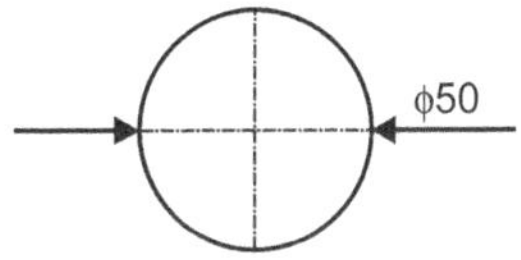

(d)

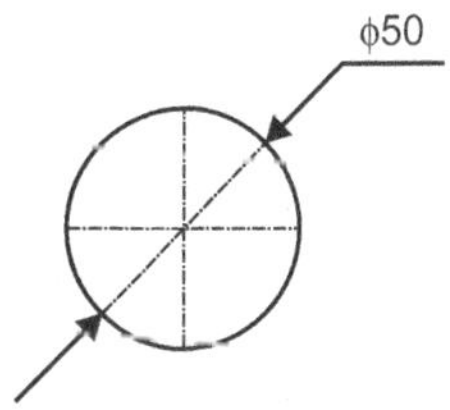

KEY Electrician 1st Year, Set–2

1. (a)	**2.** (d)	**3.** (b)	**4.** (c)	**5.** (c)	**6.** (b)	**7.** (c)	**8.** (d)	**9.** (c)	**10.** (a)
11. (b)	**12.** (b)	**13.** (c)	**14.** (b)	**15.** (c)	**16.** (b)	**17.** (a)	**18.** (c)	**19.** (a)	**20.** (d)
21. (c)	**22.** (b)	**23.** (b)	**24.** (b)	**25.** (d)	**26.** (a)	**27.** (d)	**28.** (b)	**29.** (c)	**30.** (a)
31. (b)	**32.** (b)	**33.** (b)	**34.** (d)	**35.** (b)	**36.** (b)	**37.** (b)	**38.** (a)	**39.** (a)	**40.** (b)
41. (d)	**42.** (c)	**43.** (c)	**44.** (b)	**45.** (b)	**46.** (a)	**47.** (c)	**48.** (b)	**49.** (a)	**50.** (b)

ALL INDIA TRADE TEST

Electrician Trade Theory

Latest Question Bank Set–3

First Year

Trade Theory

1. What electrical quantities are related with the Ohm's law?
   (a) Current, resistance and power
   (b) Current, voltage and resistivity
   (c) Current, voltage and resistance
   (d) Voltage, resistance and current density
2. What is the SI unit of resistivity?
   (a) ohm / cm (b) ohm / cm^2
   (c) ohm - metre (d) ohm / metre
3. Which is an application of the series circuit?
   (a) Voltmeter connection
   (b) Lighting circuits in home
   (c) Shunt resistor in ammeter
   (d) Multiplier resistor of a voltmeter
4. What is the effect of the circuit, if points 'ab' are shorted as shown in the figure?

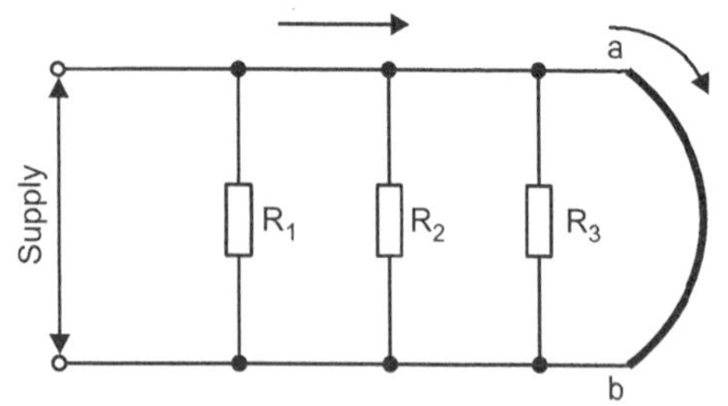

   (a) Circuit resistance will become zero
   (b) Same current will flow in all branches
   (c) Supply voltage will increase in each branch
   (d) Each branch current is equal to total current
5. What is the value of resistance of an open circuit?
   (a) Zero (b) Low
   (c) Medium (d) Infinity
6. Which resistor carries the lowest current in a parallel circuit having the values of 50 Ω, 220 Ω, 450 Ω and 560 Ω if connected with the supply?
   (a) 50 Ω (b) 220 Ω
   (c) 450 Ω (d) 560 Ω
7. What is the unit of capacitance?
   (a) mho (b) henry
   (c) farad (d) coulomb
8. Which factor affects the polarity of the electromagnet?
   (a) Length of the coil
   (b) Direction of current
   (c) Strength of current
   (d) Strength of the magnetic field
9. What is the unit of permeance?
   (a) ampere - turns
   (b) weber / ampere turns
   (c) ampere turns / weber
   (d) weber / square metre
10. Which rule is applied to find the direction of magnetic fields in a solenoid coil?
   (a) Cork screw rule
   (b) Right hand palm rule
   (c) Fleming's left hand rule
   (d) Fleming's right hand rule

11. What is the name of the part marked as 'X' in B.H curve as shown in the figure?

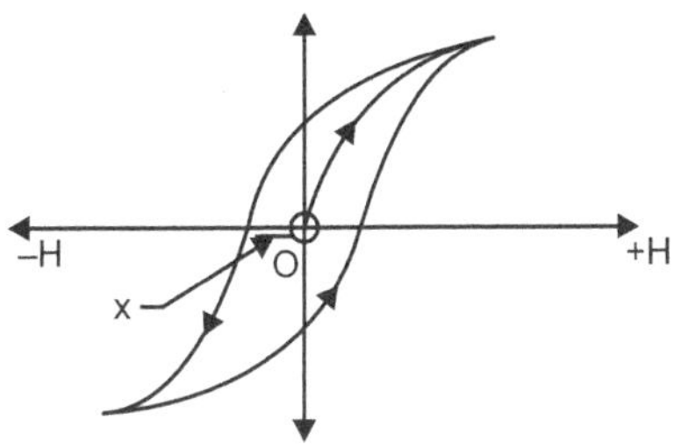

(a) Coercivity
(b) Saturation point
(c) Flux density
(d) Residual magnetism

12. What is the unit of Reluctance?
(a) weber / $metre^2$
(b) weber / metre
(c) ampere turns / weber
(d) ampere turns / $metre^2$

13. Which defines that the flux density is always lagging behind the magnetising force?
(a) Hysteresis
(b) Magnetic intensity
(c) Magnetic induction
(d) Rosidual magnetism

14. What is the SI unit of Flux density?
(a) tesla (b) weber
(c) weber/metre (d) ampere-turns

15. Which factor changes the permeability of a material?
(a) Length
(b) Flux density
(c) Diameter
(d) Cross sectional area

16. What is the effect of inductance if the distance between the turns increases?
(a) Increases (b) Decreases
(c) Becomes zero (d) Remains same

17. Which is an application of series circuit?
(a) Fuse in circuit
(b) Voltmeter connection
(c) Electrical lamp in homes
(d) Shunt resistor in ammeter

18. Which is inversely proportional to the resistance of a conductor?
(a) Length
(b) Resistivity
(c) Temperature
(d) Area of cross section

19. Which material is the paramagnetic substance?
(a) Water (b) Copper
(c) Bismuth (d) Graphite

20. Where the air capacitors are used?
(a) In VHF unit (b) In oscillator
(c) In loudspeaker (d) In radio receiver

21. Which is the diamagnetic substance?
(a) Wood (b) Nickel
(c) Platinum (d) Manganese

22. Which electrical quantity is directly proportional to the eddy current?
(a) Voltage (b) Current
(c) Frequency (d) Resistance

23. Which property of magnet is illustrated as shown in the figure?

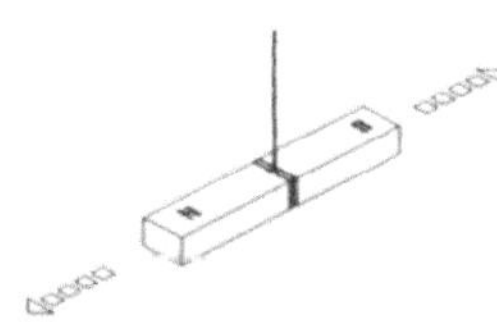

(a) Directive property
(b) Induction property
(c) Saturation property
(d) Poles existing property

24. Which is a paramagnetic substance?
(a) Air (b) Steel
(c) Glass (d) Water

25. When will the no current flow through the galvanometer in the wheat stone bridge?
(a) In balanced condition
(b) In unbalanced condition
(c) In short-circuited condition
(d) In open-circuited condition

26. What is the reading of the galvanometer in wheat stone bridge at balanced condition?
(a) High deflection (b) Low deflection
(c) Nill deflection (d) Oscillate

27. What is the change of resistance value of the conductor as its diameter is doubled?

(a) Increases to two times
(b) Decreases to four times
(c) Decrease to half of the value
(d) No change in value of resistance

28. Calculate the resistance value of the resistor by colour coding method.

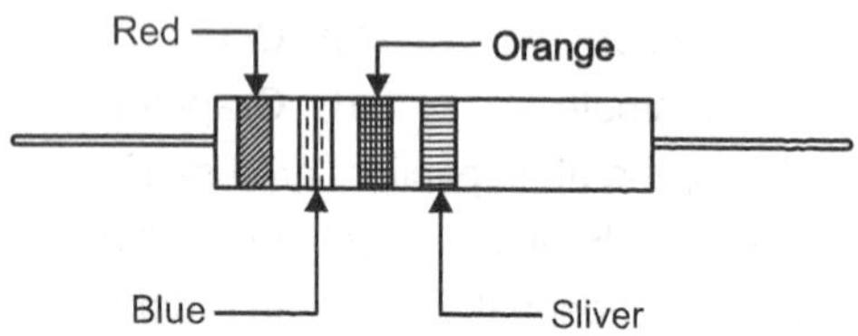

(a) $23 \times 103\ \Omega \pm 5\%$
(b) $26 \times 103\ \Omega \pm 10\%$
(c) $32 \times 104\ \Omega \pm 10\%$
(d) $37 \times 104\ \Omega \pm 5\%$

29. Why the ohmmeter is graduated with non-linear scale?

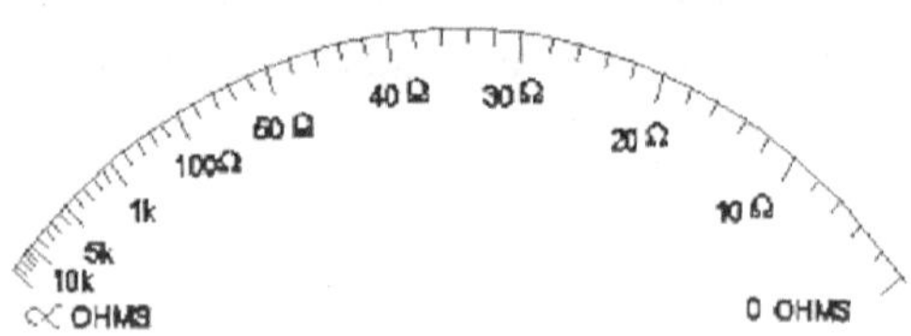

(a) Voltage is directly proportional to resistance
(b) Current is inversely proportional to resistance
(c) Resistance is inversely proportional to the square of current
(d) Voltage is directly proportional to the square of the current

30. Calculate the value of resistance by colour coding method.

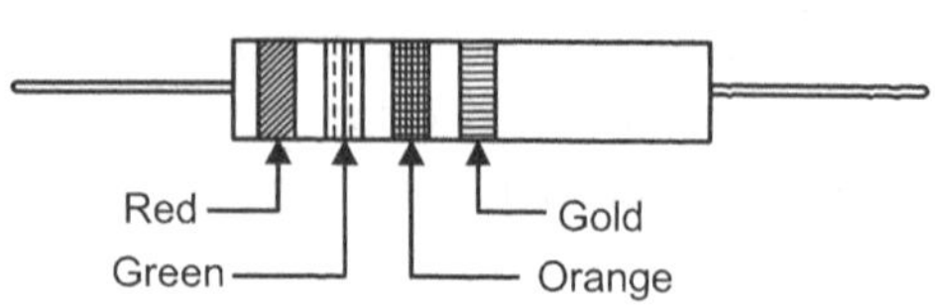

(a) $22 \times 103 \pm 10\%$ (b) $23 \times 104 \pm 10\%$
(c) $25 \times 103 \pm 5\%$ (d) $36 \times 104 \pm 5\%$

31. Which effect of the electric current takes place in the neon lamp?

(a) Heating effect
(b) Magnetic effect
(c) Chemical effect
(d) Gas ionization effect

32. What is the unit of insulation resistance?

(a) ohm (b) kilo ohm
(c) milli ohm (d) mega ohm

33. Which type of soldering flux is used for soldering aluminium conductors?

(a) Tallow (b) Kynal flux
(c) Zinc chloride (d) Sal ammonia

34. What is the name of the part marked 'X' in UG cables as shown in the figure?

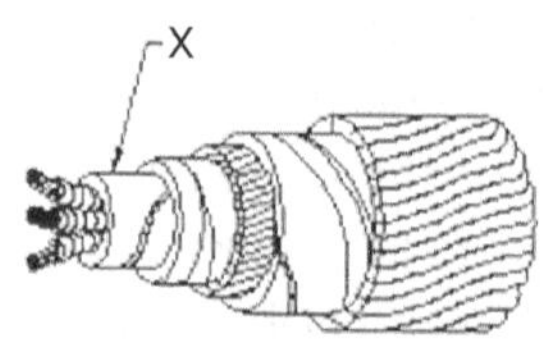

(a) Serving (b) Bedding
(c) Armouring (d) Lead sheath

35. What is the name of the joint as shown in the figure?

(a) Married joint
(b) Scarfed joint
(c) Western union joint
(d) Britannia straight joint

36. What is the use of Britannia 'T' joint?

(a) Extending the length of the lines
(b) Inside and outside wiring installation
(c) Mechanical stress not required on conductor
(d) Tapping the service connection from overhead lines

37. What is the name of the part marked 'X' of the belted U.G cable as shown in the figure?

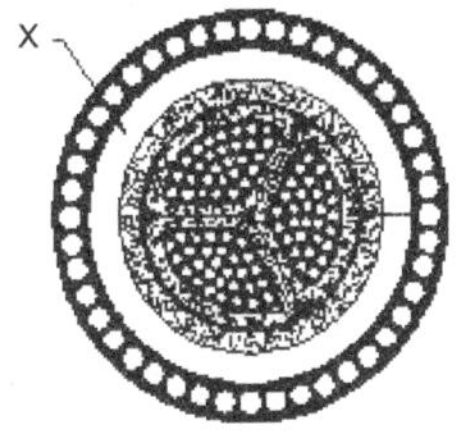

(a) Jute filling (b) Armouring
(c) Lead sheath (d) Paper insulation

38. What is the purpose of the bedding insulation in the U.G. cable?
(a) Protect the cable from mechanical injury
(b) Protect the cable from moisture and gases
(c) Protect armouring from atmospheric condition
(d) Protect the metallic sheath against corrosion

Workshop Calculation and Science

39. What are the two classifications of system of units?
(a) British and Metric
(b) Gravitational and non-gravitational
(c) Fundamental and derived
(d) Metric and International

40. What are fundamental units?
(a) Length, Mass, Volume
(b) Length, Mass, Time
(c) Length, Mass, Area
(d) Length, Pressure, Volume

41. What denotes letter 'M' in MKS system?
(a) Mile (b) Moter
(c) Millimeter (d) Micron

42. How many millimetres are there in 1 inch?
(a) 2.54 mm (b) 25.4 mm
(c) 24.5 mm (d) 2.45 mm

43. Convert decimal 0.000659 to fraction?
(a) $\frac{659}{1000}$ (b) $\frac{659}{10000}$
(c) $\frac{659}{100000}$ (d) $\frac{659}{1000000}$

44. What is the square root of 529?
(a) 13 (b) 23
(c) 33 (d) 43

Engineering Drawing

45. Identify the conventional symbol of material?

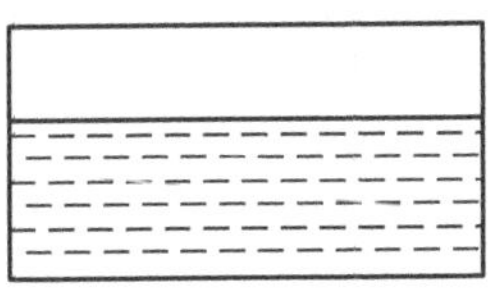

(a) Water (b) Concrete
(c) Asbestos (d) Glass

46. Identify the conventional symbol of material?

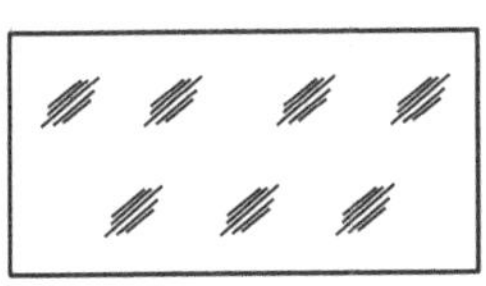

(a) Concrete (b) Steel
(c) Wood (d) Glass

47. Identify the conventional symbol of material?

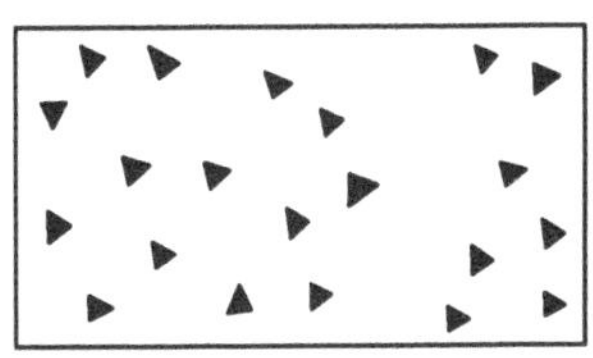

(a) Steel (b) Wood
(c) Glass (d) Concrete

48. What is the trimmed size of A4 sheet?
(a) 841 × 1189 (b) 594 × 841
(c) 210 × 297 (d) 148 × 210

49. Which drawing dimension is correct as per IS standard?

(a)

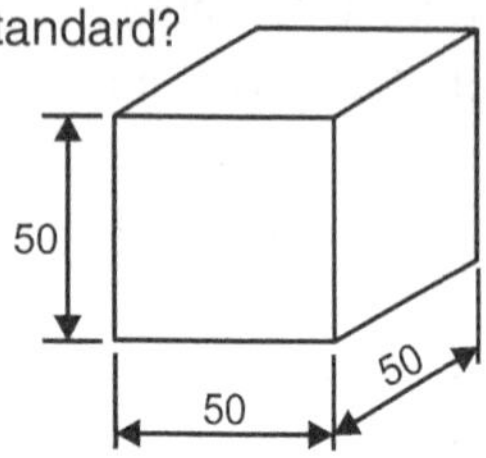

(b)

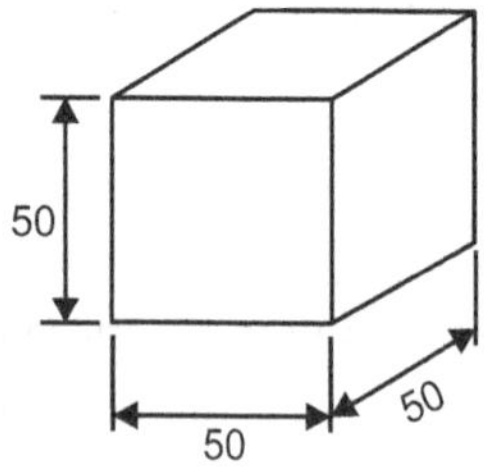

(c)

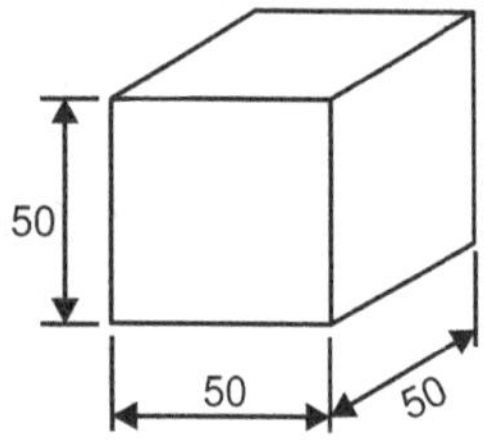

(d)

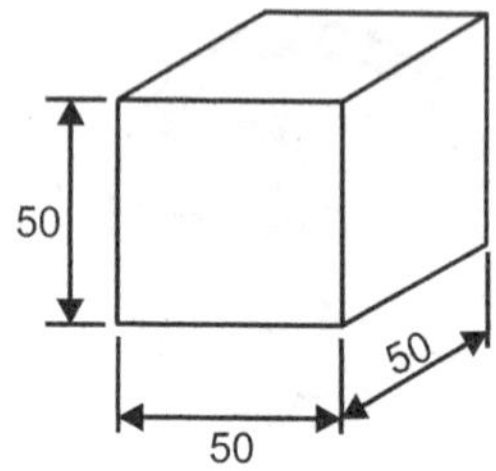

50. Identify the name of prism?

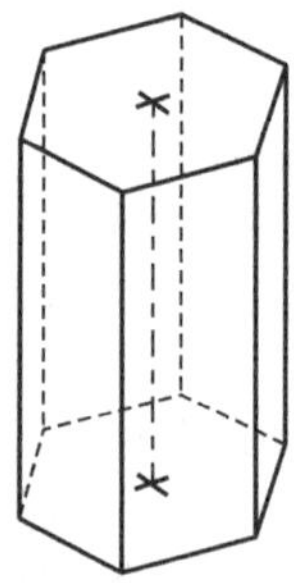

(a) Square prism
(b) Pentagonal prism
(c) Triangular prism
(d) Hexagonal prism

KEY Answer Electrician 1st Year, Set–3

1. (c)	**2.** (c)	**3.** (d)	**4.** (a)	**5.** (d)	**6.** (d)	**7.** (c)	**8.** (b)	**9.** (b)	**10.** (b)
11. (a)	**12.** (c)	**13.** (a)	**14.** (a)	**15.** (b)	**16.** (b)	**17.** (a)	**18.** (d)	**19.** (b)	**20.** (d)
21. (a)	**22.** (c)	**23.** (a)	**24.** (a)	**25.** (a)	**26.** (c)	**27.** (b)	**28.** (b)	**29.** (b)	**30.** (c)
31. (d)	**32.** (d)	**33.** (b)	**34.** (d)	**35.** (a)	**36.** (d)	**37.** (c)	**38.** (d)	**39.** (c)	**40.** (b)
41. (b)	**42.** (b)	**43.** (d)	**44.** (b)	**45.** (a)	**46.** (d)	**47.** (d)	**48.** (c)	**49.** (a)	**50.** (d)

ALL INDIA TRADE TEST
Electrician Trade Theory

Latest Question Bank Set–1

Second Year

Trade Theory

1. What is the name of the part marked as 'X' in DC generator as shown in the figure?

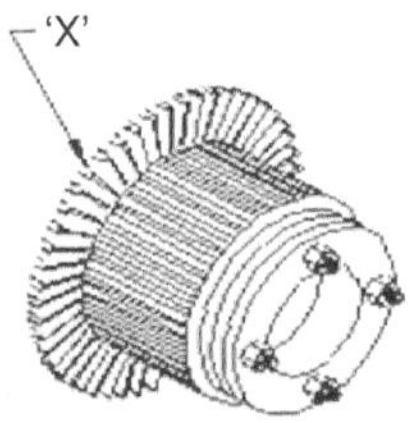

(a) Armature core
(b) Brush
(c) Commutator raisor
(d) Commutator segment

2. What is the name of the D.C generator as shown in the figure?

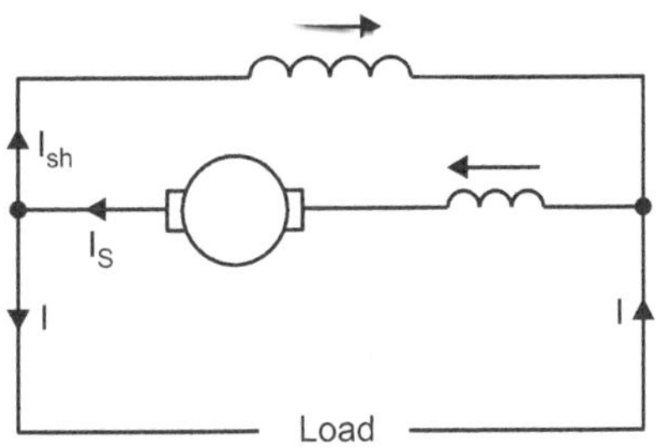

(a) Differential long shunt compound
(b) Differential short shunt compound
(c) Cumulative long shunt compound
(d) Cumulative short shunt compound

3. Which rule is used to find the direction of induced emf in D.C generator?
(a) Cork screw rule
(b) Right hand palm rule
(c) Fleming's left hand rule
(d) Fleming's right hand rule

4. Which formula is used to calculate the generated emf in D.C generator?
(a) Generated emf = $\frac{\phi ZN}{60}$ volt
(b) Generated emf = $\frac{\phi ZN}{60} \times \frac{A}{P}$ volt
(c) Generated emf = $\frac{\phi ZN}{60} \times \frac{P}{A}$ volt
(d) Generated emf = $\frac{\phi ZN}{60 \times \varphi} \times \frac{P}{A}$ volt

5. What is the name of the part marked 'X' in DC generator as shown in the figure?

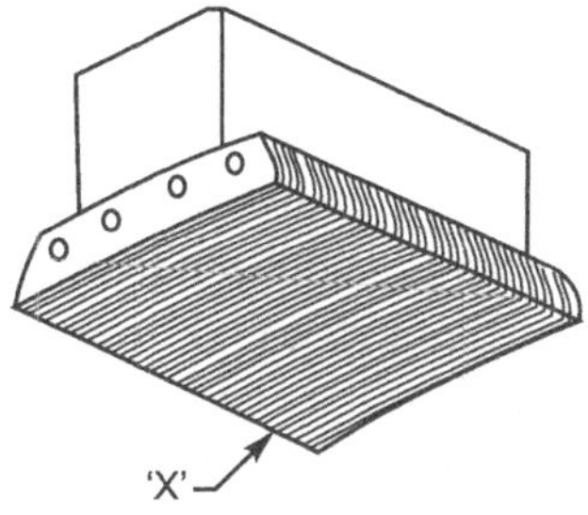

(a) Yoke (b) Pole coil
(c) Pole core (d) Pole shoe

6. What is the name of the D.C generator as shown in the circuit?

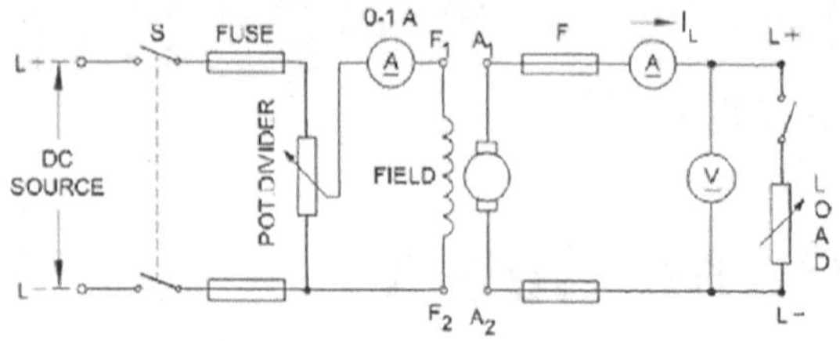

(a) Shunt generator
(b) Series generator
(c) Compound generator
(d) Separately excited generator

7. Which energy is converted into electrical energy by the generator?
(a) Heat (b) Kinetic
(c) Chemical (d) Mechanical

8. What is the name of the D.C generator according to the field as shown in the figure?

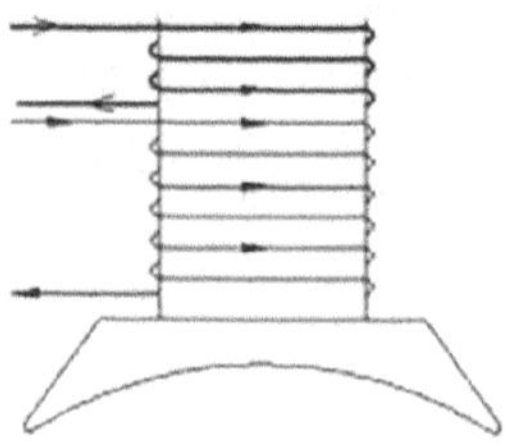

(a) Series generator
(b) Shunt generator
(c) Differential compound generator
(d) Cumulative compound generator

9. What is the working principle of D.C generator?
(a) Cork screw rule
(b) Fleming's left hand rule
(c) Fleming's right hand rule
(d) Faraday's laws of electromagnetic induction

10. What is the formula for dynamically induced emf?
(a) BLV volts (b) BL sin θ volts
(c) BLV sin θ volts (d) BLV cos θ volts

11. Which rule is used to find direction of magnetic field of the solenoid?
(a) Cork screw rule
(b) Right hand palm rule
(c) Fleming's left hand rule
(d) Fleming's right hand rule

12. What is the name of the part of DC generator as shown in the figure?

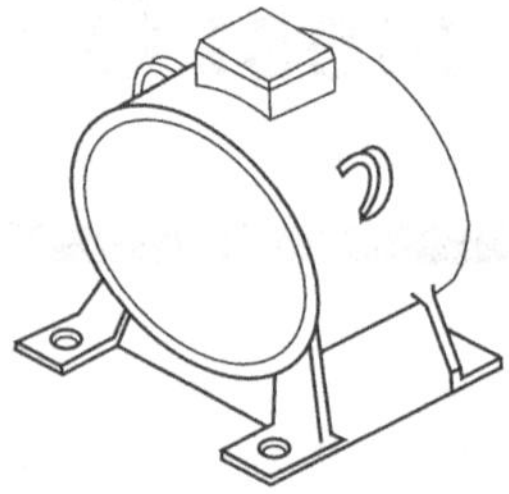

(a) Stator (b) Pole core
(c) Pole shoes (d) Yoke (or) frame

13. How many parallel paths in duplex lap winding of a 4 pole DC generator?
(a) 4 (b) 6
(c) 8 (d) 12

14. What is the name of the part of DC generator as shown in the figure?

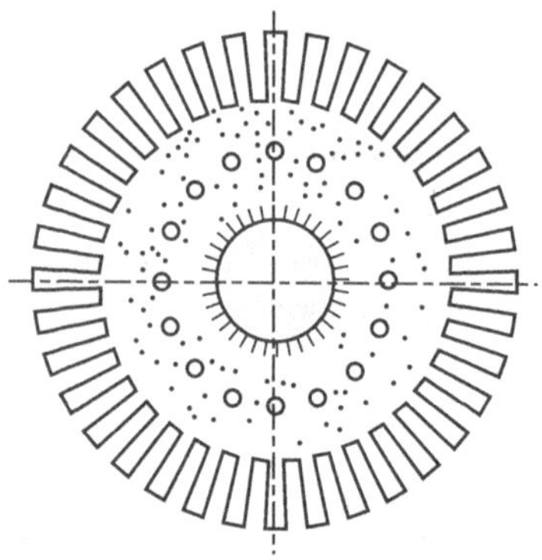

(a) Side end plates
(b) Pole shoe lamination
(c) Commutator segment
(d) Armature core lamination

15. How the interpoles are connected in a DC generator?
(a) In series with armature
(b) In parallel with armature
(c) In series with shunt field
(d) In parallel with shunt field

16. Which instrument is used to test the armature winding for short and open circuit?

(a) Tong Tester (b) Series test lamp
(c) External Growler (d) Megger

17. What is the name of the speed control method of DC motor as shown in the figure?

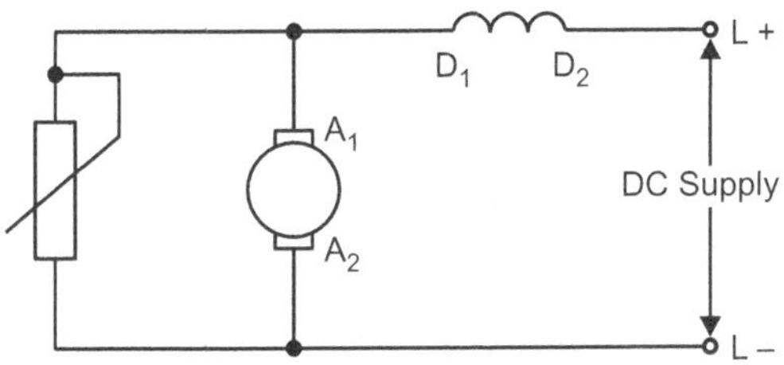

(a) Field diverter method
(b) Field tapping method
(c) Voltage control method
(d) Armature diverter method

18. Which formula is used to calculate the speed of DC motor?

(a) $N = K\frac{E_b}{\varphi}$ (b) $N = K\frac{\varphi}{E_b}$

(c) $N = K\frac{E_b \cdot \varphi}{120}$ (d) $N = K\frac{E_b \cdot \varphi}{60}$

19. Which rule is used to determine the direction of rotation of armature in D.C motor?

(a) Right hand grip rule
(b) Right hand palm rule
(c) Fleming's left hand rule
(d) Fleming's right hand rule

20. What is the name of the D.C motor as shown in the figure?

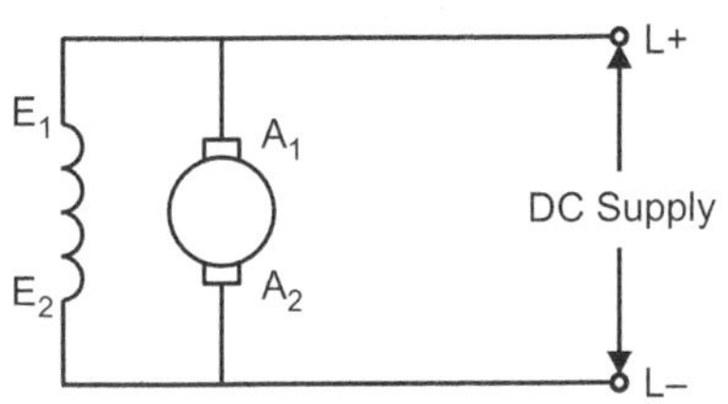

(a) D.C shunt motor
(b) D.C series motor
(c) D.C differential compound motor
(d) D.C cumulative compound motor

21. Which rule indicates the direction of current in armature conductors in D.C. Motor?

(a) Right hand grip rule
(b) Right hand palm rule
(c) Fleming's left hand rule
(d) Fleming's right hand rule

22. What is the formula to calculate the armature current taken by the D.C shunt motor ?

(a) $I_a = \frac{V}{R_a}$ (b) $I_a = \frac{E_b}{R_a}$

(c) $I_a = \frac{V - E_b}{R_a}$ (d) $I_a = \frac{V + E_b}{R_a}$

23. What is the type of the DC motor as shown in the diagram?

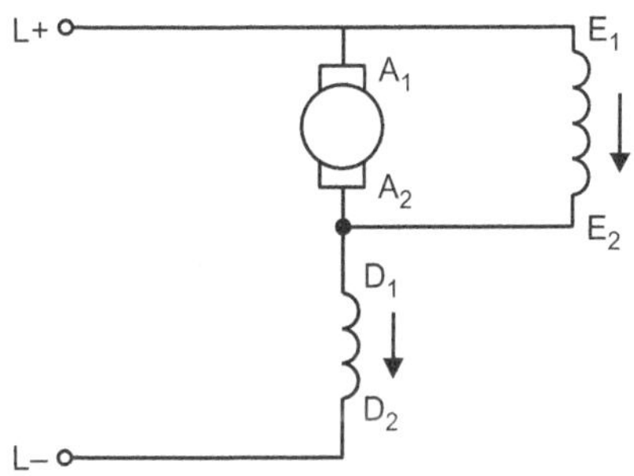

(a) Shunt motor
(b) Series motor
(c) Long shunt compound motor
(d) Short shunt compound motor

24. What is the formula to calculate the back EMF in a DC motor?

(a) $E_b = \frac{ZNP}{\varphi\ 60\ A}$

(b) $E_b = \frac{NP}{Z\ \varphi\ 60\ A}$

(c) $E_b = \frac{\varphi\ ZNP}{60\ A}$

(d) $E_b = \frac{60\ A\ \varphi}{ZNP}$

25. What is the name of the equipment as shown in the figure?

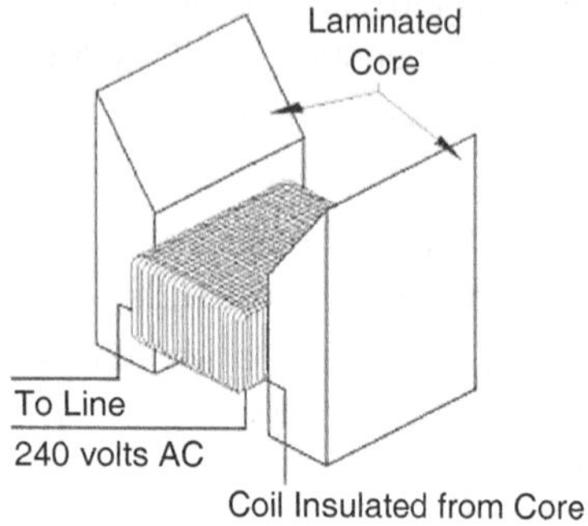

(a) Megger
(b) Earth resistance tester
(c) Internal growler
(d) External growler

26. What is the name of winding, if coil pitch is less than pole pitch?
(a) Full pitch winding
(b) Over pitch winding
(c) Long chorded winding
(d) Short chorded winding

27. What is the purpose of series resistor connected with holding coil in a D.C four point starter?
(a) Limit the current in holding coil
(b) Increase the current in holding coil
(c) Increase the voltage in holding coil
(d) Decrease the voltage in field coil

28. Why the starters are required to start the D.C motors?
(a) Regulate the field voltage
(b) Reduce the armature current
(c) Control the armature reaction
(d) Increase the armature current

29. Which insulating material belongs to class 'B' insulation?
(a) Cotton (b) Bamboo
(c) Fiberglass
(d) Leatheroid paper

30. What is the permissible temperature value of class 'F' insulation?
(a) 90°C (b) 105°C
(c) 120°C (d) 155°C

31. What is the name of the speed control method as shown in the circuit?

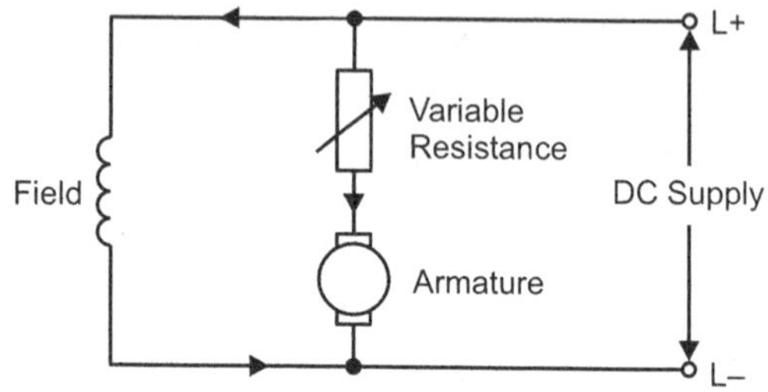

(a) Field control method
(b) Armature control method
(c) Field tapping control method
(d) Field diverter control method

32. Which is used to insulate the winding leads of a motor?
(a) Cotton tape
(b) Fibre sleeve
(c) Pressphan paper
(d) Leatheroid paper

33. Which is used to increase the insulation property of the insulating material used in the winding?
(a) Enamel paint (b) Petrol
(c) Diesel (d) Varnish

34. What is the name of the test as shown in the figure?

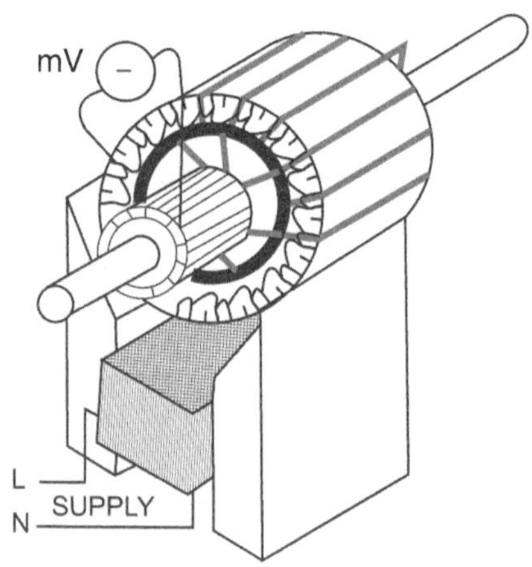

(a) Test for Grounded coil
(b) Test for Shorted coil
(c) Test for Open coil
(d) Drop test

35. How many number of parallel paths are in a wave wounded 6 pole DC machine?

(a) 2 (b) 4

(c) 6 (d) 8

36. What is the name of the test as shown in the figure?

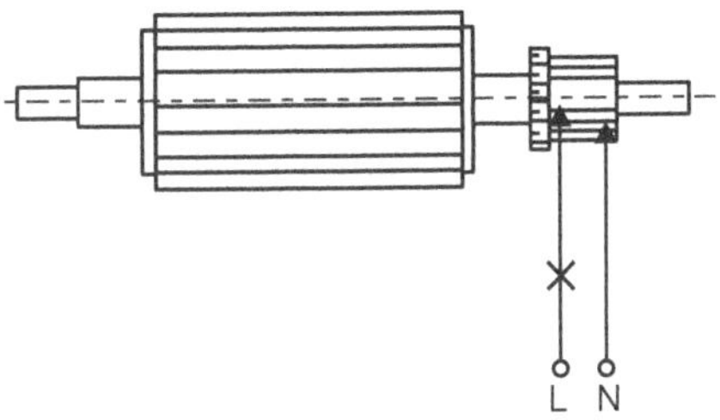

(a) Open circuit test

(b) Shorted commutator test

(c) Grounded commutator test

(d) Drop test

37. What is the name of the winding if the end lead of coil 1 is connected to the beginning lead of the adjust coil 2 through the commutator segment?

(a) Simplex lap winding

(b) Duplex lap winding

(c) Simplex wave winding

(d) Duplex wave winding

38. What is the name of the folded edges of the slot liner?

(a) Overhang insulation

(b) Coil separator

(c) Shaft insulation

(d) Cuffing

Workshop Calculation and Science

39. What is the expanded form of S.P?

(a) Selected Price (b) Special Price

(c) Selling Price (d) Super Price

40. What is denoted as 'I'?

(a) Principal (b) Interest

(c) Rate (d) Year

41. How the 'Principal' is denoted in simple interest calculation?

(a) 'P' (b) 'I'

(c) 'R' (d) 'n'

42. How the years is denoted in simple interest calculations?

(a) P (b) I

(c) n (d) r

43. What is the term used for the method of calculating various quantities and expenditure on a particular job or process?

(a) Estimation (b) Drawing

(c) Specification (d) Plan

44. What is the name of a booklet, the rates of various terms are indicated?

(a) Price bank (b) Price bunch

(c) Price tag (d) Price catalogue

Engineering Drawing

45. Which is protective device?

(a) Fuse (b) Resistor

(c) Magnet (d) Capacitor

46. Which part regulates the heat in automatic iron?

(a) Heating element

(b) Pressure plate

(c) Thermostatic switch

(d) Asbestos sheet

47. Which is the static machine that increase or decrease the AC voltage?

(a) Transformer (b) Generator

(c) Alternator (d) Induction motor

48. What is the material of line insulator?

(a) Copper (b) Aluminium

(c) Porcelain clay (d) Rubber

49. Identify the wave form.

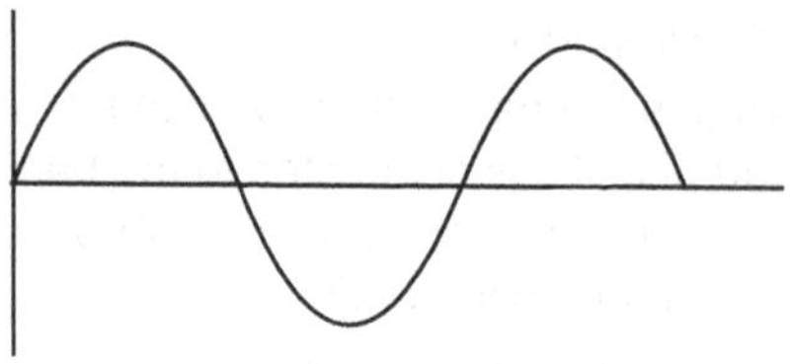

(a) Square wave (b) Sine wave
(c) Triangle wave (d) Saw-tooth wave

50. What is the name of the symbol?

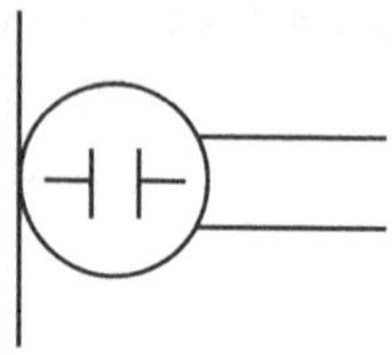

(a) Microphone
(b) Dynamic microphone
(c) Condenser microphone
(d) Loud speaker

KEY Electrician 2nd Year, Set–1

1. (c)	**2.** (a)	**3.** (d)	**4.** (c)	**5.** (d)	**6.** (d)	**7.** (d)	**8.** (d)	**9.** (d)	**10.** (c)
11. (b)	**12.** (d)	**13.** (c)	**14.** (d)	**15.** (a)	**16.** (c)	**17.** (d)	**18.** (a)	**19.** (c)	**20.** (a)
21. (c)	**22.** (c)	**23.** (d)	**24.** (c)	**25.** (d)	**26.** (d)	**27.** (a)	**28.** (b)	**29.** (c)	**30.** (d)
31. (b)	**32.** (b)	**33.** (d)	**34.** (a)	**35.** (a)	**36.** (b)	**37.** (a)	**38.** (d)	**39.** (c)	**40.** (b)
41. (a)	**42.** (c)	**43.** (a)	**44.** (d)	**45.** (a)	**46.** (c)	**47.** (a)	**48.** (c)	**49.** (b)	**50.** (c)

ALL INDIA TRADE TEST

Electrician Trade Theory

Latest Question Bank Set–2

Second Year

Trade Theory

1. Which method is used to improve the insulation resistance in DC generator?
   (a) Replacing the brushes frequently
   (b) Keeping the machine in idle
   (c) Running the machine with overload
   (d) Blowing hot air into the machine

2. What is the necessity of the residual magnetism in a self excited DC generator?
   (a) Build up the voltage
   (b) Reduce the field current
   (c) Reduce the armature current
   (d) Maintain the constant output voltage

3. Which are the two points that the brush contact resistance measured in D.C machines?
   (a) Resistance between the opposite brushes
   (b) Resistance between the brush and brush holder
   (c) Resistance between the brush and commutator
   (d) Resistance between the brush and armature

4. What is the reason for the DC generator fails to build up voltage?
   (a) Field resistance is below the critical resistance
   (b) Armature resistance is less
   (c) Field resistance is above the critical resistance
   (d) Prime mover is running at above rated speed

5. What is the name of generator, if its field is connected in parallel with armature?
   (a) Shunt generator
   (b) Series generator
   (c) Long shunt compound generator
   (d) Separately excited generator

6. Why the terminal voltage of the DC shunt generator decreases if the load increases?
   (a) Because of armature reaction effect
   (b) Due to decrease in armature resistance
   (c) Because of decrease in brush voltage drop
   (d) Due to increase in shunt field inductance

7. Which type of D.C Generator works in absence of residual magnetism?
   (a) Shunt generator
   (b) Series generator
   (c) Compound generator
   (d) Separately excited generator

8. What is the name of the generator as shown in the circuit?

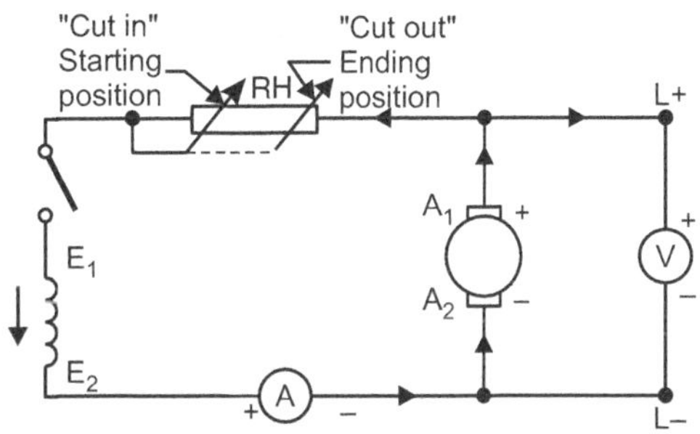

(a) DC shunt generator
(b) DC compound generator
(c) DC series generator
(d) Separately excited DC generator

9. What is the name of the generator as shown in the circuit?

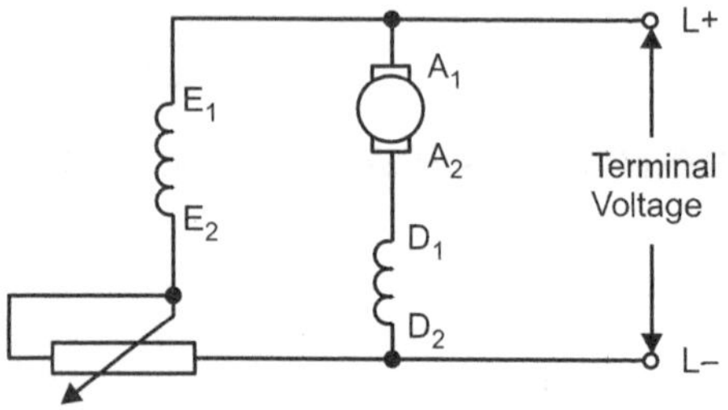

(a) DC shunt generator
(b) DC series generator
(c) Short shunt compound Generator
(d) Long shunt compound Generator

10. What is the formula for shunt field current R_{sh} of a DC shunt generator if V_T = Terminal voltage E_g = generated EMF?

(a) $I_{sh} = V_T / R_{sh}$ (b) $I_{sh} = E_g / R_{sh}$
(c) $I_{sh} = V_T \times R_{sh}$ (d) $I_{sh} = E_g \times R_{sh}$

11. What is the formula for the efficiency of a D.C generator?

(a) $\frac{\text{Input + Losses}}{\text{Input}}$ (b) $\frac{\text{Output}}{\text{Input – Losses}}$

(c) $\frac{\text{Output}}{\text{Input + Losses}}$ (d) $\frac{\text{Input}}{\text{Output}}$

12. Which loss is called as copper loss?

(a) Constant loss (b) Variable loss
(c) Friction loss (d) Windage loss

13. What is the condition for the maximum efficiency of a DC generator?

(a) Eddy current loss is equal to hysteresis loss
(b) Variable losses is less than constant losses
(c) Variable losses is more than constant losses
(d) Variable losses is equal to constant losses

14. What is the name of the symbol as shown in figure?

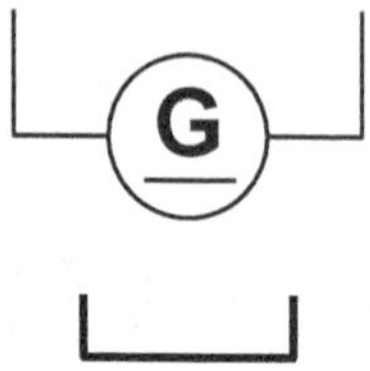

(a) DC 2 wire permanent magnet generator
(b) DC 2 wire series generator
(c) DC 2 wire electro magnet generator
(d) DC 2 wire compound generator

15. What is the formula to calculate the back emf of a D.C motor?

(a) $E_b = \frac{V}{I_a R_a}$ volts
(b) $E_b = V \times I_a R_a$ volts
(c) $E_b = V - I_a R_a$ volts
(d) $E_b = V + I_a R_a$ volts

16. Which voltage drop is indicated in the portion marked as 'X' as shown in the figure?

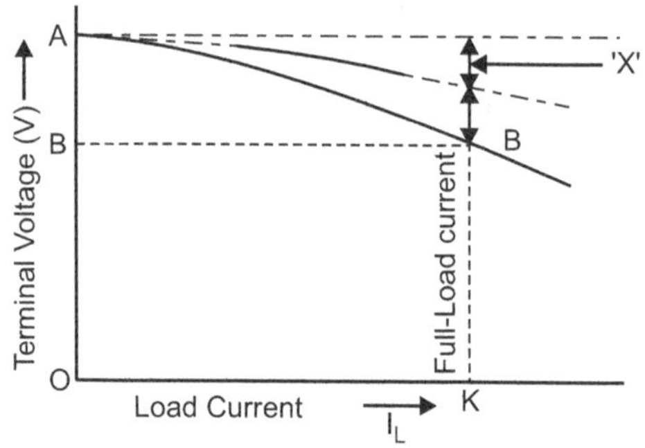

(a) Series field voltage drop
(b) Armature voltage drop
(c) Armature reaction drop
(d) Shunt field voltage drop

17. Why the armature core of a DC generator is laminated?

(a) To reduce the copper loss
(b) To reduce the friction loss
(c) To reduce the hysteresis loss
(d) To reduce the eddy current loss

18. Why the armature resistance of a D.C generator is kept very low?
(a) To reduce the armature current
(b) To reduce the armature voltage drop
(c) Helps to solve Boolean Algebra
(d) To reduce the temperature of armature

19. Why the D.C generator should run in one direction only?
(a) To protect the brushes from damage
(b) To protect the residual magnetism
(c) To avoid short circuit in the armature
(d) To avoid over loading of the generator

20. Why the compensating winding is provided in the large DC generators?
(a) To neutralize the cross-magnetizing effect of armature reaction
(b) To reduce the commutation effect
(c) To neutralize the demagnetizing effect of armature reaction
(d) To decrease the efficiency of generator

21. What is the purpose of pole shoe in DC generator?
(a) Increase the air gap
(b) Increase the field strength
(c) Minimize the magnetic losses
(d) Spread out flux uniformly in the air gap

22. What is the function of split rings in DC generator?
(a) Maintain constant voltage
(b) Collects the current unidirectionally
(c) Reduces the voltage drop at bruches
(d) Increases the terminal voltage than rated

23. Which materials are used to make the brushes in generator?
(a) Steel and graphite
(b) Carbon and graphite
(c) Cast iron and graphite
(d) Aluminium and graphite

24. Why the DC generators are loosing their residual magnetism?
(a) Flashing of field
(b) Running without load continuously
(c) Continuous running without break
(d) Change of direction of rotation very often

25. Which type of DC generator is used for long distance distribution lines?
(a) Shunt generator
(b) Differential short shunt compound generator
(c) Differential compound generator
(d) Cumulative compound generator

26. Which type of D.C generator is used for arc welding?
(a) Shunt generator
(b) Series generator
(c) Differential compound generator
(d) Cumulative compound generator

27. What is the property of wave winding in D.C generator?
(a) Low current low voltage
(b) High current low voltage
(c) Low current high voltage
(d) High current high voltage

28. What is the purpose of using resistance wire used in the commutator in D.C generator as shown in the figure?

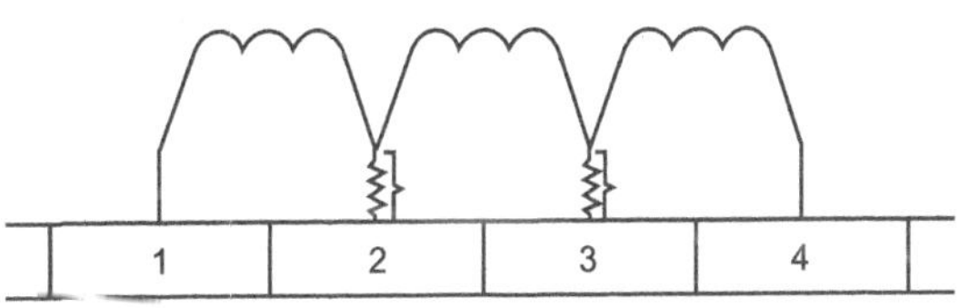

(a) To maintain constant stage
(b) To reduce the voltage drop
(c) To increase the statically induced emf
(d) For smooth reversal of current direction

29. Why solid pole shoes are used in D.C generator?
(a) To reduce the copper loss
(b) To increase the residual magnetism
(c) To decrease the residual magnetism
(d) To reduce the reluctance of magnetic path

30. Which metal is used to make the yoke of a large capacity DC generator?
(a) Cast iron (b) Soft iron
(c) Aluminium (d) Rolled Steel

31. Which type of D.C motor is used for constant speed drives?
    (a) DC series motor
    (b) DC shunt motor
    (c) Differential long shunt compound motor
    (d) Differential short shunt compound motor

32. Which DC motor is suitable for heavy starting torque and high rate of acceleration?
    (a) DC series motor
    (b) DC shunt motor
    (c) DC differential compound motor
    (d) DC cumulative compound motor

33. Which is the most effective method of balancing armature?
    (a) Static balancing
    (b) Dynamic balancing
    (c) Attached with counter balancing
    (d) Plugged with lead weight balancing

34. Which material is used for starting resistance of DC starters?
    (a) Eureka
    (b) Nichrome
    (c) Manganin
    (d) Constantine

35. Which type of armature winding is illustrated as shown in the figure?

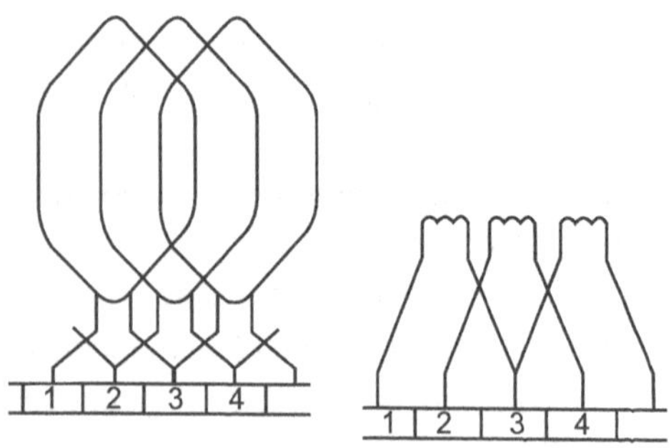

    (a) Duplex lap winding
    (b) Triplex lap winding
    (c) Simplex lap winding
    (d) Quadruplex lap winding

36. Which growler test for armature is illustrated as shown in the figure?

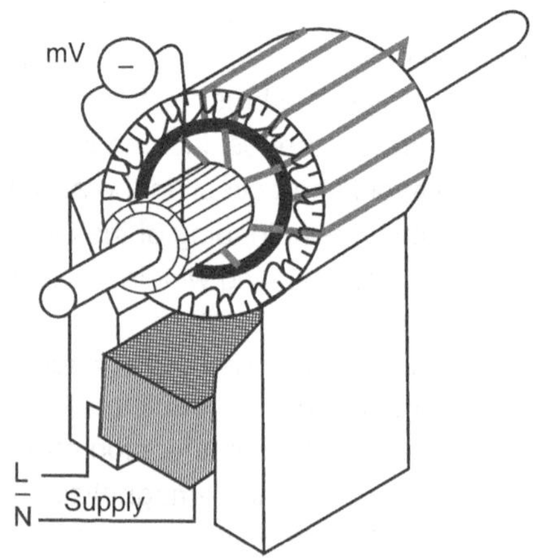

    (a) Open coil test
    (b) Grounded coil test
    (c) Shorted coil test
    (d) Shorted commutator test

37. Why commutators are sparking heavily?
    (a) Incorrect brush position
    (b) Incorrect field connection
    (c) Incorrect direction of rotation
    (d) Incorrect armature connection

38. What is the action to assist the induced emf in a running D.C motor?
    (a) Assist the applied voltage
    (b) Opposes the applied voltage
    (c) Increases the armature current
    (d) Decreases the armature current

Workshop Calculation and Science

39. What is the formula for $(a-b)^2$?
    (a) $a^2 - 2ab + b^2$
    (b) $a^2 + 2ab + b^2$
    (c) $a^2 - 2ab - b^2$
    (d) $-a^2 - 2ab - b^2$

40. What is the value of $a \times a^2 \times a^3 \times a^4$?

(a) a^7 (b) a^8

(c) a^9 (d) a^{10}

41. Which type of stress?

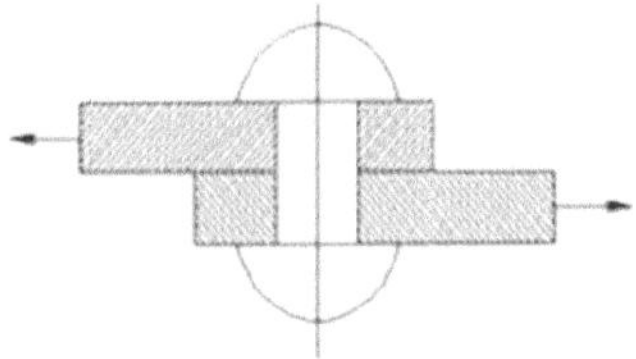

(a) Tensile stress

(b) Compressive stress

(c) Shear stress

(d) Torsional stress

42. What is the unit of strain?

(a) kg/cm^2

(b) mewton/metre^2

(c) metre

(d) No unit

43. Which symbol is used to express change in length?

(a) L (b) dl

(c) l (d) e

44. What is the expanded form of S.P?

(a) Selected Price (b) Special Price

(c) Selling Price (d) Super Price

Engineering Drawing

45. Identify the given symbol

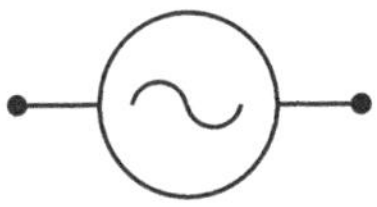

(a) AC (b) DC

(c) ohms (d) Inductance

46. Choose the symbol for main fuse board without switches (P).

(a) 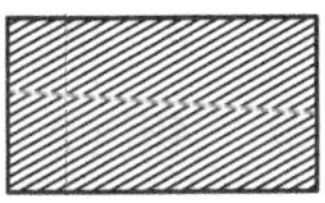 (b) 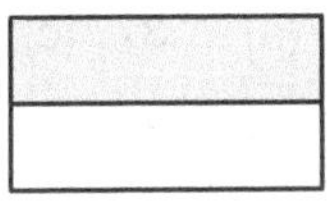

(c) 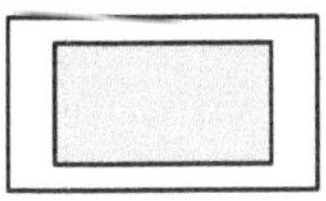 (d) 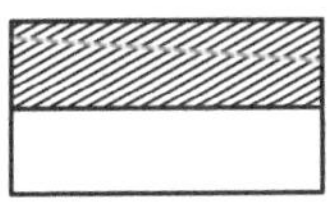

47. Identify the symbol used for microphone (general) in electrical and electronic circuits.

(a) 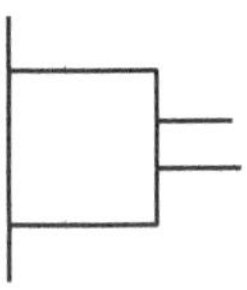 (b) 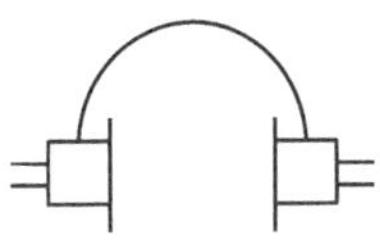

(c) 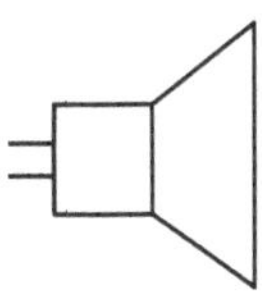 (d) 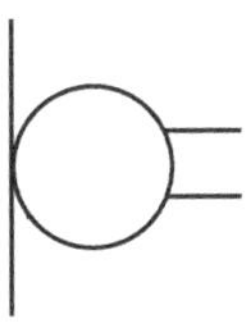

48. Identify the symbol for direct and alternating current.

(a) (b) 

(c) 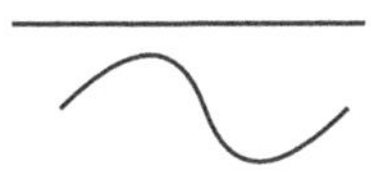 (d) 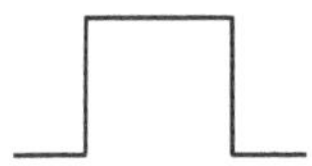

49. Which is the single pole MCB?

(a) 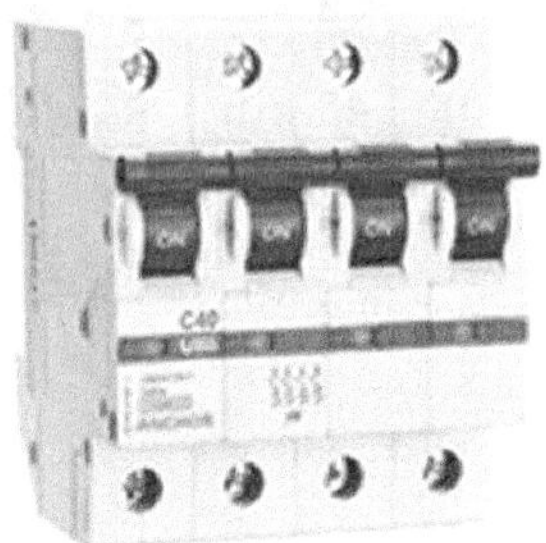

(b) 

(c) 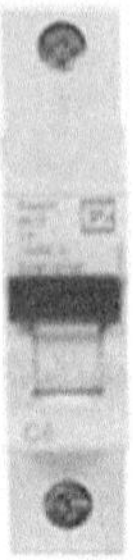

(d) 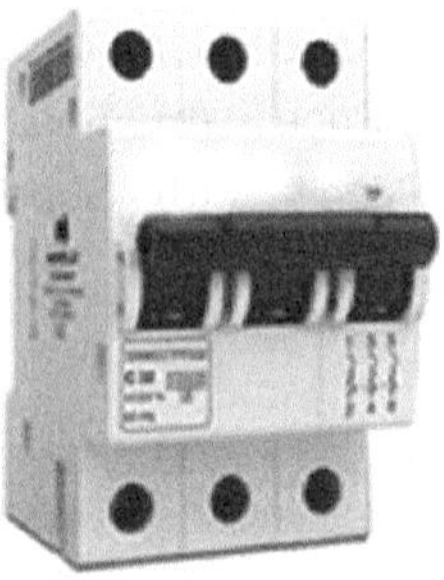

50. Which is the NOT gate in the following picture?

(a) 

(b) 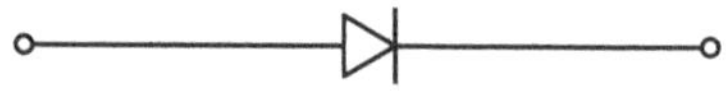

(c) 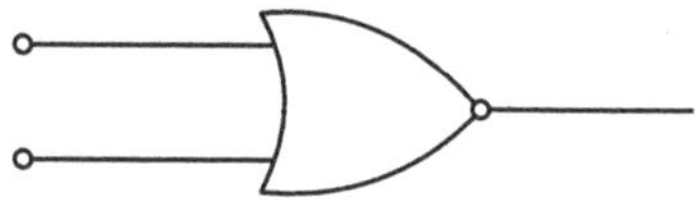

(d) 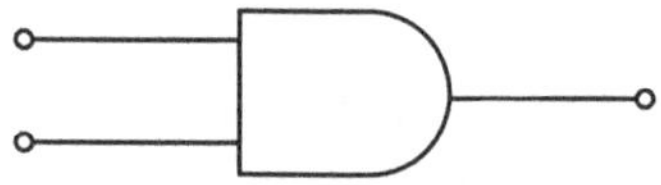

KEY Electrician 2nd Year, Set–2

1. (d)	**2.** (a)	**3.** (c)	**4.** (c)	**5.** (a)	**6.** (a)	**7.** (d)	**8.** (a)	**9.** (d)	**10.** (a)
11. (c)	**12.** (b)	**13.** (d)	**14.** (a)	**15.** (c)	**16.** (c)	**17.** (d)	**18.** (b)	**19.** (b)	**20.** (c)
21. (d)	**22.** (b)	**23.** (b)	**24.** (d)	**25.** (d)	**26.** (c)	**27.** (c)	**28.** (d)	**29.** (d)	**30.** (d)
31. (b)	**32.** (a)	**33.** (b)	**34.** (a)	**35.** (a)	**36.** (a)	**37.** (c)	**38.** (b)	**39.** (a)	**40.** (d)
41. (c)	**42.** (d)	**43.** (b)	**44.** (c)	**45.** (a)	**46.** (c)	**47.** (d)	**48.** (c)	**49.** (c)	**50.** (a)

ALL INDIA TRADE TEST
Electrician Trade Theory

Latest Question Bank Set–3

Second Year

Trade Theory

1. Which motor has this characteristics curve as shown in the figure?

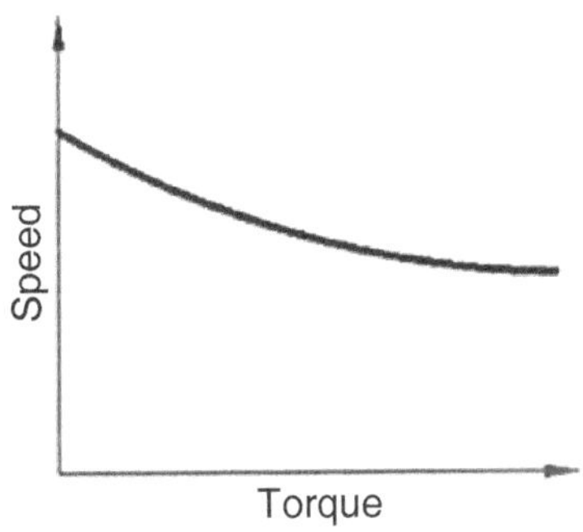

(a) Series motor
(b) Shunt motor
(c) Cumulative compound motor
(d) Differential compound motor

2. Which type of speed control of D.C series motor is shown in the figure?

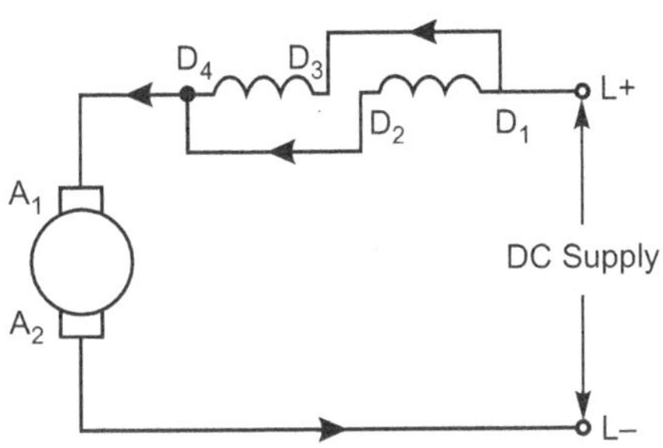

(a) Field parallel method
(b) Field diverter method
(c) Field tapping method
(d) Armature diverter method

3. Which DC Motor is designed to work with the full load limits?
(a) Shunt motor
(b) Series motor
(c) Cumulative compound motor
(d) Differential compound motor

4. Where D.C compound motors are preferred?
(a) Constant load requirements
(b) Constant speed requirements
(c) High starting torque requirements
(d) Constant speed under varying load requirements

5. Which type of DC motor is used for sudden application of heavy loads?
(a) Shunt motor
(b) Series motor
(c) Differential compound motor
(d) Cumulative compound motor

6. Which speed control method is used in food mixture motors?
(a) Voltage control method
(b) Field diverter control method
(c) Armature diverter method
(d) Series field tapping method

7. What is the purpose of tapes in winding?
(a) Insulate slots
(b) Bind the coils
(c) Wrap the conductor
(d) Insulate exposed conductors

8. What is the operation in rewinding process?

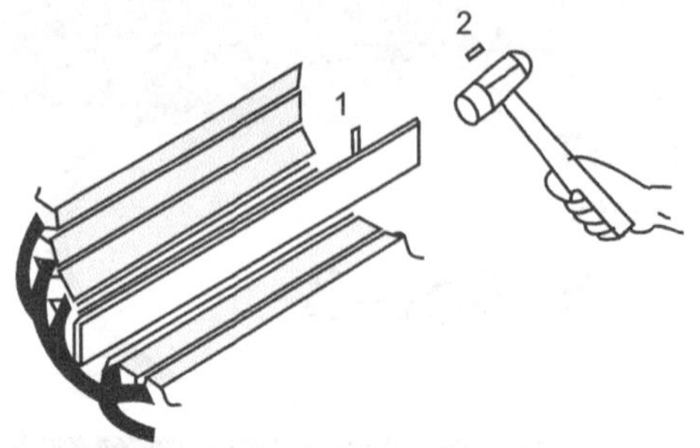

(a) Cleaning of slots
(b) Removing of winding
(c) Removing of wedges
(d) Cutting of winding wire

9. Which type of armature winding is illustrated as shown in the figure?

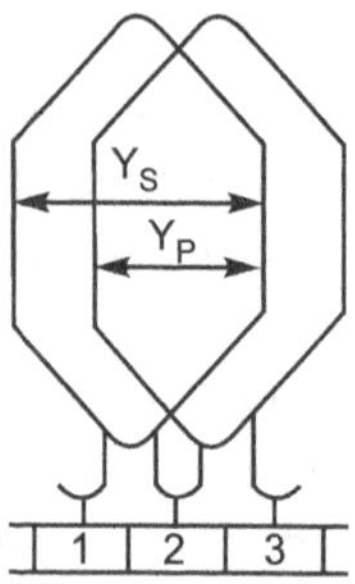

(a) Triplex wave winding
(b) Duplex wave winding
(c) Progressive lap winding
(d) Retrogressive lap winding

10. Which type of test is illustrated for the armature after rewound?

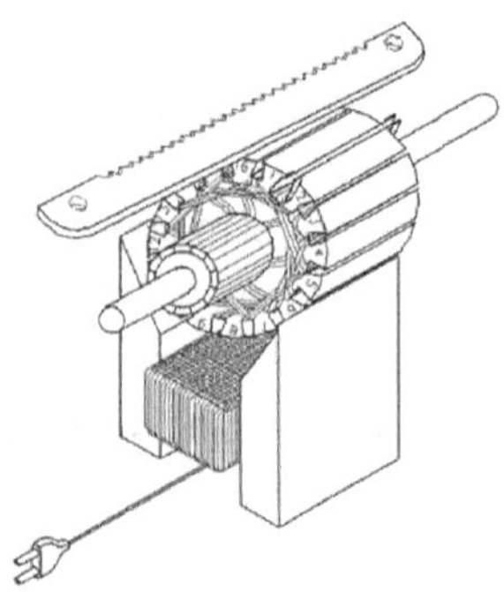

(a) Open coil test
(b) Shorted coil test
(c) Voltage drop test
(d) Grounded coil test

11. Which type of winding wire is used to wind submersible pump motors?
(a) PVC covered type
(b) Terylene thread type
(c) Super enamelled type
(d) Double cotton covered type

12. Why it is better to change the direction of armature current to change the direction of rotation of DC compound motor?
(a) To increase the rated speed
(b) To maintain the motor characteristics
(c) To increase the output power
(d) To increase the efficiency

13. What happens if the starting resistance of four point starter opens while DC compound motor is running?

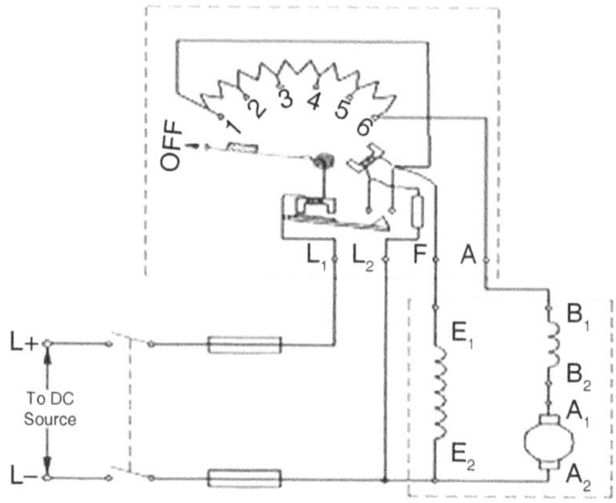

(a) Motor stopped
(b) Runs at slow speed
(c) Runs at very high speed
(d) Runs at reverse direction

14. How the opposite polarity of adjacent poles of a 4 pole DC motor is obtained?
(a) By increasing the number of turns in coil
(b) By decreasing the number of turns in coil
(c) By making the current flow in same direction
(d) By making the current flow in opposite direction

15. Why shunt field coil is connected in series with holding coil in D.C three point starter?
(a) Increase the holding coil current
(b) Decrease the holding coil current
(c) Protect the shunt field from over current
(d) Protect the motor in case of open in shunt field

16. Which method of speed control offers the speed below the rated speed of DC shunt motor?
(a) Field control method
(b) Connecting additional winding in series with field
(c) Armature control method
(d) Connecting additional resistance in series with field

17. Which method of speed control offers the speed below the rated speed of DC series motor?
(a) Field diverter method
(b) Tapped field method
(c) Connecting additional winding in series with field
(d) Armature diverter method

18. Why the series field is short circuited at the time of starting in the differential compound motor?
(a) To reduce the starting current
(b) To decrease the back EMF
(c) To decrease the speed of motor
(d) To maintain the proper direction of rotation

19. Which DC motor can be operated at constant speed under varying load?
(a) Differential long shunt compound motor
(b) Cumulative long shunt compound motor
(c) Differential short shunt compound motor
(d) Series motor

20. How the no volt coil is connected in a three point starter with DC shunt motor?
(a) Directly connected to the supply
(b) Connected in series with the armature
(c) Connected in parallel with the armature
(d) Connected in series with the shunt field

21. Which speed control method is applied to obtain both below normal and above normal speed in DC motor?
(a) Field control method
(b) Armature control method
(c) Tapped field speed control method
(d) Ward Leonard speed control method

22. Why the holding coil of a 3 point starter is connected in series with shunt field?
(a) To limit the load current
(b) To run motor at low voltage
(c) To hold the handle firmly
(d) To protect the motor if the field opens

23. What is the better method to change the DOR of a compound motor without changing its characteristics?
(a) Change the direction of armature current
(b) Change the direction of shunt field current
(c) Change the direction of series field current
(d) Change the direction of both the armature and shunt field current

24. What is the purpose of no volt coil in 3 point starter?
(a) To improve the torque
(b) To increase the field current
(c) To increase the back emf
(d) To disconnect the motor if power fails

25. Which speed control system provides a smooth variation of speed from zero to above normal?
(a) Field control
(b) Armature control
(c) Field diverter control
(d) Ward-Leonard system control

26. Which type of DC armature winding has the front pitch (Y_F) greater than back pitch (Y_B)?
(a) Lap winding
(b) Wave winding
(c) Progressive winding
(d) Retrogressive winding

27. Calculate the average pitch (Y_A) for retrogressive wave winding, if the number of armature conductor = 14, number of slots = 7 and number of poles = 2

(a) 4 (b) 6
(c) 8 (d) 14

28. Why the rewound armature must be preheated before varnishing?
(a) Dry out the moisture
(b) Help for quick drying of varnish
(c) Make easy to penetrate varnish inside
(d) Maintain uniform spreading of varnishing

29. What is the name of rule as shown in figure?

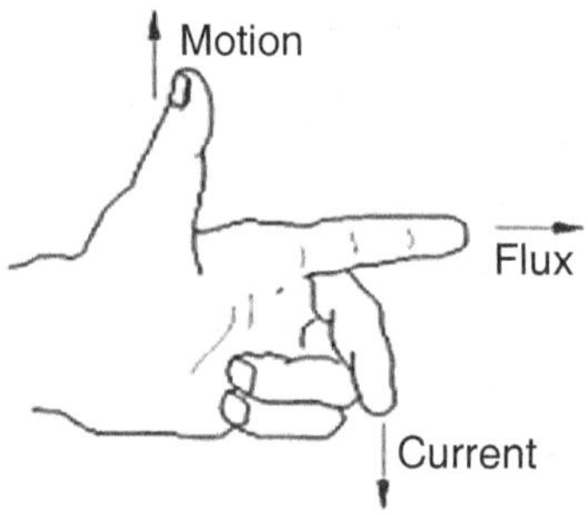

(a) Fleming's right hand rule
(b) Palm rule
(c) Fleming's left hand rule
(d) Thumb rule

30. Which is inversely proportional to the speed of DC motor?
(a) Field flux
(b) Applied voltage
(c) Armature resistance
(d) Load current

31. Which type of voltage is induced dynamically in a D.C generator?
(a) Pulsating voltage
(b) Oscillating voltage
(c) Alternating voltage
(d) Direct current voltage

32. What is the purpose of slot marked as 'X' as shown in the figure?

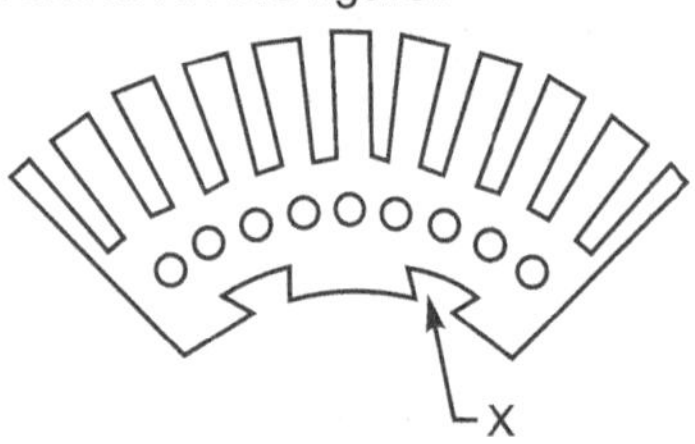

(a) To fix the key way
(b) To make air circulation
(c) For lubrication purpose
(d) For easy removal from shaft

33. What is the purpose of field coils in D.C generator?
(a) To increase the flux in air gap
(b) To decrease the magnetizing current
(c) To magnetize the poles to produce coil flux
(d) To increase the reluctance of magnetic path

34. Which metal is used to make the pole core of large DC machine?
(a) Soft iron (b) Cast iron
(c) Cast steel (d) Stainless steel

35. Why the pole core is laminated in DC generator?
(a) To reduce the friction loss
(b) To reduce the windage loss
(c) To reduce the hysteresis loss
(d) To reduce the eddy current loss

36. Which type of DC generator is used for electroplating process?
(a) Shunt generator
(b) Series generator
(c) Differential compound generator
(d) Over Compounded Cumulative generator

37. What is the name of the compound generator, if the shunt field is connected in parallel with armature?
(a) Long shunt compound generator
(b) Shunt generator
(c) Separately excited generator
(d) Short shunt compound generator

38. How does the magnetic flux of the field system completed the circuit in a generator?
(a) Through the field coil and field core
(b) Through the field core, armature core and yoke
(c) Through the laminated pole core
(d) Through the armature winding and armature core

Workshop Calculation and Science

39. What is the purpose of a lubricant?
(a) To increase the pressure
(b) To increase friction
(c) To reduce friction
(d) To reduce pressure

40. Which is used to reduce the friction in machine parts?
(a) Kerosene (b) Petrol
(c) Water (d) Lubricants

41. What is the value of $14x + 3y + 25x + 2y$?
(a) $17x + 27y$ (b) $16x + 28y$
(c) $39x + 5y$ (d) $44xy$

42. What is the formula for $a^m \times a^n$?
(a) a^{m+n} (b) a^{m-n}
(c) a^{mn} (d) $n.a^m$

43. Which is the formula for $\frac{a^m}{a^n}$?
(a) a^{m+n} (b) a^{m-n}
(c) $a^{m \times n}$ (d) $(a^m)^n$

44. What is the expanded form of $(a + b)^2$?
(a) $a^2 + 2ab + b^2$ (b) $a^2 - 2ab + b^2$
(c) $a^2 + 2ab - b^2$ (d) $-a^2 - 2ab + b^2$

Engineering Drawing

45. Which device is used to start and stop a motor?
(a) Rotor (b) Starter
(c) Stator (d) Slip ring

46. What is the name of the gate

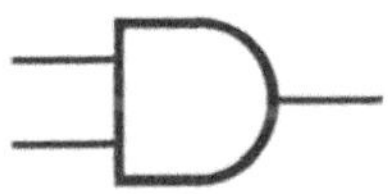

(a) NOT Gate (b) Ex-NOR Gate
(c) OR Gate (d) AND Gate

47. Which is photo diode?

(a) 

(b) 

(c) 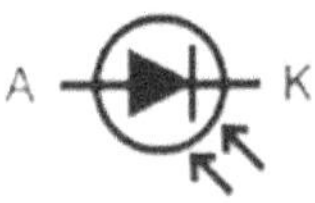

(d) 

48. Which of the following is AC ammeter?

(a) 

(b) 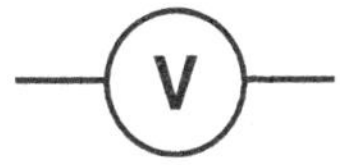

(c) 

(d) 

49. Which is a symbol of DC motor?

(a) 

(b) 

(c) 

(d) 

50. Voltage less than 250 volt is called______.
(a) LT (Low Tension)
(b) HT (High Tension)
(c) MT (Medium Tension)
(d) EHT (Extra High Tension)

KEY Electrician 2nd Year, Set–3

1. (c)	**2.** (a)	**3.** (d)	**4.** (d)	**5.** (d)	**6.** (d)	**7.** (c)	**8.** (c)	**9.** (c)	**10.** (b)
11. (a)	**12.** (b)	**13.** (a)	**14.** (d)	**15.** (d)	**16.** (c)	**17.** (d)	**18.** (d)	**19.** (b)	**20.** (d)
21. (d)	**22.** (d)	**23.** (a)	**24.** (d)	**25.** (d)	**26.** (d)	**27.** (b)	**28.** (a)	**29.** (c)	**30.** (a)
31. (c)	**32.** (a)	**33.** (c)	**34.** (c)	**35.** (d)	**36.** (a)	**37.** (d)	**38.** (b)	**39.** (c)	**40.** (d)
41. (c)	**42.** (a)	**43.** (b)	**44.** (a)	**45.** (b)	**46.** (d)	**47.** (c)	**48.** (a)	**49.** (a)	**50.** (a)

FIRST YEAR JULY 2023

All India Trade Test (CBT)

Electrician Trade Theory

NSQF Level – 4

Trade Theory

1. Why distribution transformers are normally connected as primary in delta and secondary in star?
   - (a) To maintain constant voltage
   - (b) To avoid over loading
   - (c) To easy distribution of 3 phase 4 wire system
   - (d) To reduce transformer losses
2. Why A.C is required to measure the earth resistance by using earth resistance tester?
   - (a) Avoid electrolytic emf interference
   - (b) Regulate the current
   - (c) Increase the voltage drop
   - (d) Decrease the voltage drop
3. What is the type of test in domestic wiring installation?

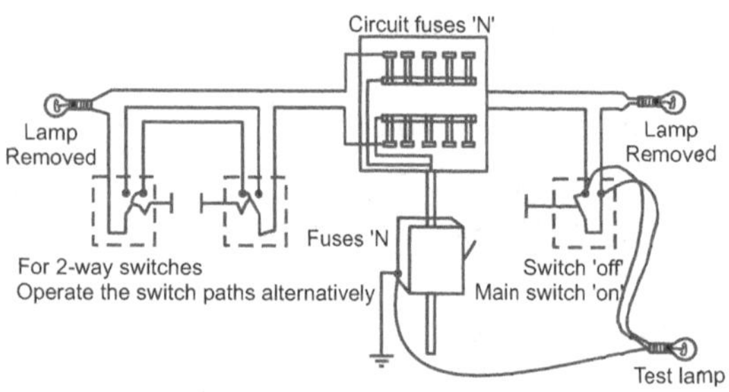

   - (a) Continuity (or) open circuit test
   - (b) Insulation resistance test between conductors and earth
   - (c) Polarity test
   - (d) Insulation resistance test between conductors
4. What is the minimum cross-sectional area of copper conductor for flexible cord as per BIS?
   - (a) 1 mm^2
   - (b) 0.5 mm^2
   - (c) 2.5 mm^2
   - (d) 1.5 mm^2
5. What is the composition of steel and silicon steel in transformer core?
   - (a) Steel 90% and silicon 10%
   - (b) Steel 97% and silicon 3%
   - (c) Steel 93% and silicon 7%
   - (d) Steel 95% and silicon 5%
6. What is the frequency of AC supply in India?
   - (a) 50 Hz
   - (b) 60 Hz
   - (c) 75 Hz
   - (d) 25 Hz
7. Which type of A.C single phase motor is used in food mixer?
   - (a) Split phase motor
   - (b) Shaded pole motor
   - (c) Universal motor
   - (d) Repulsion motor
8. What is the purpose of phase sequence meter?
   - (a) To indicate the incorrect phase sequence of 3 phase
   - (b) To protect motor against short circuit fault
   - (c) To ensure the correct phase sequence of 3 phase system
   - (d) To control the speed of 3 phase motor

9. What is the use of die stock set?
(a) Cut internal threads on cylindrical pipe
(b) Cut external threads on cylindrical pipe
(c) Cut internal threads on rectangular pipe
(d) Cut external threads on square pipe

10. Which rule is applied to find the direction of magnetic fields in a solenoid coil?
(a) Fleming's left hand rule
(b) Cork screw rule
(c) Right hand palm rule
(d) Fleming's right hand rule

11. How many electrons are there in the valence shell of a copper atom?
(a) 2 (b) 1
(c) 18 (d) 8

12. What is the S.I. unit of specific resistance?
(a) ohm/metre^2 (b) ohm-metre
(c) ohm/cm (d) micro ohm/cm^2

13. Which material is used to make heating element?
(a) Nichrome (b) Copper
(c) Silver (d) Aluminium

14. What is the name of the soldering method?

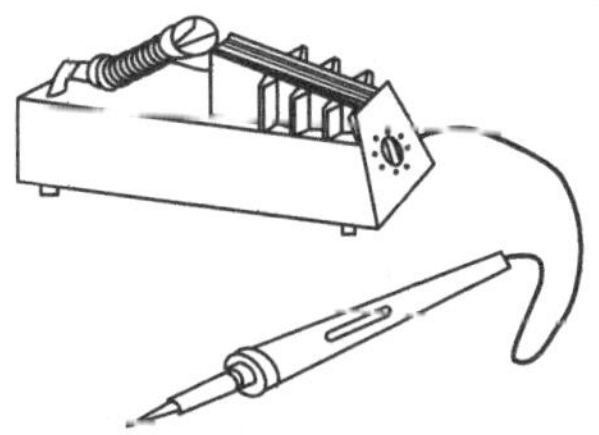

(a) Soldering with soldering gun
(b) Soldering with blow lamp
(c) Temperature controlled soldering
(d) Dip soldering

15. Which property of magnet is illustrated?

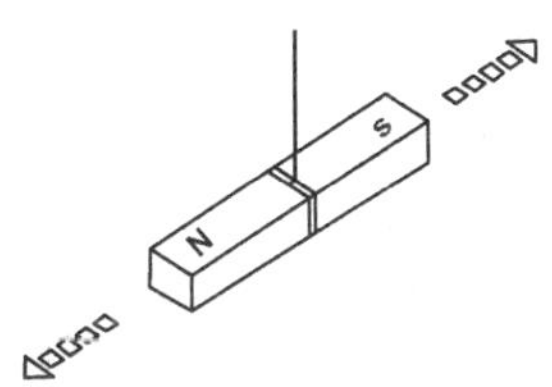

(a) Directive property
(b) Induction property
(c) Poles existing property
(d) Saturation property

16. What is the minimum size of copper earth continuity conductor used in single phase domestic wiring as per BIS?
(a) 3 sq.mm (b) 2.5 sq.mm
(c) 1.5 sq.mm (d) 3.5 sq.mm

17. Which file is used for sharpening the blunt teeth of a tenon saw?
(a) Square file (b) Half round file
(c) Round file (d) Triangular file

18. What is the output voltage of lithium cell?
(a) 2.5 V (b) 1.2 V
(c) 1.5 V (d) 1.8 V

19. What is the reason of lamp glowing dim and motor running slow in a domestic wiring circuit?
(a) Short circuit between conductors
(b) Open circuit in the earth conductor
(c) High value series resistance fault
(d) Open circuit in the neutral line

20. Which type of soldering flux is used for soldering galvanised iron?
(a) Sal ammonia (b) Zinc chloride
(c) Rosin (d) Hydrochloric acid

21. Which type of emf is induced in an ideal two winding transformer?
(a) Self induced emf
(b) Dynamically induced emf
(c) Statically induced emf
(d) Mutually induced emf

22. Which quantity is measured by an electrodynamo type instrument?
(a) Resistance (b) Current
(c) Power (d) Voltage

23. What is the unit of luminous flux?
(a) lux (b) lumen/m^2
(c) candela (d) lumen

24. Which factor affects the polarity of the electromagnet?
(a) Direction of current
(b) Length of the coil
(c) Strength of the magnetic field
(d) Strength of current

25. What is the maximum permissible load for a power sub circuit as per I.E. rules?
(a) 3000 watt (b) 800 watt
(c) 2000 watt (d) 1500 watt

26. Which device converts sunlight into electrical energy?
(a) Light emitting diode
(b) Liquid crystal diode
(c) Photo voltaic cell
(d) Light dependent resistor

27. What is the name of the tool?

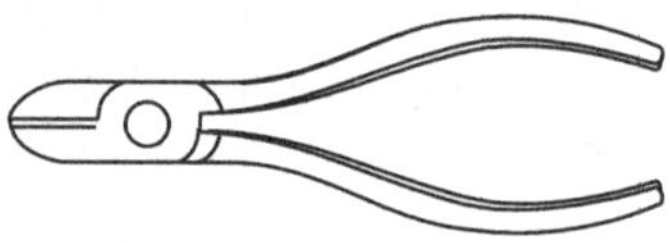

(a) Diagonal cutting pliers
(b) Combination pliers
(c) Crimping tool
(d) Wire stripper

28. Where system earthing is done?
(a) Domestic wiring installation
(b) Generating station
(c) Electroplating installation
(d) Small industrial installation

29. What is the fusing factor for rewireable fuse?
(a) 1.4 (b) 2.1
(c) 2.5 (d) 1.1

30. What is the expansion of ECC?
(a) Earth Conductor Continuity
(b) Earth Carrying Conductor
(c) Earth Continuity Conductor
(d) Earth Continuity Cable

31. Which location the service connection supply leads to be connected at consumer main board?
(a) IC cut out
(b) Main switch
(c) Distribution board
(d) Energy meter

32. Which quantity is rotating at a constant angular velocity?
(a) Vector quantity (b) Phasor quantity
(c) Scalar quantity (d) Algebraic quantity

33. Which test is conducted to locate the faults in U.G. cables?
(a) Insulation resistance test
(b) External growler test
(c) Break down voltage test
(d) Loop test

34. Which parameter is the cause for loading effect on measuring instruments?
(a) High sensitivity
(b) Low influence error
(c) Low accuracy
(d) Low sensitivity

35. What is the formula for resonance frequency?
(a) $\frac{1}{2\pi LC}$ (b) $2\pi\sqrt{LC}$
(c) $\frac{1}{2\sqrt{LC}}$ (d) $\frac{1}{2\pi\sqrt{LC}}$

36. Which instrument is an example of an integrating instrument?
(a) Energy meter
(b) Tangent galvanometer
(c) AC voltmeter
(d) DC voltmeter

37. Why tree system of wiring most suitable for Multistoried building?
(a) Constant voltage distribution
(b) Offers minimum voltage drop
(c) Easy in fault finding with many fuses
(d) Easy load balancing

38. What is the phase displacement in a single phase AC circuit?
(a) 120° (b) 270°
(c) 180° (d) 90°

WORKSHOP CALCULATION AND SCIENCE

39. What is called mass per unit volume of a substance?
(a) Density (b) Mass
(c) Volume (d) Weight

40. What is the S.I unit of heat?
(a) British thermal unit
(b) calorie
(c) centigrade heat unit
(d) joule

41. What is the side of a square whose area is 625mm^2?

(a) 30 mm (b) 25 mm
(c) 15 mm (d) 20 mm

42. What are fundamental units?

(a) Length, Pressure, Volume
(b) Length, Mass, Volume
(c) Length, Mass, Area
(d) Length, Mass, Time

43. What are the two classifications of system of units?

(a) Metric and International
(b) Fundamental and derived
(c) Gravitational and non-gravitational
(d) British and Metric

44. What denotes letter 'M' in MKS system?

(a) Meter (b) Mile
(c) Micron (d) Millimeter

ENGINEERING DRAWING

45. Which single stroke inclined letter 'A' is correct as per IS standard?

(a) 
7.5

(b) 
7.5

(c) 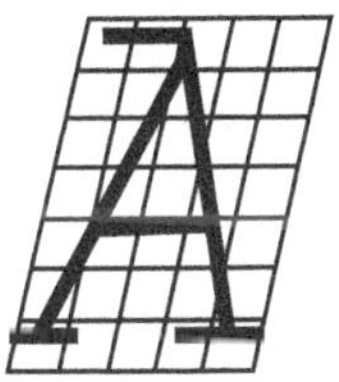
7.5

(d) 
7.5

46. Identify the name of prism?

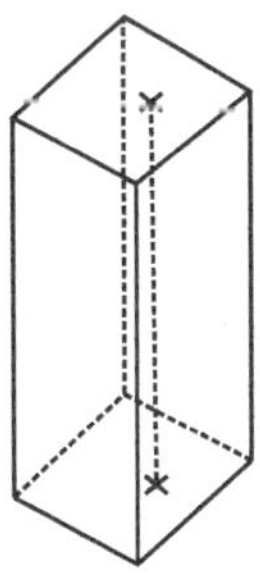

(a) Square prism (b) Triangular prism
(c) Hexagonal prism (d) Rectangle prism

47. Which drawing dimension is correct as per standard drawing practice?

(a) 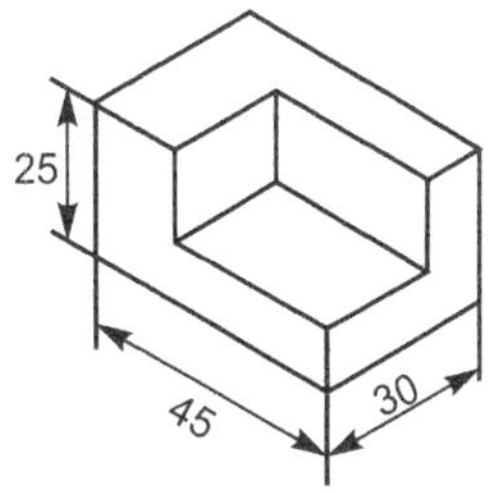

(b) 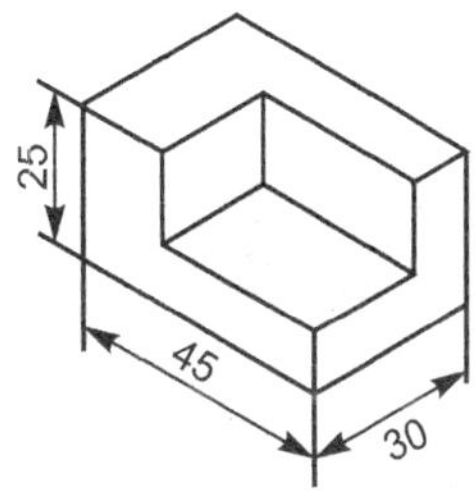

(c) 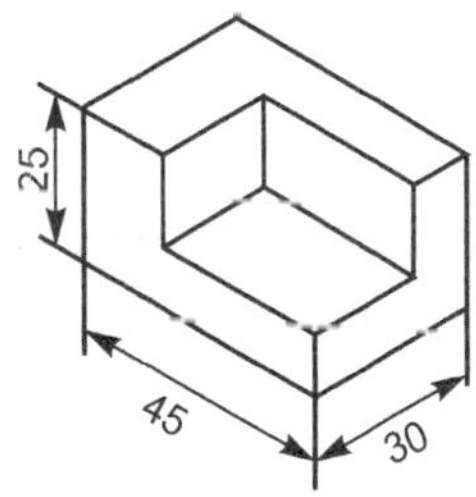

(d) 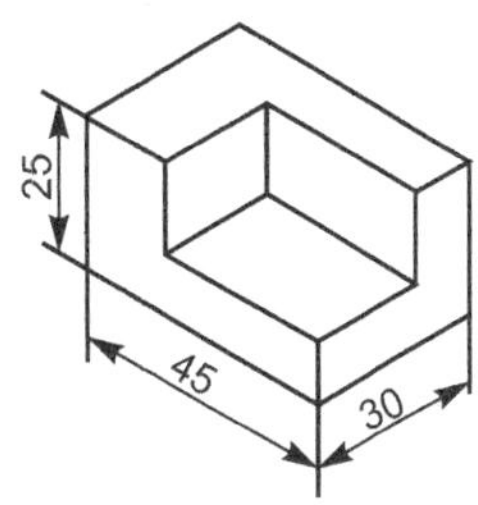

48. Identify the name of tool?

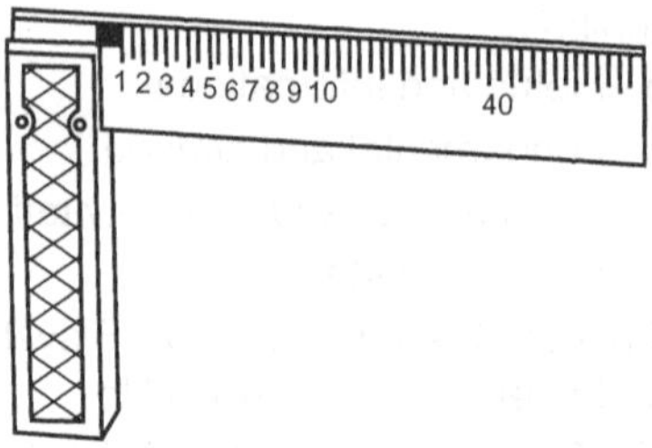

(a) Feeler gauge (b) Try square
(c) Steel rule (d) Surface gauge

49. Identify the name of tool?

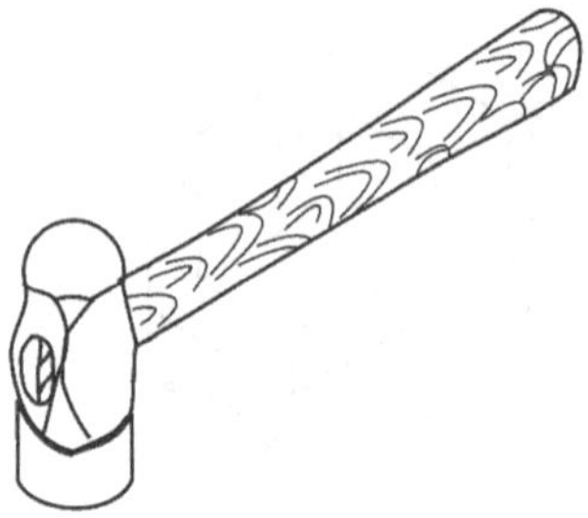

(a) Sledge hammer
(b) Ball pein hammer
(c) Straight pein hammer
(d) Cross pein hammer

50. Identify the name of tool?

(a) Staright pein hammer
(b) Cross pein hammer
(c) Mallet
(d) Ball pein hammer

KEY First Year July 2023

1. (c)	**2.** (a)	**3.** (c)	**4.** (b)	**5.** (c)	**6.** (a)	**7.** (c)	**8.** (c)	**9.** (b)	**10.** (c)
11. (b)	**12.** (b)	**13.** (a)	**14.** (c)	**15.** (a)	**16.** (c)	**17.** (d)	**18.** (a)	**19.** (c)	**20.** (d)
21. (d)	**22.** (c)	**23.** (d)	**24.** (a)	**25.** (a)	**26.** (c)	**27.** (a)	**28.** (b)	**29.** (a)	**30.** (c)
31. (d)	**32.** (b)	**33.** (d)	**34.** (d)	**35.** (d)	**36.** (a)	**37.** (c)	**38.** (d)	**39.** (a)	**40.** (d)
41. (b)	**42.** (d)	**43.** (b)	**44.** (a)	**45.** (b)	**46.** (a)	**47.** (d)	**48.** (b)	**49.** (b)	**50.** (c)

SECOND YEAR JULY 2023

All India Trade Test (CBT)

Electrician Trade Theory

NSQF Level – 4

Trade Theory

1. Which is proportional to the torque in D.C motor?
   (a) Armature current (b) Back e.m.f
   (c) Terminal voltage (d) Field current
2. What is the purpose of field coils in D.C generator?
   (a) To increase the reluctance of magnetic path
   (b) To decrease the magnetizing current
   (c) To increase the flux in air gap
   (d) To magnetize the poles to produce coil flux
3. Name the type of DC motor.

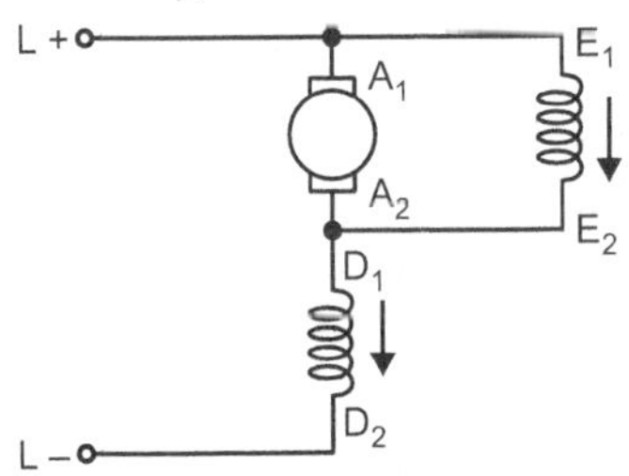

   (a) Series motor
   (b) Short shunt compound motor
   (c) Shunt motor
   (d) Long shunt compound motor
4. Which relay hold their contacts in position after power is cutoff?
   (a) Latching relay (b) Current relay
   (c) Voltage relay (d) Read relay
5. What is the criteria to select the contactor?
   (a) Place of use of the contactor
   (b) Type of load connected
   (c) Type of supply
   (d) Supply voltage and load
6. Where D.C compound motors are preferred?
   (a) Constant load requirements
   (b) High starting torque requirements
   (c) Constant speed under varying load requirements
   (d) Constant speed requirements
7. What is the use of 'G' channels in control panel?
   (a) For fixing terminal connectors
   (b) For fixing instruments
   (c) For fixing contactors
   (d) For fixing relays
8. Which type of DC motor is used for sudden application of heavy loads?
   (a) Differential compound motor
   (b) Series motor
   (c) Shunt motor
   (d) Cumulative compound motor
9. What is the name of line insulator?

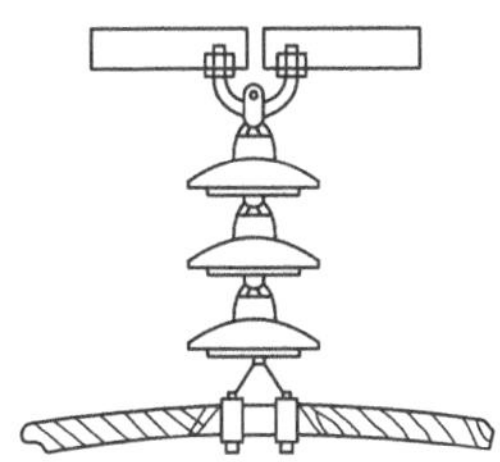

   (a) Disc type insulator
   (b) Suspension type insulator
   (c) Shackle type insulator
   (d) Pin type insulator

10. Which type of winding wire is used to wind submersible pump motors?

(a) Super enamelled type
(b) PVC covered type
(c) Double cotton covered type
(d) Terylene thread type

11. What is the formula to calculate emf equation of an alternator?

(a) $E = 1.11\ K_d\ K_c\ F\ \phi_m$
(b) $E = 4.44\ K_d\ K_c\ T\ \phi_m$
(c) $E = 2.22\ K_d\ K_c\ F\ \phi_m$
(d) $E = 4.44\ K_d\ K_c\ FT\ \phi_m$

12. Which type of output transformer is used in automatic voltage stabilizer?

(a) Ring core type transformer
(b) Auto transformer
(c) Ferrite core type transformer
(d) Static transformer

13. What is the function of split rings in DC generator?

(a) Reduces the voltage drop at brushes
(b) Maintain constant voltage
(c) Collects the current unidirectionally
(d) Increases the terminal voltage than rated

14. Why D.C supply is necessary for synchronous motor operation?

(a) Start the motor initially
(b) Reduce the losses
(c) Run the motor with overload
(d) Run the motor at synchronous speed

15. What is the purpose of connecting L_1 through C_1 to the transistor base?

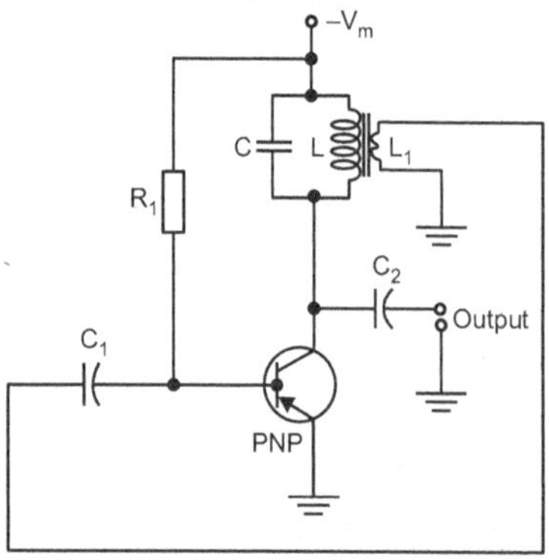

(a) Provides transistor biasing voltage
(b) Provides DC supply
(c) Provides positive feed back
(d) Provides negative feedback

16. Which type of winding has more space for cooling?

(a) Between overhanging coil and rotor
(b) Between overhanging coils
(c) Between overhanging coil and wedge
(d) Between overhanging coils and yoke

17. Which device protects from overload and short circuit in a panel board?

(a) Miniature circuit breaker
(b) Thermal overload relay
(c) Time delay relay
(d) Isolating switch

18. Which rule is used to find the direction of induced emf in an alternator?

(a) Fleming's left hand rule
(b) Cork screw rule
(c) Fleming's right hand rule
(d) Right hand palm rule

19. What is the name of the coil winding?

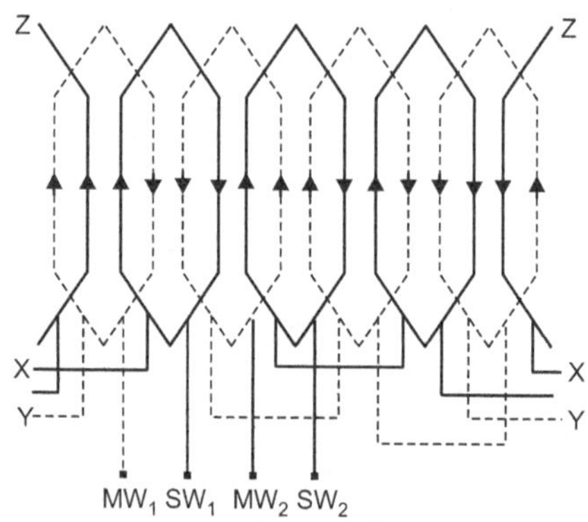

(a) Diamond mesh shaped coil winding
(b) Distributed coil winding
(c) Mesh shaped coil winding
(d) Concentric coil winding

20. What is the name of the A.C motor starter?

(a) DOL starter
(b) Semi automatic star delta starter
(c) Auto transformer starter
(d) Fully automatic star delta starter

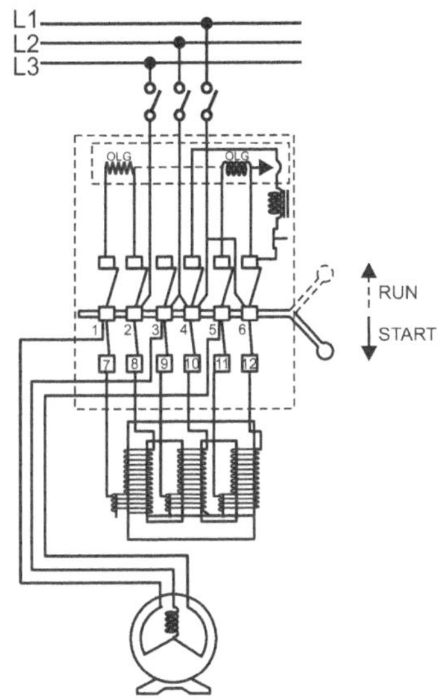

21. Which is the conventional power generation?

(a) Thermal power generation
(b) Solar power generation
(c) Tidal power generation
(d) Wind power generation

22. Name the part marked as 'X' of the winding machine?

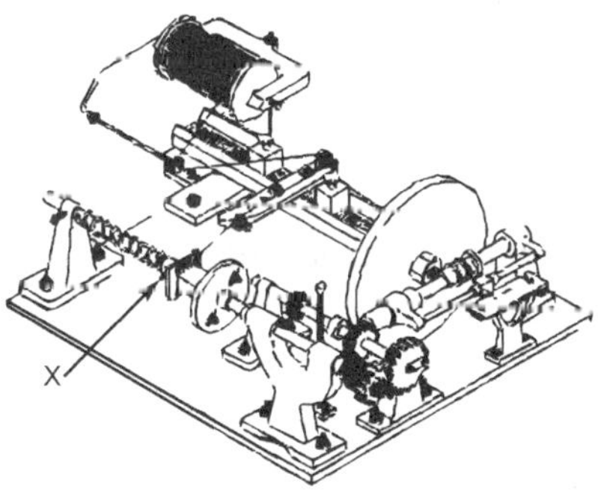

(a) Wire feed (b) Wire guides
(c) Spool carrier (d) Mandrel

23. Which instrument is used to test armature winding for short and open circuit?

(a) Digital multimeter
(b) Internal Growler
(c) Tong Tester
(d) External Growler

24. What is the name of the accessory used in control panel wiring?

(a) Terminal connector
(b) Lugs
(c) Grommet
(d) Thimble

25. What is the type of A.C voltage stabilizer?

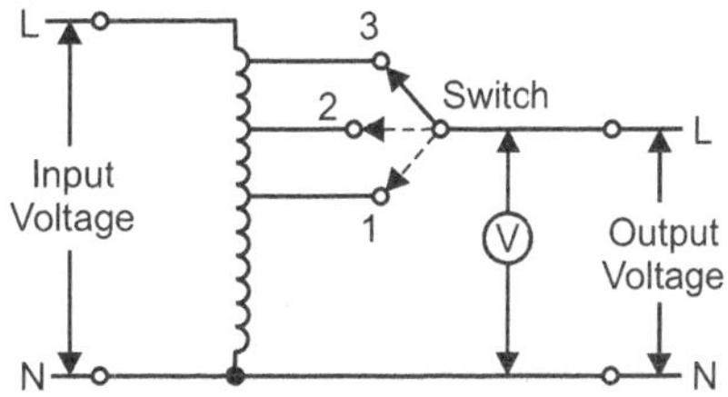

(a) Automatic voltage stabilizer
(b) Manual stepped voltage stabilizer
(c) Servo voltage stabilizer
(d) Constant voltage transformer stabilize

26. Which loss of 3 phase induction motor is determined by blocked rotor test?

(a) Eddy current loss
(b) Hysteresis loss
(c) Friction loss
(d) Copper loss

27. Which formula is used to calculate the speed of DC motor?

(a) $N = \frac{E_b}{\varphi}$ (b) $N = \frac{\varphi}{E_b}$
(c) $N = \frac{E_b \varphi}{120}$ (d) $N - \frac{E_b \varphi}{60}$

28. Which motor is having half coil winding?

(a) Washing machine (b) Mixer
(c) Ceiling fan (d) Grinder

29. Why DC generators are loosing their residual magnetism?

(a) Heavy short circuit in load
(b) Continuous running without break
(c) Running without load continuously
(d) Change of direction of rotation very often

30. Which speed is called as synchronous speed in 3 phase induction motor?

(a) Full load speed
(b) Rotating magnetic field speed
(c) No load speed
(d) Relative speed between stator and rotor

31. What is the characteristic property of base material in a transistor?
(a) Lightly doped and very thin
(b) Heavily doped and very thin
(c) Heavily doped and very larger
(d) Lightly doped and very larger

32. What is the name of the converter?

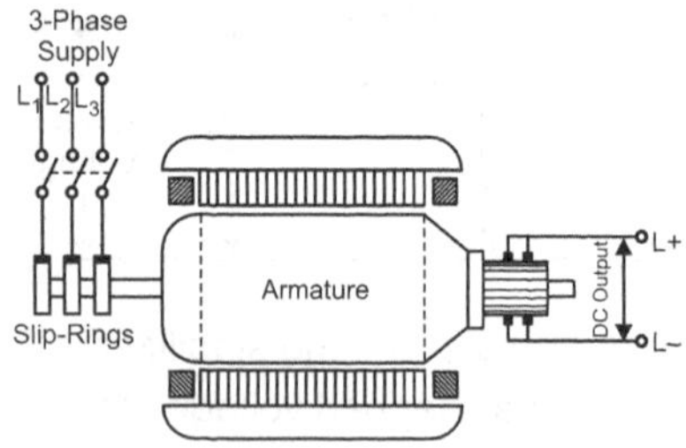

(a) Motor-Generator set
(b) Rotary converter
(c) Mercury arc rectifier
(d) Metal rectifier

33. What is the name of the part marked 'X' in DC generator?

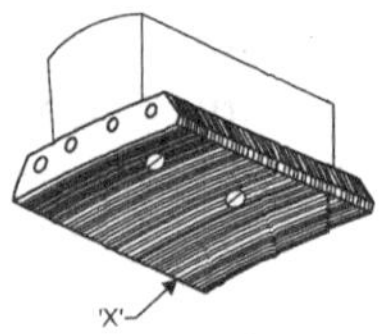

(a) Pole shoe
(b) Pole tip
(c) Pole coil
(d) Pole core

34. Which type of test is illustrated for the armature after rewound?

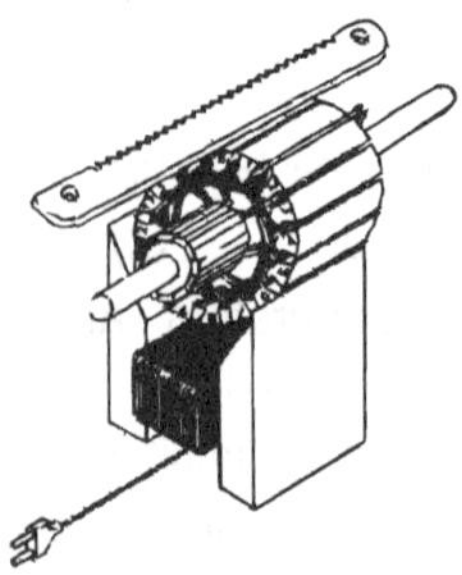

(a) Shorted coil test
(b) Open coil test
(c) Voltage drop test
(d) Grounded coil test

35. What is the name of amplifier?

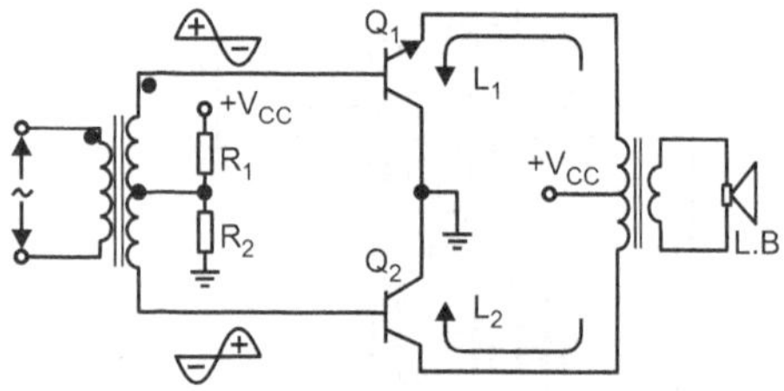

(a) Common emitter amplifier
(b) Common collector amplifier
(c) Class AB push pull amplifier
(d) Class B push pull amplifier

36. Which power modulator is used in the electric drive system?
(a) Cyclo converters
(b) Phase sequence indicator
(c) Servo controlled voltage stabilizer
(d) Frequency multiplier

37. What refers coil in AC winding?
(a) Number of turns connected in series
(b) Number of turns under two similar poles
(c) Number of turns under two dissimilar poles
(d) Number of turns connected in parallel

38. Which device controls the operations in sequential control systems?
(a) Control transformer
(b) Relays
(c) Timer (d) Contactor

WORKSHOP CALCULATION AND SCIENCE

39. What is the name of a booklet, the rates of various terms are indicated?
(a) Price bank (b) Price tag
(c) Price bunch (d) Price catalogue

40. What is the value of 625^{0}?
(a) 25 (b) 0
(c) 525 (d) 1

41. What is the value of $a \times a^2 \times a^3 \times a^4$?
(a) a^9 (b) a^7
(c) a^8 (d) a^{10}

42. How the years is denoted in simple interest calculations?
(a) I (b) r
(c) P (d) n

43. Which hand book is referred by machine engineer?

(a) Oxford Dictionary
(b) Mark standard
(c) CRC
(d) Parry's cheorikal

44. What is denoted as 'I'?

(a) Principal (b) Interest
(c) Year (d) Rate

ENGINEERING DRAWING

45. Choose the symbol for main fuse board without switches (P).

(a) 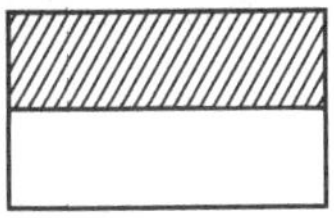 (b) 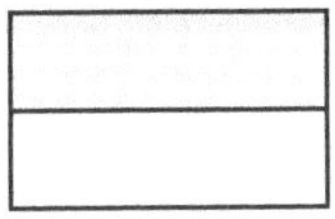
(c) 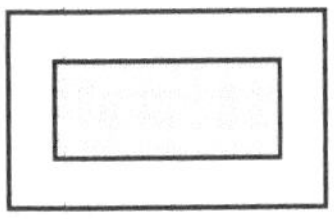 (d) 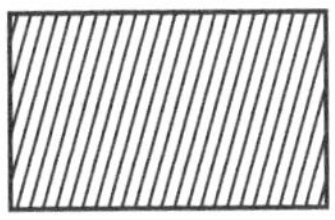

46. Identify the given symbol

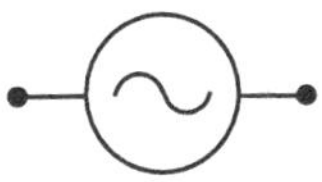

(a) ohms (b) Inductance
(c) AC (d) DC

47. Identify the symbol of battery.

(a)  (b) 

(c) 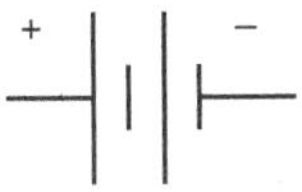 (d) 

48. Which is protective device?

(a) Fuse (b) Resistor
(c) Capacitor (d) Magnet

49. Which is photo diode?

(a) 

(b) 

(c) 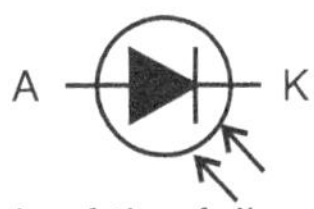

(d) 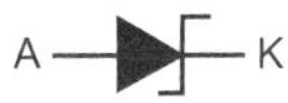

50. Which of the following is AC ammeter?

(a) 

(b) 

(c) 

(d) 

KEY Second Year July 2023

1. (a)	**2.** (d)	**3.** (b)	**4.** (a)	**5.** (d)	**6.** (c)	**7.** (a)	**8.** (d)	**9.** (b)	**10.** (b)
11. (d)	**12.** (b)	**13.** (c)	**14.** (d)	**15.** (c)	**16.** (d)	**17.** (a)	**18.** (c)	**19.** (a)	**20.** (c)
21. (a)	**22.** (d)	**23.** (d)	**24.** (c)	**25.** (b)	**26.** (d)	**27.** (a)	**28.** (c)	**29.** (d)	**30.** (b)
31. (a)	**32.** (b)	**33.** (a)	**34.** (a)	**35.** (d)	**36.** (a)	**37.** (a)	**38.** (c)	**39.** (d)	**40.** (d)
41. (d)	**42.** (d)	**43.** (b)	**44.** (b)	**45.** (c)	**46.** (c)	**47.** (c)	**48.** (a)	**49.** (c)	**50.** (b)

FIRST YEAR AUGUST 2024

All India Trade Test (CBT)

Electrician Trade Theory

NSQF Level – 4

Trade Theory

1. What is the effect of inductance if the distance between the turns increases?
   (a) Remains same (b) Increases
   (c) Becomes zero (d) Decreases
2. Which type of joint is used in overhead lines for high tensile strength?
   (a) Married joint
   (b) Western union joint
   (c) Rat-tail joint
   (d) Scarfed joint
3. What is the name of the diagram as shown in the figure?

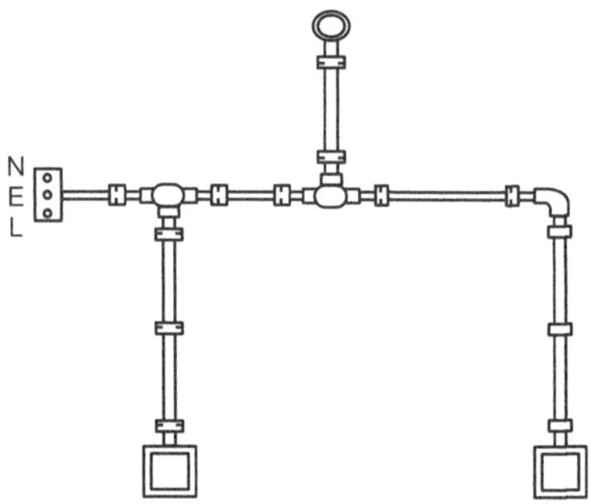

   (a) Layout plan (b) Wiring diagram
   (c) Schematic diagram
   (d) Installation plan
4. Where the Iron Clad Double Pole (ICDP) main switch is used?
   (a) Large industrial installations
   (b) Single phase domestic installations
   (c) Three phase power circuit installations
   (d) Large commercial installations
5. What is the formula for the equivalent resistance (R_T) of the three resistors R_1, R_2 & R_3 are connected in parallel?
   (a) $R_T = \frac{1}{R_1} + \frac{1}{R_2} + \frac{1}{R_3}$
   (b) $R_T = R_1 + R_2 + R_3$
   (c) $R_T = \frac{1}{R_1 + R_2 + R_3}$
   (d) $R_T = \frac{1}{\frac{1}{R_1} + \frac{1}{R_2} + \frac{1}{R_3}}$
6. What is the unit for Quantity of electricity?
   (a) ampere / second (b) volt / second
   (c) coulomb (d) mho
7. Which part is losing its weight during the electrolysis?
   (a) Separator (b) Anode
   (c) Electrolyte (d) Cathode
8. When the no current will flow through the galvanometer in the wheat stone bridge?
   (a) In short-circuited condition
   (b) In open-circuited condition
   (c) In unbalanced condition
   (d) In balanced condition
9. Which conductors are used for the over-head distribution lines?
   (a) Bare conductors
   (b) Insulated solid conductors

(c) Two core cable
(d) Insulated conductors

10. Which is employed in circuit breaker for the thermal overload protection?
(a) No volt coil
(b) Bimetallic strip
(c) Magnetic repulsion
(d) Mechanical vibration

11. Which is the position to use the instrument provided with gravity control?
(a) Horizontal position
(b) Vertical position
(c) Any position
(d) Inclined position

12. Calculate the electrical energy in unit consumed by 500W lamp for 5 hours.
(a) 1.5 unit (b) 2.5 unit
(c) 1.0 unit (d) 0.5 unit

13. How to determine the core loss in a transformer?
(a) By ratio test
(b) By impulse test
(c) By short circuit test
(d) By open circuit test

14. What is the name of meter as shown in the figure?

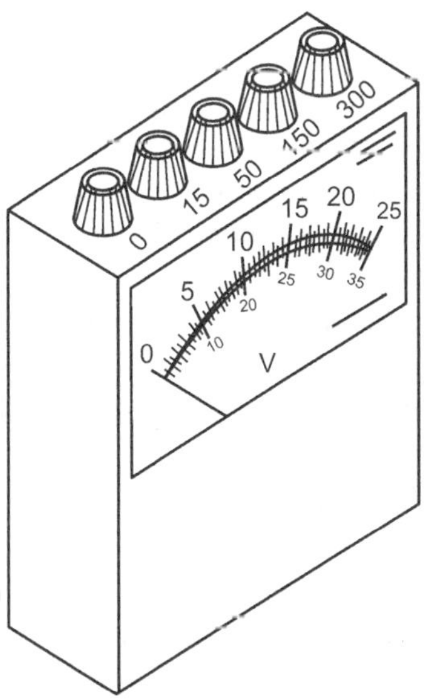

(a) AC and DC multirange voltmeter
(b) AC multirange ammeter
(c) DC multirange voltmeter
(d) AC and DC multirange ammeter

15. What is an application of 'L' series type MCB's?
(a) Halogen lamps (b) Air conditioners
(c) Geysers (d) Locomotives

16. Which type of fire extinguisher is used for fire on electrical equipment?
(a) Gas cartridge water type
(b) Foam type
(c) Stored pressure water type
(d) Halon type

17. Which 3 phase system requires the artificial neutral to measure the phase voltage?
(a) 3 wire delta connected system
(b) 3 wire star connected system
(c) 4 wire star connected system
(d) Scott connected system

18. Which material is used to make the core of power transformer?
(a) Hot rolled steel
(b) Copper alloy
(c) Soft iron
(d) Cold rolled grain oriented steel

19. Which is an application of the series circuit?
(a) Lighting circuits in home
(b) Voltmeter connection
(c) Shunt resistor in ammeter
(d) Multiplier resistor of a voltmeter

20. What is the advantage of the stranded conductor over the solid conductor?
(a) Less voltage drop
(b) Cost is less
(c) More insulation resistance
(d) More flexible

21. What is the standard frequency of AC supply in India?
(a) 75 Hz (b) 25 Hz
(c) 50 Hz (d) 60 Hz

22. Why the system earthing Is different from the equipment earthing?
(a) It is associated with non-current carrying conductors

(b) It protects human only
(c) It is associated with current carrying conductors
(d) It protects equipment only

23. What is the voltage drop in resistor 'R_2' in the series circuit?

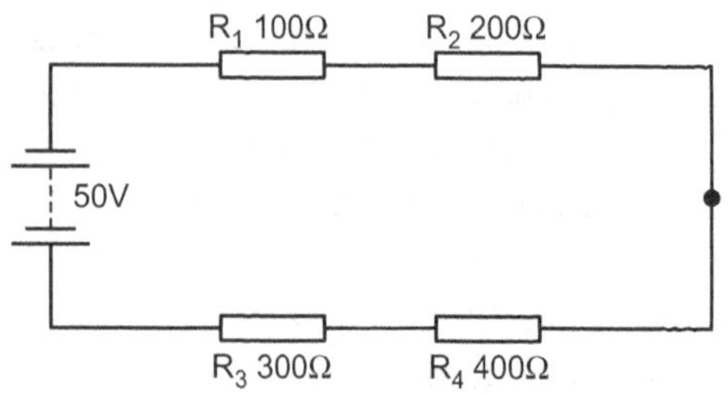

(a) 20 volt (b) 10 volt
(c) 15 volt (d) 5 volt

24. Which device converts the sunlight into electrical energy?
(a) Light emitting diode
(b) Liquid crystal diode
(c) Light dependent resistor
(d) Photo voltaic cell

25. Which apparatus is used to check the charging condition of the battery?
(a) Voltmeter
(b) High rate discharge tester
(c) Hydrometer
(d) Multimeter

26. Which electrical quantity is unable to measure directly by the multimeter?
(a) AC voltage (b) Power
(c) DC voltage (d) Resistance

27. Which is the indication of a moisture contained transformer oil is placed in the crackle test?
(a) A crackle sound is heard
(b) A humming noise is heard
(c) A flash over occurs
(d) A sizzle noise is heard

28. Which factor changes the permeability of a material?
(a) Flux density
(b) Length
(c) Diameter
(d) Cross sectional area

29. Which power measurement method is used for both balanced and unbalanced loads?
(a) Voltmeter and Ammeter method
(b) Single Wattmeter method
(c) 2 Wattmeter method
(d) Energy meter method

30. Where the service mains are to be connected in the domestic installation?
(a) Main switch
(b) Distribution board
(c) Energy meter
(d) IC cut out

31. What is the name of the wiring diagram?

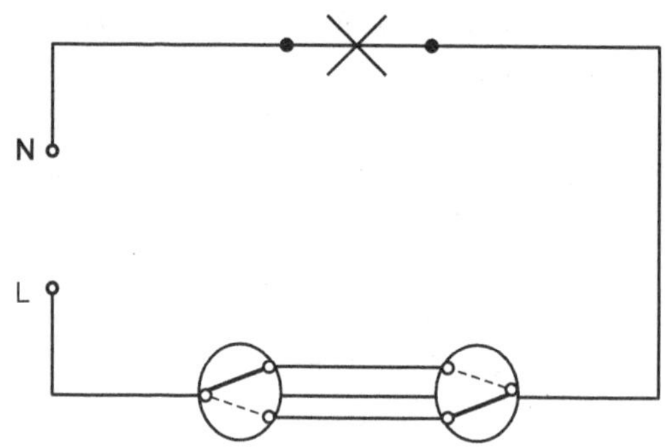

(a) Staircase wiring (b) Hostel wiring
(c) Godown wiring (d) Tunnel wiring

32. What is the SI unit of resistivity?
(a) ohm / metre (b) ohm / cm^2
(c) ohm / cm (d) ohm - metre

33. What will happen to the value of earth resistance if length of the earth pipe is increased?
(a) Infinity (b) Remains same
(c) Increases (d) Decreases

34. What is the function of the rotary switch in food mixer?
(a) For reverse rotation
(b) Overload protection
(c) Speed selection
(d) Over current protection

35. What is the unit of insulation resistance?

(a) milli ohm (b) mega ohm

(c) milo ohm (d) ohm

36. Which type of motor is used in the wet grinder?

(a) Universal motor

(b) Shaded pole motor

(c) Capacitor start induction run motor

(d) Repulsion motor

37. What is the name of the part marked 'X' as shown in the figure?

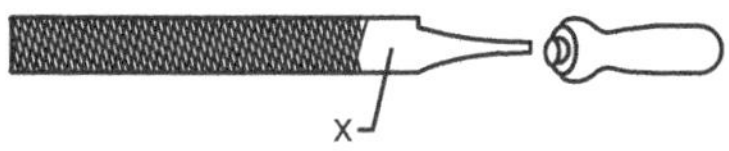

(a) Point (b) Tang

(c) Edge (d) Heel

38. Which wiring installation uses the system earthing?

(a) Domestic wiring

(b) Godown wiring

(c) Substations

(d) Commercial wiring

WORKSHOP CALCULATION AND SCIENCE

39. Which is equal to sin θ?

(a) $\frac{\text{Opposite side}}{\text{Hypotenuse}}$ (b) $\frac{\text{Hypotenuse}}{\text{Adjacent side}}$

(c) $\frac{\text{Hypotenuse}}{\text{Opposite side}}$ (d) $\frac{\text{Adjacent side}}{\text{Hypotenuse}}$

40. What is the square root of 529?

(a) 43 (b) 23

(c) 33 (d) 13

41. What are the two classifications of system of units?

(a) Metric and International

(b) British and Metric

(c) Fundamental and derived

(d) Gravitational and non-gravitational

42. What is the side of a square whose area is 625 mm^2?

(a) 20 mm (b) 15 mm

(c) 30 mm (d) 25 mm

43. What are fundamental units?

(a) Length, Mass, Volume

(b) Length, Mass, Area

(c) Length, Mass, Time

(d) Length, Pressure, Volume

44. Convert decimal 0.000659 to fraction?

(a) $\frac{659}{1000}$ (b) $\frac{659}{100000}$

(c) $\frac{659}{10000}$ (d) $\frac{659}{1000000}$

ENGINEERING DRAWING

45. Identify the conventional symbol of material?

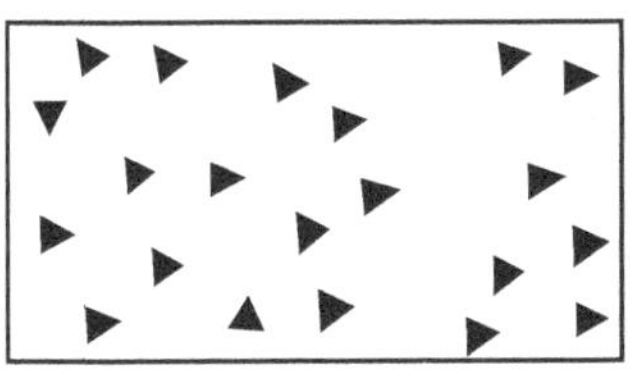

(a) Glass (b) Wood

(c) Concrete (d) Steel

46. Identify the conventional symbol of material?

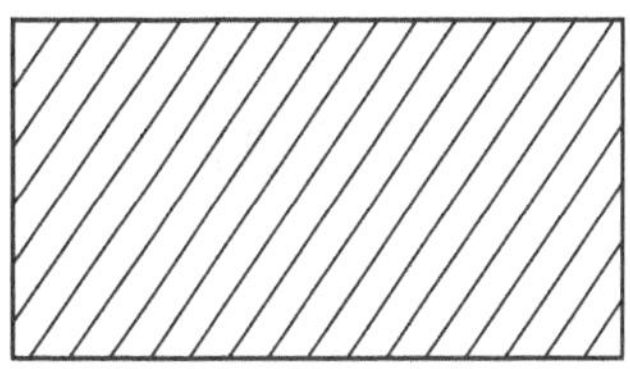

(a) Porcelain (b) Steel

(c) Lead (d) Glass

47. Which drawing dimension is correct as per IS standard?

(a)

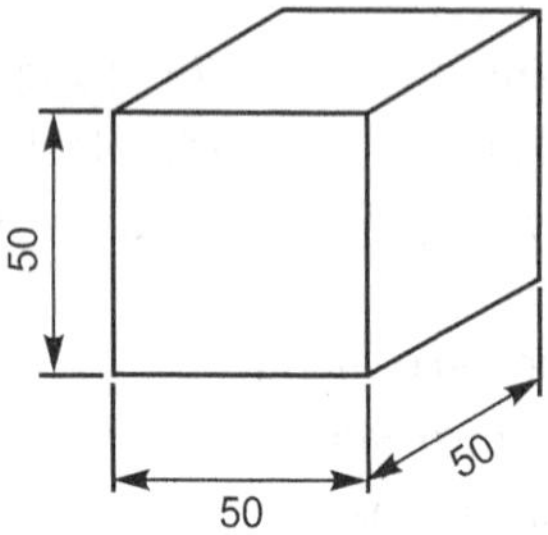

(b)

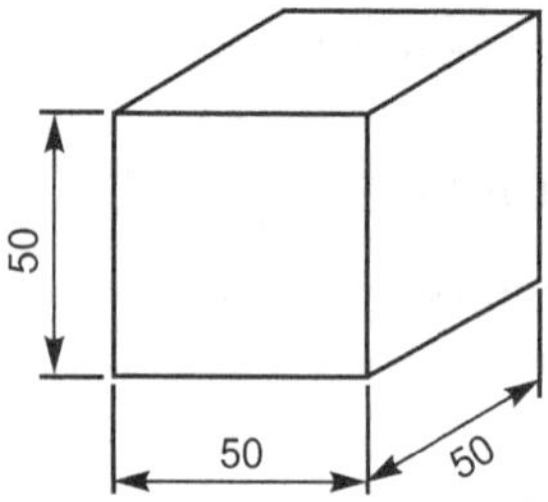

(c)

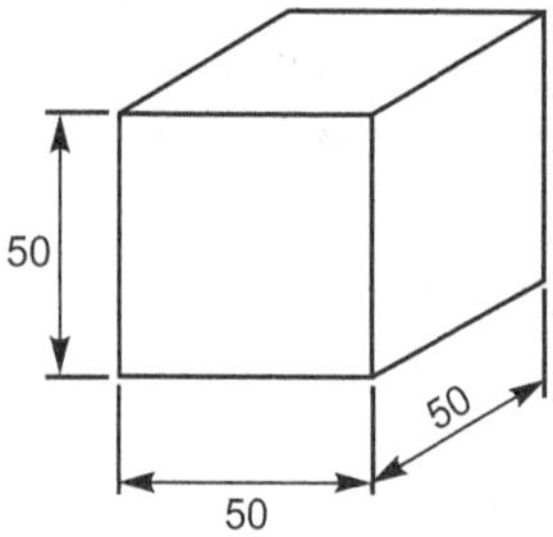

(d)

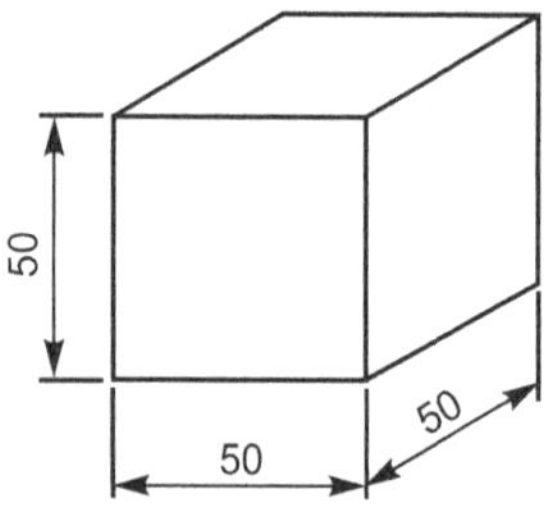

48. Identify the line 'b'?

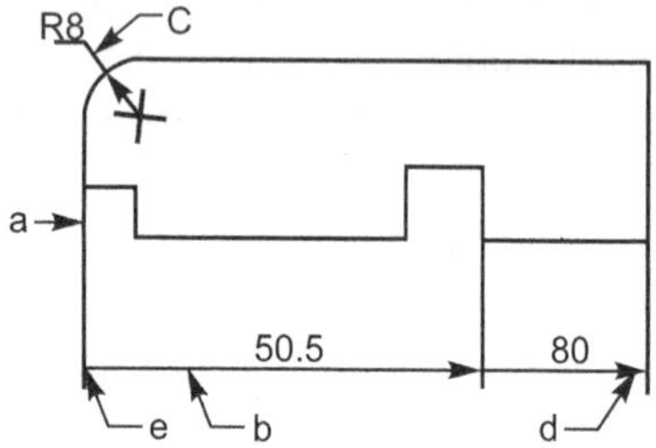

(a) Extension line
(b) Termination of dimension line
(c) Dimension line
(d) Leader line

49. What is the trimmed size of A4 sheet?

(a) 594 × 841 (b) 210 × 297
(c) 148 × 210 (d) 841 × 1189

50. Identify the conventional symbol of material?

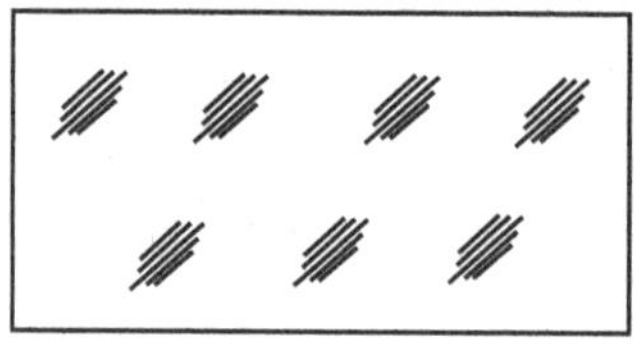

(a) Wood (b) Glass
(c) Steel (d) Concrete

KEY First Year August 2024

1. (d)	**2.** (b)	**3.** (d)	**4.** (b)	**5.** (a)	**6.** (c)	**7.** (b)	**8.** (d)	**9.** (a)	**10.** (b)
11. (b)	**12.** (b)	**13.** (d)	**14.** (a)	**15.** (c)	**16.** (d)	**17.** (a)	**18.** (d)	**19.** (d)	**20.** (d)
21. (c)	**22.** (c)	**23.** (b)	**24.** (d)	**25.** (b)	**26.** (b)	**27.** (a)	**28.** (a)	**29.** (c)	**30.** (c)
31. (a)	**32.** (d)	**33.** (d)	**34.** (c)	**35.** (b)	**36.** (c)	**37.** (d)	**38.** (c)	**39.** (a)	**40.** (b)
41. (c)	**42.** (d)	**43.** (c)	**44.** (d)	**45.** (c)	**46.** (b)	**47.** (d)	**48.** (c)	**49.** (b)	**50.** (b)

SECOND YEAR AUGUST 2024

All India Trade Test (CBT)

Electrician Trade Theory

NSQF Level – 4

1. How many parallel paths in duplex lap winding of a 4 pole DC generator?
   (a) 8 (b) 4
   (c) 12 (d) 6

2. Which type of testing of winding is illustrated as shown in the figure?

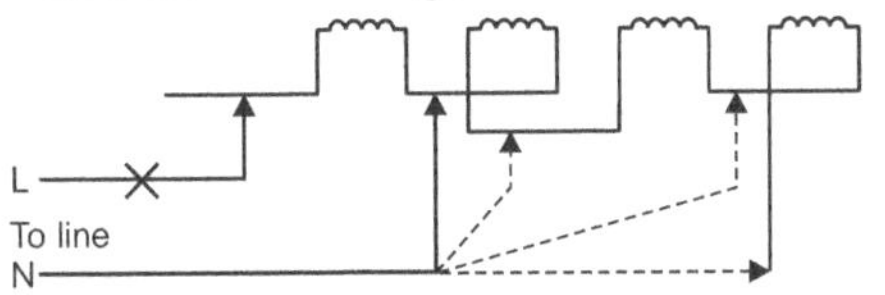

   (a) Short circuit test (b) Voltage drop test
   (c) Resistance test (d) Polarity test

3. Which circuit, the limit switches are used?
   (a) Domestic power circuits
   (b) Battery Charging Circuit
   (c) Lift circuits
   (d) Street lighting

4. Which device Is used to test stator winding short and open fault?
   (a) Digital multimeter
   (b) Internal Growler
   (c) Tong Tester
   (d) External Growler

5. Which type of DC motor is used for sudden application of heavy loads?
   (a) Differential compound motor
   (b) Series motor
   (c) Shunt motor
   (d) Cumulative compound motor

6. Which device is controling the operations in the sequential control systems?
   (a) Timer (b) Relays
   (c) Control transformer
   (d) Contactor

7. What is the full form of VVVFD?
   (a) Variable Value Voltage Frequency Drive
   (b) Voltage Value Variable Frequency Drive
   (c) Variable Voltage Variable Frequency Drive
   (d) Value Variable Voltage and Frequency Drive

8. Which material is used for the damper winding?
   (a) Cast iron bar (b) Copper bar
   (c) Stainless steel bar (d) Silicon steel bar

9. What is the purpose of slot marked as 'X' as shown in the figure?

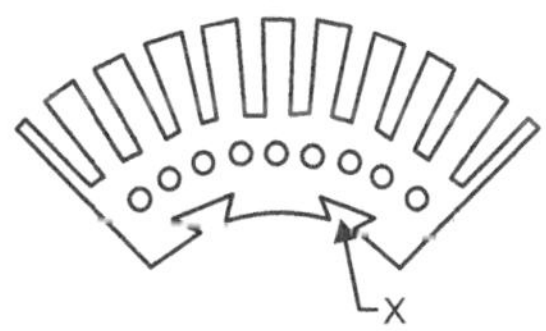

   (a) To make air circulation
   (b) For lubrication purpose
   (c) For easy removal from shaft
   (d) To fix the key way

10. What is the name of the AC single phase motor as shown in the diagram?

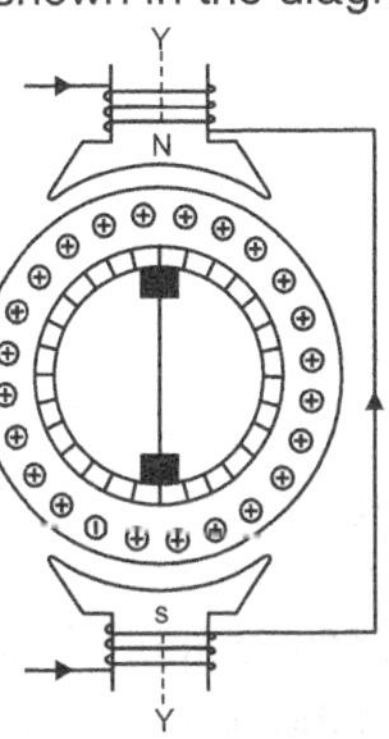

(a) Shaded pole motor
(b) Permanent capacitor motor
(c) Universal motor
(d) Repulsion motor

11. Why negative feedback is required in amplifier circuits?
(a) To reduce the distortion
(b) To increase the output voltage gain
(c) To increase the output current gain
(d) To increase the amplification factor

12. Which type of line insulator is used for terminating on corner post?
(a) Suspension insulator
(b) Shackle insulator
(c) Strain insulator
(d) Pin insulator

13. What is the function of IGBT in AC drive?
(a) Stabilize the output voltage from the rectifier
(b) Controls the power delivered to the motor
(c) Converts incoming A.C. into D.C.
(d) Smoothening incoming A.C. supply

14. Which formula is used to calculate EMF/ phase in a ideal alternator?
(a) $E = 4.44\ _{\phi}\ FT$ (b) $E = \frac{\varphi\ FT}{4.44}$
(c) $E = \frac{\varphi\ FT}{2.22}$ (d) $E = 2.22\ _{\phi}\ FT$

15. What is the synchronous speed of an A.C. 3 phase induction motor having 6 poles at a frequency of 50 hertz?
(a) 800 rpm (b) 1200 rpm
(c) 1440 rpm (d) 1000 rpm

16. Which is proportional to the torque in D.C. motor?
(a) Back e.m.f
(b) Armature current
(c) Field current
(d) Terminal voltage

17. Which relay holds their contacts in the position after power is cutoff?
(a) Voltage relay (b) Read relay
(c) Current relay (d) Latching relay

18. What is the use of synchroscope?
(a) Adjust the output voltage
(b) Adjust the phase sequence
(c) Adjust the supply frequency
(d) Indicate the correct instant for paralleling

19. What is the name of the speed control method of AC single phase motor as shown in the diagram?

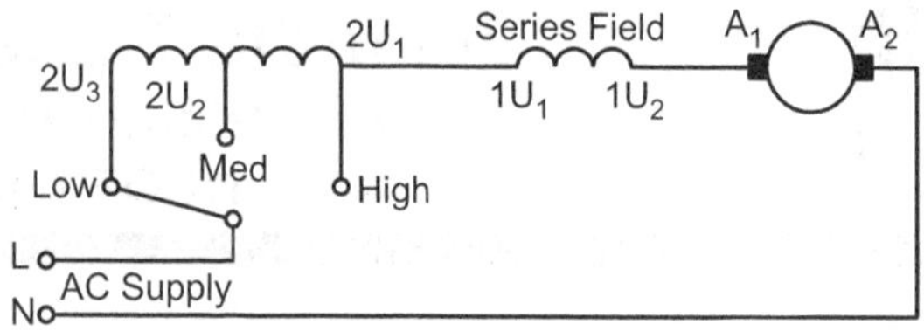

(a) Tapped field method
(b) Centrifugal switch method
(c) Field diverter method
(d) Voltage control method

20. When all the 3 lamps become dark in the dark lamp method of parallel operation of two alternators:
(a) Terminal voltages are equal
(b) Voltage and frequency are equal
(c) Voltage and power rating are equal
(d) Frequency are same in both alternator

21. What is the name of the speed control method of DC motor as shown in the figure?

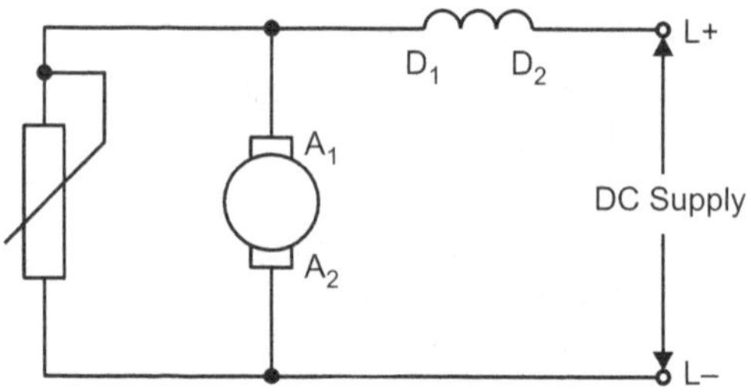

(a) Field diverter method
(b) Field tapping method
(c) Armature diverter method
(d) Voltage control method

22. What is the effect in internal resistance of a discharged cell?
(a) Decrease (b) Increase
(c) Remains same (d) Becomes zero

23. Why the armature resistance of a D.C generator is kept very low?
(a) Helps to solve Boolean Algebra
(b) To reduce the armature current
(c) To reduce the armature voltage drop
(d) To reduce the temperature of armature

24. Calculate the voltage and ampere/hour, if four cells rated as 1.5 V and 8 A.H are in parallel?

(a) 1.5 V and 32 AH (b) 6 V and 32 AH
(c) 6 V and 8 AH (d) 3 V and 16 AH

25. What is the formula to calculate the back emf of a D.C. motor?

(a) $E_b = V + I_a R_a$ volts
(b) $E_b = V \times I_a R_a$ volts
(c) $E_b = V - I_a R_a$ volts
(d) $E_b = \frac{V}{I_a R_a}$ volts

26. Which schedule of maintenance the resistance of earth connection of an induction motor is to be examined?

(a) Monthly (b) Yearly
(c) Daily (d) Weekly

27. Why a feedback network is used in the oscillator as shown in the circuit?

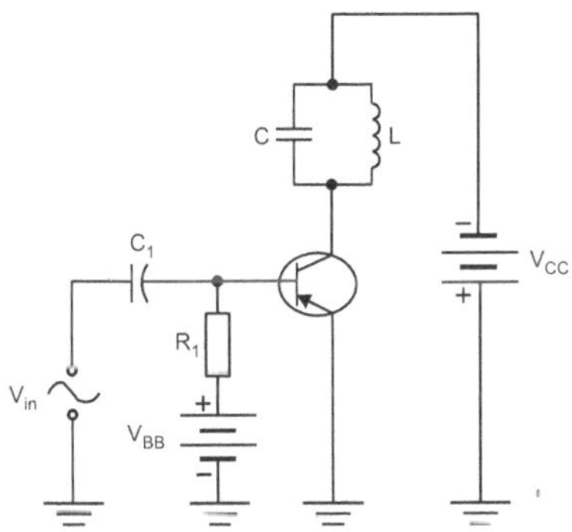

(a) To cancel second harmonic distortion
(b) To phase shift the signal by 60°
(c) To cancel noise distortion
(d) To phase shift the signal by 180°

28. Which is an active component?

(a) Transistor (b) Inductor
(c) Resistor (d) Capacitor

29. Which instrument is used to measure insulation resistance of a 3 phase induction motor?

(a) Shunt type ohmmeter
(b) Series type ohmmeter
(c) Multimeter
(d) Megger

30. Why most of the semi conductor devices are made by silicon compared to germanium?

(a) High current carrying capacity
(b) High barrier voltage
(c) High resistance range
(d) High thermal conductivity

31. What is the main function of Uni Junction Transistor (UJT)?

(a) Relaxation oscillator
(b) Microphone input device
(c) Loud speaker amplifier
(d) Broadcast transmitter

32. What is the name of the wiring accessory used in control panel wiring as shown in the figure?

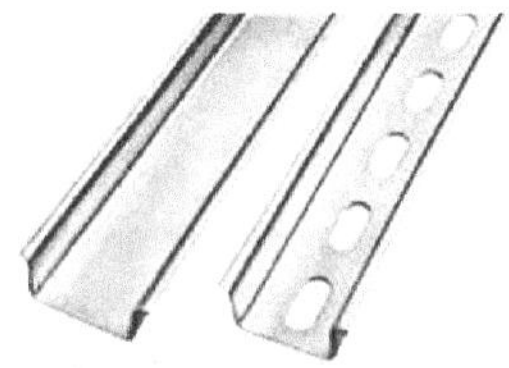

(a) DIN rails (b) Grommets
(c) Race ways (d) G channel

33. What is the name of the substation that all equipment of substation is installed within the station building?

(a) Outdoor substation
(b) Indoor substation
(c) Plinth mounted substation
(d) Pole mounted substation

34. Which material is used for starting resistance of DC starters?

(a) Eureka (b) Manganin
(c) Constantine (d) Nichrome

35. How the voltage is received in the rotor of induction motor?

(a) Direct connection to rotor from supply
(b) Direct connection from stator
(c) By the transformer action of stator and rotor
(d) Due to back emf produced in stator

36. Which type of D.C. motor is used for constant speed drives?

(a) Differential long shunt compound motor
(b) Differential short shunt compound motor

(c) DC series motor
(d) DC shunt motor

37. What is the property of wave winding in D.C. generator?
(a) High current high voltage
(b) Low current low voltage
(c) Low current high voltage
(d) High current low voltage

38. What is the phase displacement between winding in 3 phase motor?
(a) 180° (b) 120°
(c) 90° (d) 360°

WORKSHOP CALCULATION AND SCIENCE

39. What is the unit of strain?
(a) kg/cm^2 (b) $newton/metre^2$
(c) metre (d) No unit

40. What is the formula for $(a - b)^2$?
(a) $a^2 - 2ab - b^2$ (b) $a^2 - 2ab + b^2$
(c) $-a^2 - 2ab - b^2$ (d) $a^2 + 2ab + b^2$

41. What is the expanded form of $(a + b)^2$?
(a) $a^2 - 2ab + b^2$ (b) $a^2 + 2ab - b^2$
(c) $a^2 + 2ab + b^2$ (d) $-a^2 - 2ab + b^2$

42. How the years is denoted in simple interest calculations?
(a) r (b) n
(c) P (d) I

43. What is the formula for $a^m \times a^n$?
(a) a^{m-n} (b) a^{m+n}
(c) a^{mn} (d) $n.a^m$

44. What is the name of a booklet, the rates of various terms are indicated?
(a) Price tag (b) Price bank
(c) Price bunch (d) Price catalogue

ENGINEERING DRAWING

45. Which is protective device?
(a) Magnet (b) Capacitor
(c) Resistor (d) Fuse

46. Which device is used to start and stop a motor?
(a) Rotor (b) Slip ring
(c) Stator (d) Starter

47. Which of the following is AC ammeter?
(a) 

(b) 

(c) 

(d) 

48. Which is a symbol of DC motor?
(a) 

(b) 

(c) 

(d) 

49. Which part regulates the heat in automatic iron?
(a) Heating element
(b) Asbestos sheet
(c) Thermostatic switch
(d) Pressure plate

50. Which is ball peen hammer?
(a) 
(b) 
(c) 
(d) 

KEY Second Year August 2024

1. (a)	**2.** (c)	**3.** (c)	**4.** (b)	**5.** (d)	**6.** (a)	**7.** (c)	**8.** (b)	**9.** (d)	**10.** (d)
11. (a)	**12.** (b)	**13.** (b)	**14.** (a)	**15.** (d)	**16.** (b)	**17.** (d)	**18.** (d)	**19.** (a)	**20.** (b)
21. (c)	**22.** (b)	**23.** (c)	**24.** (a)	**25.** (c)	**26.** (b)	**27.** (d)	**28.** (a)	**29.** (d)	**30.** (b)
31. (a)	**32.** (a)	**33.** (b)	**34.** (a)	**35.** (c)	**36.** (d)	**37.** (c)	**38.** (b)	**39.** (d)	**40.** (b)
41. (c)	**42.** (b)	**43.** (b)	**44.** (d)	**45.** (d)	**46.** (d)	**47.** (b)	**48.** (c)	**49.** (c)	**50.** (b)